THEORY OF FLEXIBLE SHELLS

E.L. AXELRAD

Institut für Mechanik
Universität der Bundeswehr München
Fed. Rep. Germany

1987

NORTH-HOLLAND – AMSTERDAM · NEW YORK · OXFORD · TOKYO

ISBN: 0 444 87954 4

Publishers:
ELSEVIER SCIENCE PUBLISHERS B.V.
P.O. Box 1991
1000 BZ Amsterdam
The Netherlands

Sole distributors for the U.S.A. and Canada:
ELSEVIER SCIENCE PUBLISHING COMPANY, INC.
52 Vanderbilt Avenue
New York, N.Y. 10017
U.S.A.

Library of Congress Cataloging-in-Publication Data

Axelrad, E. L. (Ernest L.), 1927–
Theory of flexible shells.

(North-Holland series in applied mathematics and
mechanics ; v. 28)
Bibliography: p. 385
1. Shells (Engineering)   2. Elastic plates and
shells. I. Title. II. Series.
TA660.S5A94   1986   624.1'7762   86-11475
ISBN 0-444-87954-4 (U.S.)

PRINTED IN THE NETHERLANDS

To A.I. Lur'e
one of those few, whom so many
owe so much in mechanics

PREFACE

Engineers who must cope with problems of thin-walled structures and their research colleagues have fundamental monographs and textbooks on shell theory available. But there is next to nothing on "flexible shells" there. What indeed is meant by this term?

The almost exclusive concern of the shell-theory manuals are shells designed for strength and *stiffness*. Analysis of these shells is based predominantly on two special branches of shell theory, namely, the membrane and the Donnell–Koiter theories. These are theories of stressed states which vary, respectively, slowly or markedly with *both* surface coordinates.

It is less well known that there is a complementary class of shells and a corresponding branch of the shell theory, which occupies the ground between the above classic domains. These are *flexible shells*, designed for maximum elastic displacements.

Flexible shells—curved tubes, Bourdon pressure elements, bellows, expansion joints, corrugated diaphragms, twisted tubes—have been used in industry for many decades. Their underrepresentation in shell-theory books is partly due to the striking diversity of shapes and problems involved (cf. the schemes in Fig. 12 on p. 72). By contrast great effort has been invested in solving individual problems. (Curved tubes alone were treated in 1911–1985 in over 250 articles.)

The experience thus gained facilitated the perception of the immanent feature of small-strain deformation rendering large displacements: such deformation varies slowly and is nearly membrane-like along *one* of the surface coordinates while along the other it can vary intensively and involves substantial wall-bending. This feature constitutes the foundation for the specialized theory of flexible shells.

The prediction of stability and the analysis of postbuckling can be adequately served by the Donnell–Koiter theory also for flexible shells. However, in contrast to the "stiff" shells, stability and post-buckling are not the sole, not even the main objects of nonlinear analysis of flexible shells. It requires the specialized theory just mentioned.

For the buckling analysis, the flexibility causes substantial complication of perturbing both the shape and the stress state by large precritical displacements. Effective aid is provided here by the *local-buckling* approach.

The general theory is stated in the vector form. This formulation is as far as possible coordinate free, facilitating physical insights essential for complicated (flexible shell) problems. Transition to the component form is most direct for the orthogonal coordinates sufficient for all problems considered. There is no need of "exposing the gears of a machine for grinding out the workings of a tensor" (J.G. Simmonds). The equivalence of the vector formulation to the tensor one is briefly traced (p. 67).

Throughout the book the intrinsic theory is used. The evaluation of displacements has been avoided also for kinematical edge conditions. This mitigates an obstacle in the intrinsic formulation, one of the causes of the recent revival of the displacement approach. This approach has been the object of research of many a distinguished mechanician. Remarkable progress has been achieved. But the crucial for applications complexity of the equations could not be overcome, not even at the price of drastic restrictions on rotations. The intrinsic approach retains its advantages. These are absolute by large displacements (cf. §8.3, Ch. 3).

For both periodic and boundary-value problems, the trigonometric-series solution in matrix form[1] is widely applied in the book. It remains one of the most suitable methods, even more so with the advent of computers. Useful in reducing the partial differential equations to the ordinary ones, it leads in some cases to closed-form solutions. Of course, the series method does not replace powerful numerical tools like direct integration or finite elements with their computer programs. On the contrary, such numerical studies are complemented by a more articulate theory. An advance assessment of eventual results can be particularly helpful in solving a nonlinear problem.

---

[1] Another way is opened by Fast Fourier Transforms. It proved effective in numerical modelling of semiconductor devices described by highly nonlinear partial differential equations.

The book contains five chapters. Chapter 1 presents the general nonlinear theory of elastic thin shells. Despite its conciseness it includes enough basic theory to make the book self-contained. The formulation is unconventional in regard to its use of a vector description of the shell geometry and of strain, including nonlinear compatibility equations.

Chapter 2 starts with formulation of the main problems and the hypothesis of flexible shells (§1). This is followed by the analysis of the Saint-Venant problems, resulting in the nonlinear Reissner equations (§2), the linear Schwerin–Chernina system of the axisymmetric shells under wind-type loading and of curved tubes with forces applied at the ends (§3). Next (§4), the general nonlinear shell equations of equilibrium and compatibility are integrated and simplified with the aid of the hypothesis of flexible shells. This leads to a solution system extending the Reissner and Schwerin–Chernina equations to nonsymmetric problems. The analysis also shows the flexible-shell hypothesis to be tantamount to the semi-momentless model. This model gives (in §5) another, Vlasov-type resolving system. The rest of Chapter 2 is concerned with the solution methods of the basic problems.

Chapters 3–5 are devoted to applications of the theory.

More space is allotted (Chapter 3) to the problems of tubes and torus shells, which in recent years (since 1973), have acquired prominent significance in energy technology. Besides the analysis of "long" tubes, the nonlinear flexure and buckling is considered for prescribed boundary conditions. Buckling under external pressure is discussed for tubes and for torus shells including those with ribs.

The nonlinear flexure of open-section beams is considered in Chapter 4.

Chapter 5 treats flexible shells of revolution including the cross-bending of seamless bellows and the contact effects in welded bellows due to overloading.

Three remarks on the manner of presentation are in order:

(1) The discussion is concerned mainly with isotropic homogeneous shells. The possibility of extending the results to orthotropic and layered shells and to include the effects of thermal expansion is indicated in Chapter 1, §7 and illustrated by examples in Chapter 3, §12 and Chapter 4, §4.

(2) The book contains many graphs and some tables as aids to the

evaluation of stresses and displacements. However, the main aim of the author is to present a system of methods, rather than a list of recipes.

(3) In Physics, "the more complicated the system, the more simplified must necessarily be its theory" (Ya.I. Frenkel). Shells are complicated systems; flexible shells are among the more intricate of these. Their analysis is dominated, indeed shaped, by the stratagem of trimming all that projects beyond the limits of accuracy of the thin-shell theory.

In preparing the book the author has drawn on his book *Flexible Shells* published in 1976 in Russian.

The book reflects the progress achieved by efforts of many investigators. It grew out of the author's work on problems of interest to several industrial organizations and from lecture courses given in Leningrad, Darmstadt and Munich.

The help of friends and associates in programming and computation is gratefully acknowledged in relevant places of the book.

The author's sincere thanks for valuable discussion and suggestions are due to Professor F. Emmerling. Appreciation for many improvements is expressed to Professor W. Stadler who first read the text and to Dr. W. Hübner.

The author is grateful to Professor W.T. Koiter for his support.

TABLE OF CONTENTS

CHAPTER 1

FOUNDATIONS OF THIN-SHELL THEORY

CHAPTER 1

FOUNDATIONS OF THIN-SHELL THEORY

A closed shell is a body bounded by two surfaces whose overall dimensions are much greater than distance between the surfaces—the thickness of the shell wall. If a shell is not closed (as a ball is) there is yet another bounding surface—the edge surface.

Shell theory is a branch of Mechanics of Deformable Bodies. It is a practice-oriented, engineering theory. Providing a two-dimensional representation to three-dimensional problems, the shell theory makes the analysis immensely more tractable. The reduction is achieved in a way similar to that of the beam theory. The theory deals with variables defined only on the reference surface (lying mostly in the middle of the wall-thickness). The accuracy of the shell theory depends on the shell being sufficiently thin and also on the mechanical properties of the shell material and on the load distribution. Only elastic shells conforming to Hooke's law are considered in the following.

There are several approaches to the formulation of the shell theory. In this book the theory is derived on the basis of hypotheses. This axiomatic method goes back to the ground-breaking work of G. Kirchhoff (on plates) and to the first work on shells by ARON [1]. It is well suited for an engineering theory. Its relative simplicity and explicit form are particularly valuable for the nonlinear case. However, the evaluation of the bounds of applicability of the theory and of its errors is not possible within the theory's framework set by the hypotheses. The verification is achieved when the shell theory is evolved from a three-dimensional theory. After the classic works of Cauchy and Poisson (on plates), this approach was developed in the investigations of A.I. Lur'e, E. Reissner, A.L. Gol'denveiser, F. John and others. This is the main way of verifying an axiomatically evolved shell theory.

A third approach to shell theory consists of the consideration of a shell as a two-dimensional continuum supporting moment stresses. The idea of constructing a model of a shell independent of a three-dimensional theory and its simplifications is due to E. and F. COSSERAT in 1909. This approach is exhaustively presented in the work of NAGHDI [48].

A comprehensive analysis of the development of the foundations of shell theory may be found in the review of KOITER and SIMMONDS [155].

1. Shell geometry

The form of a shell is defined in a way similar to the description of the shape of a beam in terms of the shape of its axis. The shell form is determined by that of the reference surface and by the thickness of the shell, which may be variable along the shell. (For nonhomogeneous material of the shell it is expedient to choose the reference surface lying not in the middle of the shell thickness (present chapter, §6).)

Correspondingly, this section is concerned mainly with the theory of surfaces. All the information needed to obviate reference to the special literature is presented. No derivation is omitted or based on any fact beyond the scope of college mathematics.

1.1. Coordinates

A point of the shell (M in Fig. 1) is defined by its distance ζ from the reference surface and by the coordinates of the point m of this surface, lying on a ζ-axis normal to the surface and passing through M. The reference surface is described in terms of the curvilinear coordinates ξ and η. Each pair of values of ξ and η determines the position vector $r(\xi, \eta)$ of a single point m of the reference surface.

Points of the surface corresponding to a fixed value of ξ constitute a line—a coordinate line η or, in short η-line. Every point of the surface can be found as an intersection of a ξ-line and an η-line. Obviously, all points of the reference surface correspond to $\zeta = 0$. A point M is thus defined by the three coordinates ξ, η and ζ. When the reference surface coincides with the middle surface, the two surfaces bounding the shell wall have equations $\zeta = \pm h/2$. The shell-wall thickness may be variable, defined by $h = h(\xi, \eta)$.

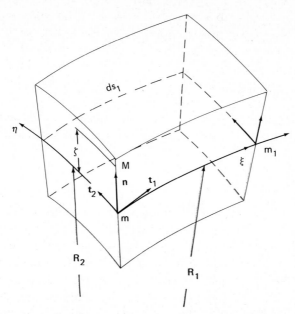

FIG. 1. Reference surface and curvilinear coordinates.

The vector equation of the middle surface $r = r(\xi, \eta)$ also determines the directions of the coordinate lines and the differentials of their arc lengths ds_1 and ds_2 at each point of the surface.

An element of the ξ-line connecting points $m(\xi, \eta)$ and $m_1(\xi + d\xi, \eta)$ shown in Fig. 2 is determined by the tangent vector $dr = (\partial r/\partial \xi)\, d\xi$. The introduction of the unit vectors t_1 and t_2 tangent to the coordinate lines yields

$$\frac{\partial r}{\partial \xi} = at_1 , \quad a = \left| \frac{\partial r}{\partial \xi} \right| , \qquad \frac{\partial r}{\partial \eta} = bt_2 , \quad b = \left| \frac{\partial r}{\partial \eta} \right| . \tag{1.1}$$

Thus, the lengths of the elements of coordinate lines are determined by

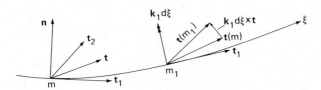

FIG. 2. Element of ξ-line ($\eta =$ const.) and tangent vectors t_i.

the parameters a and b,

$$ds_1 = a\, d\xi , \qquad ds_2 = b\, d\eta . \tag{1.2}$$

Usually, and also in what follows, the surface coordinates are chosen to be orthogonal at every point. For orthogonal coordinates the unit vectors t_1 and t_2 are connected with each other and with the unit vector $t_3 = n$, normal to the middle surface, by the formulas

$$t_1 \times t_2 = n , \qquad t_2 \times n = t_1 , \qquad n \times t_1 = t_2 ,$$

$$t_i \cdot t_j = \begin{cases} 1, & i = j \\ 0, & i \neq j . \end{cases} \tag{1.3}$$

The undeformed shape of a shell is a particular case of its deformed (under the applied load) shape. In what follows, the values of parameters decribing the geometry of a shell *after* deformation will be denoted by an asterisk. The shear deformation of the reference surface makes the ξ- and η-lines (their unit tangent vectors t_1^* and t_2^*) nonorthogonal to each other even when they are orthogonal before the deformation. This makes in the following some relations of the nonorthogonal coordinates unavoidable.

Relations (1.1) and (1.2) remain valid for the r^*, t_i^*, a^*, b^* and ds_i^*, describing the *deformed* shape of the shell. During an *arbitrary* deformation of the shell, the coordinates ξ and η of any point remain unchanged.

1.2. Curvature of surfaces

Consider shape of a surface in the vicinity of a point on it. The curvature of the surface can be measured in terms of rotation of a tangent plane moving along the surface. The position of the plane is determined by the vector n, which is normal to it. To measure also rotation of the plane around the normal n, we must relate the plane at a point of the surface $m(\xi, \eta)$ to some material linear element of it. We choose for this an element of the surface $(t\, ds)$ bisecting the angle between the coordinate lines as shown in Fig. 2. Thus, the position of the tangent plane is identified in terms of the two unit vectors

$$n = t_1 \times t_2 , \qquad t = (t_1 + t_2)/|t_1 + t_2| . \tag{1.4}$$

(Clearly, the shear deformation makes an angle between two linear elements different from the value it had before deformation. Thus, the angular displacements of different linear elements at a point of the surface are not equal to each other. It may be verified that the angular displacement of the vectors n and t into a position n^* and $t^* = (t_1^* + t_2^*)/|t_1^* + t_2^*|$ is equal to the rotation at a point of a deformable body, as defined in theory of elasticity.)

We introduce two vector parameters $k_1(\xi, \eta)$ and $k_2(\xi, \eta)$ of surface curvature, defining $k_1 \, d\xi + k_2 \, d\eta$ as an angle $(d\Phi)$ between the tangent planes at points $m(\xi, \eta)$ and $m'(\xi + d\xi, \eta + d\eta)$ of the surface. The vector k_1 (or k_2) is a measure of the curvature of the surface, observed in moving along the ξ-line (or η-line).

The geometric meaning of the curvature vectors k_1 and k_2 becomes clearer with their component representation

$$\frac{k_1}{a} = \frac{n \times t_1}{R_1} + \frac{n \times t_2}{R_{12}} + \frac{n}{\rho_1}, \qquad \frac{k_2}{b} = \frac{n \times t_1}{R_{21}} + \frac{n \times t_2}{R_2} + \frac{n}{\rho_2}. \qquad (1.5)$$

According to the definitions of k_1 and a the value of $1/R_1$ is equal to that angle of rotation of the normal vector n in the plane of n and t_1 (Fig. 1), which corresponds to a unit distance along the ξ-line. This means: R_1 is the radius of curvature of the intersection line of the surface with a plane passing through n and tangent to the ξ-line. In other words, $1/R_1$ is the curvature of the normal section of the surface along the coordinate line ξ. The variable $1/R_2$ has a similar meaning for a normal section along the η-line.

The meaning of the variables R_{12} and R_{21} is illustrated in Fig. 3. For this example, the normal section curvatures $1/R_i$ are equal to zero and the coordinate lines ξ and η are locally straight. The coordinates are also orthogonal making rotation of n with t identical to that of n with t_2 and t_1. According to the definition of k_1 and Fig. 3, in this case $k_1 \, d\xi = -t_1 \, ds_1/R_{12}$. With the shift $ds_2 = b \, d\eta$ along the other coordinate line the triad n, t_1 and t_2 turns by an angle $k_2 \, d\eta = t_2 \, ds_2/R_{21}$. Determining the distance denoted in Fig. 3 by δ through each of the two angles results in the equality $R_{12} = R_{21}$ (proved in §1.3 for the general case of local shape of a surface). The variable $1/R_{12} = 1/R_{21}$ is called the *twist* of the surface.

The remaining curvature parameters ρ_i determine the radii of the

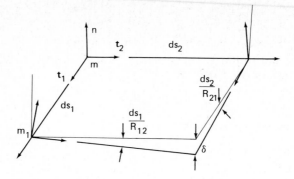

FIG. 3. Parameters $1/R_{12}$ and $1/R_{21}$ giving the twist of the surface.

in-plane or *geodetic curvatures* of the coordinate lines ξ and η. Indeed, $k_1 \, ds_1$ is an angle of rotation for the vectors n and t in turning through the distance $ds_1 = a \, d\xi$. The component $n \, ds_1/\rho_1$ indicates the rotation around the normal (in the tangent plane). When the coordinates ξ and η are orthogonal, the tangents of the coordinate lines turn by the same angles as the pair n and t, in particular by $n \, ds_1/\rho_1$. The curvature of the projection of the ξ-line on the tangent plane is thus equal to the quantity $1/\rho_1 = |n \, ds_1/\rho_1|/ds_1$.

The curvature vectors k_1 and k_2 render formulas for derivatives with respect to the coordinates ξ and η for the basis vectors.

Consider first the derivatives for an auxiliary unit vector $v(\xi, \eta)$ directed at any point of the surface at some fixed angles to n and t. The definition of k_1 means in fact that $k_1 \, d\xi$ determines the angle between the vectors v in two adjacent points $m(\xi, \eta)$ and $m_1(\xi + d\xi, \eta)$. This amounts to the relation

$$v(\xi + d\xi, \eta) = v(\xi, \eta) + k_1 \, d\xi \times v(\xi, \eta) \tag{1.6}$$

or to the formula

$$v_{,1} = \frac{\partial}{\partial \xi} v = k_1 \times v . \tag{1.7}$$

Naturally, there exists a similar formula for the other derivative: $v_{,2} = \partial v/\partial \eta = k_2 \times v$. The formulas for the derivatives remain valid for the particular cases of $v = t_i$ and $v = n$. Thus,

$$v_{,j} = k_j \times v \,, \qquad t_{i,j} = k_j \times t_i \,, \qquad n_{,j} = k_j \times n \quad (i, j = 1, 2)\,. \quad (1.8)$$

(If the angles between the vectors t_1 and t_2 vary with the coordinates ξ and η, the derivation formulas for t_1 and t_2 are, naturally, different from that for t or v [197].)

1.3. Surface-geometry relations

Four vectors describing the local shape of the surface have been introduced thus far: k_1, k_2, $r_{,1} = at_1$ and $r_{,2} = bt_2$. The four variables are related by two vector equations.

The first of the equations is obvious. Any vector function $r(\xi, \eta)$ defining a continuous smooth surface satisfies the relation

$$r_{,12} = r_{,21} \,.$$

The corresponding three scalar equations are particularly simple for the orthogonal coordinates ξ and η; they can be written with the aid of (1.1), (1.8), (1.5) and (1.3) in the form

$$\frac{a}{\rho_1} = -\frac{a_{,2}}{b} \,, \qquad \frac{b}{\rho_2} = \frac{b_{,1}}{a} \,, \qquad \frac{1}{R_{12}} = \frac{1}{R_{21}} \,. \qquad (1.9)$$

The second of the mentioned vector relations is rendered by the condition

$$v_{,12} = v_{,21} \,,$$

which must be fulfilled for the vector $v(\xi, \eta)$ having at any point of a smooth surface a fixed position in relation to the tangent plane at the point.

Applying the differentiation formulas (1.8) to the equation $v_{,12} - v_{,21} = 0$ and using the triple-product relation $A \times (B \times C) + B \times (C \times A) + C \times (A \times B) = 0$ transforms the left-hand side of the equation to a product of a differential expression and the vector v. Since v is arbitrary, the cofactor must be equal to zero, resulting in

$$k_{1,2} - k_{2,1} + k_1 \times k_2 = 0 \,. \qquad (1.10)$$

This equation is equivalent to three scalar ones which for orthogonal coordinates ξ and η are

$$\left(\frac{b}{R_{21}}\right)_{,1} - \left(\frac{a}{R_1}\right)_{,2} = \frac{ab}{R_2\rho_1} - \frac{ab}{R_{12}\rho_2},$$

$$\left(\frac{b}{R_2}\right)_{,1} - \left(\frac{a}{R_{12}}\right)_{,2} = \frac{ab}{R_1\rho_2} - \frac{ab}{R_{21}\rho_1};$$

(1.11)

$$\left(\frac{b}{\rho_2}\right)_{,1} - \left(\frac{a}{\rho_1}\right)_{,2} = \frac{ab}{R_{12}R_{21}} - \frac{ab}{R_1R_2}.$$

(1.12)

The relations (1.11) bear the name of Codazzi, while (1.12) is that of Gauss.

It is proved in the next section that for any surface there exist orthogonal coordinates ξ and η, such that at any point of the surface $1/R_{12} = 1/R_{21} = 0$. These coordinate lines coincide with the *lines of curvature* of the surface. That is, one of the normal curvatures $1/R_1$ and $1/R_2$ is the maximum; the other, the minimum of curvature of normal sections at a point of the surface. A simple procedure for determining two (orthogonal) directions for the lines of curvature is given in present chapter, §5. However, in all the cases discussed in this book, the lines of curvature are easily determined by inspection.

A feature of the curvature line, which is useful in identifying it, follows directly from the expansions of k_i (1.5): along a line of curvature $(1/R_{12} = 0)$ the normals n turn only in the plane *tangent* to

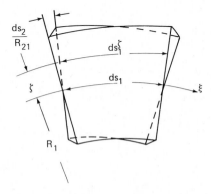

FIG. 4. Twisted shell element $(1/R_{12} \neq 0)$.

the line. An example of the opposite situation, in which an element of a shell is twisted, viz., $1/R_{12} \neq 0$, is presented in Fig. 4.

1.4. Determination of curvature. Surfaces of revolution

Consider two ways of determining the values of R_i, R_{12} and ρ_i. First, having (for the chosen orthogonal coordinates) expressions of $t_i(\xi, \eta)$, their derivatives may be found by direct differentiation. The subsequent use of (1.8), (1.5) and (1.3) yields

$$\frac{a}{R_1} = -t_{1,1} \cdot n, \qquad \frac{b}{R_2} = -t_{2,2} \cdot n,$$

$$\frac{1}{R_{12}} = \frac{1}{R_{21}} = -t_{2,1} \cdot n \frac{1}{a}. \tag{1.13}$$

Alternatively, the right-hand sides of the formulas may be $n_{,1} \cdot t_1$, $n_{,2} \cdot t_2$ and $n_{,1} \cdot t_2/a$.

Another possibility, existing in many cases, is the direct determination of the vector curvature parameters k_1 and k_2 from their geometric definitions, by inspection.

Let us apply the two methods to a *surface of revolution*—a surface developed by a plane η-curve (*meridian* η-line in Fig. 5) rotating around an axis (denoted by z in Fig. 5). In this rotation, each point of the η-curve constitutes a *parallel circle*, a ξ-line. Take $b = $ const., making $b\eta$ equal to the length of the meridian measured from a chosen parallel $\eta = 0$. For the coordinate ξ, take the polar angle to the

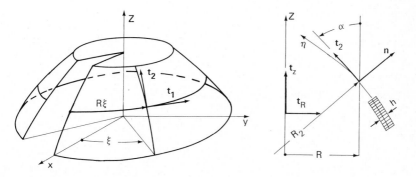

FIG. 5. Surface of revolution developed by a meridian ($\xi = $ const.).

meridian plane of a point considered. This means $s_1 = R\xi$, $a = R(\eta)$. The cylindrical coordinates R and z will be used besides ξ and η.

An inspection of Fig. 5 gives expressions for t_1, t_2 and n in terms of the unit vectors i, j and t_z of the Cartesian system x, y and z, which are constant, independent of ξ and η,

$$t_1 = j \cos \xi - i \sin \xi , \qquad t_2 = t_z \cos \alpha - t_R \sin \alpha ,$$

$$n = t_z \sin \alpha + t_R \cos \alpha , \qquad t_R = i \cos \xi + j \sin \xi ; \qquad (1.14)$$

$$b \cos \alpha = dz/d\eta , \qquad b \sin \alpha = -dR/d\eta .$$

The unit vector t_R and the angle $\alpha(\eta)$ are shown in Fig. 5.

Substitution of the expressions for t_1, t_2 and n in terms of ξ and η and of i, j and t_z into (1.13) gives

$$\frac{1}{R_1} = \frac{\cos \alpha}{R} , \qquad \frac{1}{R_2} = \frac{d\alpha}{b \, d\eta} , \qquad \frac{1}{R_{12}} = \frac{1}{R_{21}} = 0 . \qquad (1.15)$$

With $a = R$, $b = $ const., formulas (1.9) give

$$\frac{R}{\rho_1} = -\frac{dR}{b \, d\eta} = \sin \alpha , \qquad \frac{1}{\rho_2} = 0 . \qquad (1.16)$$

Clearly, $1/R_1$ and $1/\rho_1$ are components of the curvature $1/R$ of the ξ-line, which is, of course, a parallel circle (Fig. 5).

The absence of twist and of in-plane curvature of the η-line ($1/R_{12} = 0$, $1/\rho_2 = 0$) is an obvious consequence of this line lying in the symmetry plane of the local form of the surface.

The values of the curvature parameters in (1.15) and (1.16) follow more directly from values of k_1 and k_2, determined by inspection of the angles $k_1 \, d\xi$ and $k_2 \, d\eta$ between triads t_1, t_2 and n at points located on the coordinate lines at distances $R \, d\xi$ and $b \, d\eta$, respectively. Thus, Fig. 5 and definitions of k_1, k_2 and t_z result in

$$\left(-\frac{t_1}{R_{12}} + \frac{t_2}{R_1} + \frac{n}{\rho_1} \right) d\xi = t_z \frac{d\xi}{R} , \qquad \left(-\frac{t_1}{R_2} + \frac{t_2}{R_{21}} + \frac{n}{\rho_2} \right) d\eta = -t_1 \frac{d\alpha}{b} .$$

Equating the three components of both sides of these equations gives the same six formulas for R_i, R_{12}, R_{21} and ρ_i as in (1.15) and (1.16).

1.5. Local-geometry parameters for different coordinate systems

A full picture of the shape of a surface around a point can be provided by a survey of the normal curvature and the twist for *all* possible directions of the coordinate lines. Fortunately, this picture can be deduced from the values of R_1, R_2, R_{12}, R_{21}, a and b known for *one* coordinate system ξ, η. The values of the parameters $(R_1^\alpha, R_2^\alpha, \ldots)$ for any coordinate system ξ^α, η^α turned in relation to ξ, η through an angle α can easily be calculated. The corresponding formulas follow directly from relations

$$k_1^\alpha \, d\xi^\alpha = k_1 \, d\xi + k_2 \, d\eta \,, \qquad dr = t_1^\alpha a^\alpha \, d\xi^\alpha = t_1 a \, d\xi + t_2 b \, d\eta \,. \quad (1.17)$$

The variables k_1^α, t_1^α and a^α, introduced here, are the values of the k_1, t_1 and a for the coordinates ξ^α and η^α, shown in Fig. 6. The second of equations (1.17) is obvious. The first becomes so if we recall that $k_1^\alpha \, d\xi^\alpha$ is the angle between tangent planes at the points $m(\xi, \eta)$ and $m_2(\xi + d\xi, \eta + d\eta)$. The angle has the same value when the tangent plane goes from point m to m_2 not along the line ξ^α but along the component segments of the lines ξ and η (Fig. 6).

For orthogonal coordinates, there follows from (1.17),

$$\frac{k_1^\alpha}{a^\alpha} = \frac{k_1}{a}c + \frac{k_2}{b}s \,, \qquad c \equiv \cos \alpha = \frac{a \, d\xi}{a^\alpha \, d\xi^\alpha}, \qquad s \equiv \sin \alpha = \frac{b \, d\eta}{a^\alpha \, d\xi^\alpha}.$$

$$(1.18)$$

With the obvious from Fig. 6 representation

$$t_1 = t_1^\alpha c - t_2^\alpha s \,, \qquad t_2 = t_2^\alpha c + t_1^\alpha s$$

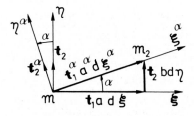

FIG. 6. Coordinates ξ, η and ξ^α, η^α.

and an expansion of k_1^α similar to that for k_1 in (1.5)

$$\frac{k_1^\alpha}{a^\alpha} = \frac{-t_1^\alpha}{R_{12}^\alpha} + \frac{t_2^\alpha}{R_1^\alpha} + \frac{n}{\rho_1^\alpha} \quad (-t_1^\alpha = n \times t_2^\alpha, t_2^\alpha = n \times t_1^\alpha),$$

equating of components on both sides of relation (1.18) results in

$$\frac{1}{R_1^\alpha} = \frac{c^2}{R_1} + \frac{cs}{R_{12}} + \frac{cs}{R_{21}} + \frac{s^2}{R_2}, \qquad \frac{1}{R_{12}^\alpha} = \frac{c^2}{R_{12}} + \frac{cs}{R_2} - \frac{cs}{R_1} - \frac{s^2}{R_{21}}.$$

$$(1.19)$$

These formulas give normal-section curvature and twist for any system of orthogonal coordinates. The extremum condition

$$\frac{\partial}{\partial \alpha} \frac{1}{R_1^\alpha} = 0 \qquad (1.20)$$

with $1/R_1^\alpha$ from (1.19) yields the angles α_1 and $\alpha_2 = \alpha_1 + \pi/2$ from the original ξ-direction to the planes of maximum and minimum normal-section curvature. The extremum condition happens to be identical to the equation $1/R_{12}^\alpha = 0$.

Thus, for any given surface there exist two mutually orthogonal directions of extremum normal-section curvature at any point of the surface. The lines passing through each point of the surface in these directions constitute a net of *lines of curvature*. When these lines are used as coordinate lines, the twist is equal to zero ($1/R_{12} = 1/R_{21} = 0$).

From the equation $\partial(1/R_1^\alpha)/\partial\alpha = 0$ or from $1/R_{12}^\alpha = 0$, we obtain

$$\tan 2\alpha_1 = \tan 2\alpha_2 = \frac{2/R_{12}}{1/R_1 - 1/R_2}, \qquad (1.21)$$

as the formula for the angles between a ξ-line and the lines of curvature at a point.

Of course, (1.19) indicate that $1/R_i$ and $1/R_{ij}$ constitute a two-dimensional second-order tensor. The same follows from the relation (1.17) for a^2, b^2 and $abt_1 \cdot t_2$.

A graphical representation of the curvatures in all possible normal sections at a point of a surface, illustrating local shape of a surface, follows directly from (1.19). It is a curve in an xy-plane with $[x \; y] =$

$\sqrt{R_1^\alpha}[c\ s]$, defined by an equation following from (1.19)

$$\frac{x^2}{R_1} + \frac{2xy}{R_{12}} + \frac{y^2}{R_2} = 1 .$$

This curve is called Dupin's indicatrix. The form of the indicatrix does not depend on the individual values of R_i and R_{12}, but on

$$\frac{1}{R_1} \cdot \frac{1}{R_2} - \frac{1}{R_{12}} \cdot \frac{1}{R_{21}} = K , \qquad \frac{1}{2}\left(\frac{1}{R_1} + \frac{1}{R_2}\right) .$$

Consequently, these two parameters (called the Gaussian and the mean curvature) are independent of the choice of the coordinate lines ξ and η. The invariance is, of course, deducible from (1.19) and the similar formulas for R_2^α and R_{21}^α.

(For $K > 0$, $K < 0$ or $K = 0$, the indicatrix is an ellipse, a hyperbola or two parallel lines. Examples of corresponding shapes are an ellipsoid, a hyperboloid and a cylinder surface.)

2. Deformation of the reference surface

2.1. Deformation parameters [1]

The local shape of the surface *after deformation* is described by new values k_1^*, k_2^*, $r_{,1}^* = a^* t_1^*$ and $r_{,2}^* = b^* t_2^*$ of its four vector parameters, its overall shape—by the new vector function r^* changed by the displacement u according to

$$r^*(\xi, \eta) = r(\xi, \eta) + u(\xi, \eta) . \qquad (1.22)$$

Each pair of coordinate values ξ, η designates some "material particle" of the surface. These (Lagrangian) coordinates of a particle remain unchanged by any deformation of the surface.

In describing the local deformation of the surface it is natural to use characteristics (strain resultants of the shell), directly reflecting the change of the four geometry parameters from k_i, $r_{,i}$ to k_i^*, $r_{,i}^*$. Introduce strain resultants κ_i and ε_i in the simplest way compatible with the requirement that the strain resultants be equal to zero at any

---

[1] Equivalence of the following vector description of strain and the classical tensor description is discussed on pp. 67–68.

point where the shell is not deformed,

$$k_1^* = k_{1R} + a\kappa_1 , \qquad k_2^* = k_{2R} + b\kappa_2 ,$$
$$r_{,1}^* = (r_{,1})_R + a\varepsilon_1 , \qquad r_{,2}^* = (r_{,2})_R + b\varepsilon_2 . \tag{1.23}$$

The values of k_{iR} and $(r_{,i})_R$ must be equal to k_i^* and $r_{,i}^*$ at any point of the surface where there is no local deformation. This means that the variables k_{iR} and $(r_{,i})_R$ describe the *initial*, undeformed local shape of the surface. They merely take into account the local *rotation* of the surface, i.e. its angular displacement, caused by a deformation of the surface as a whole. Components of the vectors k_{iR} and $(r_{,i})_R$ can be made equal to the components of the parameters k_i and $r_{,i}$ of the *initial* shape of the surface, provided a special basis is used for the k_{iR} and $(r_{,i})_R$. This basis is composed of unit vectors t_1', t_2' and n^* coinciding with the t_1, t_2 and n initially but moving with the tangent plane during any deformation of the surface[2]. This *rotated basis* remains in the same position relative to the pair n^* and t^* as was the initial basis with respect to n and t. As illustrated in Fig. 7, the angles between the tangent vectors of the coordinate lines (t_i^*) and the t_i' are equal to a half of the shear angle γ. The t_i' are easily expressed in terms of the t_i^*. In the case of small strain ($|\gamma| \ll 1$), they become

$$t_1' = t_1^* - \frac{\gamma}{2} t_2^* , \qquad t_2' = t_2^* - \frac{\gamma}{2} t_1^* . \tag{1.24}$$

Besides being the unique basis for the expansions of the k_{iR} and $(r_{,i})_R$, the vectors t_1', t_2' and n^* have the advantage of being independent of the

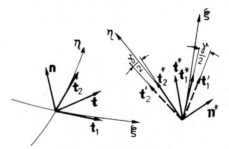

FIG. 7. Rotated basis t_i'.

[2] The use of the rotated basis is actually a counterpart to the polar decomposition (e.g., [133, p. 146] and N.A. ALUMAE, *Prikl. Mat. Mekh.* **13** (1949) 95–106).

shear deformation. In particular, they retain the orthogonality if the coordinates ξ and η are orthogonal before the deformation. The following expansions of the vectors introduced in (1.23) define the scalar parameters of the deformed–surface geometry and strain:

$$\left[\frac{k_1^*}{a} \quad \frac{k_{1R}}{a} \quad \kappa_1\right] = n^* \times t_1'\left[\frac{1}{R_1'} \quad \frac{1}{R_1} \quad \kappa_1\right] + n^* \times t_2'\left[\frac{1}{R_{12}'} \quad \frac{1}{R_{12}} \quad \tau_1\right]$$

$$+ n^*\left[\frac{1}{\rho_1'} \quad \frac{1}{\rho_1} \quad \lambda_1\right] \quad (n^* \times t_1' = t_2', \ldots) ; \quad (1.25)$$

$$(r_{,1})_R = at_1' , \qquad r_{,1}^* = a^*t_1^* , \qquad \varepsilon_1 = \varepsilon_1 t_1' + \varepsilon_{12} t_2' \quad (1\ 2\ a\ b) .$$

Expansions for the remaining parameters $k_2^*, \ldots$ are obtained from (1.25) when the indices 1, 2 and 12 and the parameter a are replaced by 2, 1, 21 and b, respectively. This is indicated by the symbol $(1\ 2\ a\ b)$.

With the definitions (1.25) and expressions of the t_i^* in terms of t_i' according to (1.24), one obtains the formulas for strain resultants

$$\kappa_1 = \frac{1}{R_1'} - \frac{1}{R_1} , \qquad \tau_1 = \frac{1}{R_{12}'} - \frac{1}{R_{12}} , \qquad \lambda_1 = \frac{1}{\rho_1'} - \frac{1}{\rho_1} \quad (1\ 2) ; \quad (1.26)$$

$$\varepsilon_1 = \frac{a^*}{a} - 1 , \qquad \varepsilon_2 = \frac{b^*}{b} - 1 , \qquad \varepsilon_{12} = \varepsilon_{21} = \frac{\gamma}{2} . \quad (1.27)$$

Similar formulas are easily written out for large strain.

The actual curvature and twist of the deformed surface $(1/R_i^*, 1/R_{12}^*)$ can be expressed with the aid of the foregoing formulas and of expansions

$$\frac{k_1^*}{a^*} = \frac{n^* \times t_1^*}{R_1^*} + \frac{n^* \times t_2^*}{R_{12}^*} + \frac{n^*}{\rho_1^*} \quad (1\ 2\ a\ b) \quad (1.28)$$

in terms of the strain resultants κ_i, τ_i and λ_i. But the expressions are not as simple as those for $1/R_i', \ldots$ from (1.26).

The formulas, derived directly from (1.28) and (1.26) for $\gamma = 0$,

$$[R_1^* \quad R_{12}^* \quad \rho_1^*] = (1 + \varepsilon_1)[R_1' \quad R_{12}' \quad \rho_1'] , \quad (1.29)$$

are used in Chapter 2, §2.

2.2. Compatibility equations

As long as the surface retains its continuity during deformation, the deformed-shape parameters k_i^* and $r_{,i}^*$ satisfy the Gauss–Codazzi equation (1.10) and $r_{,12}^* = r_{,21}^*$. This and the expressions (1.23) imply two vector compatibility equations for the four strain resultants κ_i and ε_i.

To derive the equations, we need the derivatives of k_{iR} and $(r_{,i})_R$ with respect to ξ and η. According to the definition of t_i' (as rotating together with t^*), the derivatives of n^* and t_i' are determined by (1.8), in which the k_i are now replaced with k_i^*,

$$[n^* \ t_1' \ t_2']_{,j} = k_j^* \times [n^* \ t_1' \ t_2'] . \qquad (1.30)$$

Consider the deformation of the surface at a point m. The use of the tangent plane at m as a reference plane for the displacements results in $t_i'(m) = t_i(m)$. Hence, the relations (1.23) become

$$k_1^*(m) = k_1(m) + a\kappa_1(m) , \qquad k_2^*(m) = k_2(m) + b\kappa_2(m) . \qquad (1.31)$$

With this, the derivative formulas (1.30) are transformed to

$$[n^* \ t_1' \ t_2']_{,1} = [n \ t_1 \ t_2]_{,1} + a\kappa_1 \times [n \ t_1 \ t_2] \quad (1 \ 2 \ a \ b) .$$

Now, differentiating each term in the expansions (1.25) as a product of a vector and a scalar yields

$$[k_{iR} \ (r_{,i})_R]_{,1} = [k_i \ r_{,i}]_{,1} + a\kappa_1 \times [k_i \ r_{,i}] \quad (1 \ 2 \ a \ b) . \qquad (1.32)$$

Insert the expressions (1.23) into equation (1.10), presented for the *deformed* shape of the surface as $k_{1,2}^* - k_{2,1}^* + k_1^* \times k_2^* = 0$, and into $r_{,12}^* = r_{,21}^*$. The use of the derivative formulas (1.32), equations (1.10), $r_{,12} = r_{,21}$ and finally (1.31) results in the two compatibility equations [197]

$$(a\kappa_1)_{,2} - (b\kappa_2)_{,1} + ab\kappa_1 \times \kappa_2 = 0 ,$$
$$(a\varepsilon_1)_{,2} - (b\varepsilon_2)_{,1} - ab(t_1' \times \kappa_2 - t_2' \times \kappa_1) = 0 . \qquad (1.33)$$

The equations are written here in the general form. The tangent plane at a point being considered is not necessarily a plane of reference for the displacements and thus $t_i'(m)$ may not be coincident with t_i. Clearly, a choice of a reference plane for the displacements does not influence the relations (1.33) between the local strain parameters κ_i and ε_i.

Consider small strain and decomposition (1.25) of the strain resultants in terms of the orthogonal basis t_1', t_2' and n^*.

With the differentiation formulas (1.30) and relations

$$n^* \times t_1' = t_2', \qquad n^* \times t_2' = -t_1',$$

the six scalar equations equivalent to the vector equations (1.33) may be obtained in the form

$$\frac{(b\kappa_2)_{,1}}{ab} - \frac{(a\tau_1)_{,2}}{ab} + \frac{\lambda_1}{R_{21}} + \frac{\tau_2}{\rho_1'} - \frac{\lambda_2}{R_1'} - \frac{\kappa_1}{\rho_2} = 0,$$

$$\frac{(b\tau_2)_{,1}}{ab} - \frac{(a\kappa_1)_{,2}}{ab} + \frac{\lambda_2}{R_{12}} + \frac{\tau_1}{\rho_2'} - \frac{\lambda_1}{R_2'} - \frac{\kappa_2}{\rho_1} = 0,$$

$$\frac{(b\lambda_2)_{,1}}{ab} - \frac{(a\lambda_1)_{,2}}{ab} + \frac{\kappa_1}{R_2'} + \frac{\kappa_2}{R_1} - \frac{\tau_1}{R_{21}'} - \frac{\tau_2}{R_{12}} = 0,$$

$$\frac{(b\gamma)_{,1}}{2ab} - \frac{(a\varepsilon_1)_{,2}}{ab} + \frac{\gamma}{2\rho_2'} - \frac{\varepsilon_2}{\rho_1'} = \lambda_1, \tag{1.34}$$

$$\frac{(b\varepsilon_2)_{,1}}{ab} - \frac{(a\gamma)_{,2}}{2ab} + \frac{\gamma}{2\rho_1'} - \frac{\varepsilon_1}{\rho_2'} = \lambda_2,$$

$$\tau_1 + \frac{\gamma}{2R_1'} - \frac{\varepsilon_1}{R_{21}'} = \tau_2 + \frac{\gamma}{2R_2'} - \frac{\varepsilon_2}{R_{12}'} = \tau.$$

The last of these equations will be used to express τ_1 and τ_2 in terms of the new variable τ introduced there.

For coordinates with $1/R_{12} = 0$ and consistent with the approximation of a small-strain theory

$$\tau_1 = \tau - \frac{\gamma}{2R_1}, \qquad \tau_2 = \tau - \frac{\gamma}{2R_2}. \tag{1.35}$$

The strain resultants ε_1, ε_2, γ and, as will be shown in the present chapter, §3, the values of $\kappa_1 h$, $\kappa_2 h$ and τh are of the order of magnitude of the strain components. The values of $a\lambda_1$ and $b\lambda_2$ are, according to the fourth and fifth of equations (1.34), also of that order of magnitude. These quantities will be neglected relative to unity.

After elimination of τ_1 and τ_2 and the simplification of small nonlinear terms, the first five compatibility equations may be written as

$$\frac{(b\kappa_2)_{,1}}{ab} - \frac{(a^2\tau)_{,2}}{a^2b} + \frac{R_1}{a^2b}\left(\frac{a^2\gamma}{2R_1^2}\right)_{,2} - \frac{\kappa_1}{\rho_2} - \frac{\lambda_2}{R_1'} = 0,$$

$$\frac{(b^2\tau)_{,1}}{ab^2} - \frac{(a\kappa_1)_{,2}}{ab} - \frac{R_2}{ab^2}\left(\frac{b^2\gamma}{2R_2^2}\right)_{,1} - \frac{\kappa_2}{\rho_1} - \frac{\lambda_1}{R_2'} = 0,$$

$$\frac{(b\lambda_2)_{,1}}{ab} - \frac{(a\lambda_1)_{,2}}{ab} + \frac{\kappa_1}{R_2} + \frac{\kappa_2}{R_1} = \tau^2 - \kappa_1\kappa_2, \qquad (1.36)$$

$$\frac{(b^2\gamma)_{,1}}{2ab^2} - \frac{(a\varepsilon_1)_{,2}}{ab} - \frac{\varepsilon_2}{\rho_1'} = \lambda_1,$$

$$\frac{(b\varepsilon_2)_{,1}}{ab} - \frac{(a^2\gamma)_{,2}}{2a^2b} - \frac{\varepsilon_1}{\rho_2'} = \lambda_2.$$

The in-plane curvatures $1/\rho_i$ can be expressed in terms of a and b through (1.9). The expressions of $1/\rho_i'$ (as well as R_i' and R_{ij}') in terms of the strain parameters follow from the relations (1.26) and (1.36).

The variables λ_1 and λ_2 are easily eliminated from the first three compatibility equations by using the fourth and fifth (where for the *small* strain, $\varepsilon_2/\rho_1' = \varepsilon_2/\rho_1$ and $\varepsilon_1/\rho_2' = \varepsilon_1/\rho_2$).

Note that, in contrast to the vector equations (1.33), equations (1.34) and (1.36) are valid only for the *orthogonal* coordinates ξ and η and for *small strain*. However, the displacements and the change of shape are unrestricted.

2.3. Strain-displacement relations

In the foregoing, the local-form parameters are expressed in terms of the equation of the surface $r = r(\xi, \eta)$. The same is easily done with $r^* = r + u$ for the shape *after* arbitrary displacement u. Thus there is no basic difficulty in expressing all strain resultants $(\varepsilon_i, \gamma, \kappa_i, \tau, \lambda_i)$ in terms of the displacement u. But the corresponding formulas for *large* displacements will not be presented here[3]. They are too cumbersome. (This is one reason why the displacement u is often inconvenient as a basic dependent variable in the analysis of nonlinear problems.) However, the *small* displacement relations between u and the strain parameters are simple and can be effectively used.

Consider a *small* displacement u and an angle of rotation ϑ of the tangent plane, their components in terms of an orthogonal basis being given by

$$u = ut_1 + vt_2 + wn , \qquad \vartheta = -\vartheta_2 t_1 + \vartheta_1 t_2 + \omega n . \qquad (1.37)$$

The angles of rotation ϑ_1 and ϑ_2 are, by definition, positive when directed from n towards t_1 or t_2, respectively. (The rotation ϑ brings the triad n, t_i into the position n^*, t_i' described previously. This defines ϑ as comprising the angle of rotation of the linear elements of the surface oriented towards the main directions of the deformation. The definition coincides with that of angular displacement in the theory of elasticity.)

Consider the expression of the strain resultants in terms of u and ϑ. The difference in the values of the $u(\xi, \eta)$ at the ends of a linear element of the surface $t_1 a\, d\xi$ consists of the effects of its elongation and rotation. With ε_1 defined in (1.27) as the extension and with the angle $\vartheta + \gamma n/2$ of rotation of the element $t_1 a\, d\xi$ (Fig. 7), we have

$$u(\xi + d\xi, \eta) - u(\xi, \eta) = u_{,1}\, d\xi = \varepsilon_1 t_1 a\, d\xi + (\vartheta + \gamma n/2) \times t_1 a\, d\xi .$$

This results in the expression for $\varepsilon_1 t_1 + \frac{1}{2}\gamma t_2 \equiv \varepsilon_1$. With the expression for ε_2 derived in a similar manner, we obtain

$$\varepsilon_1 = \frac{1}{a} u_{,1} + t_1 \times \vartheta , \qquad \varepsilon_2 = \frac{1}{b} u_{,2} + t_2 \times \vartheta . \qquad (1.38)$$

To express κ_i in terms of ϑ we start from the definitions (1.23) $\kappa_1 a = k_1^* - k_{1R}$. For small angular displacements ϑ, the angle between the tangent planes at the points $m(\xi, \eta)$ and $m_1(\xi + d\xi, \eta)$ is changed by the deformation from $k_1\, d\xi$ to $k_1^*\, d\xi = k_{1R}\, d\xi + d\vartheta$. This results in the formulas

$$\kappa_1 = \frac{1}{a}\vartheta_{,1} , \qquad \kappa_2 = \frac{1}{b}\vartheta_{,2} . \qquad (1.39)$$

---

[3] The displacement formulation of the theory is treated in the well-known work [188, 224, 227, 231, 232].

The conditions of integrability for (1.38) and (1.39) namely, $u_{,12} = u_{,21}$ and $\vartheta_{,12} = \vartheta_{,21}$ amount to the linear approximation of the compatibility equations (1.33), which follows from (1.33) when the term $ab\kappa_1 \times \kappa_2$ is omitted and t'_i are replaced by t_i.

Equations (1.38) are equivalent to the following formulas determining the extension and shear of the surface and the three components of angular displacement as functions of the components of displacements (for $1/R_{12} = 0$)

$$\varepsilon_1 = \frac{u_{,1}}{a} - \frac{v}{\rho_1} + \frac{w}{R_1}, \qquad \varepsilon_2 = \frac{v_{,2}}{b} + \frac{u}{\rho_2} + \frac{w}{R_2},$$

$$\gamma = \frac{a}{b}\left(\frac{u}{a}\right)_{,2} + \frac{b}{a}\left(\frac{v}{b}\right)_{,1};$$

$$\vartheta_1 = -\frac{w_{,1}}{a} + \frac{u}{R_1}, \qquad \vartheta_2 = -\frac{w_{,2}}{b} + \frac{v}{R_2},$$

$$2\omega = \frac{(bv)_{,1}}{ab} - \frac{(au)_{,2}}{ab}.$$

(1.40)

Similarly, projecting (1.39) on t_i and n gives

$$\kappa_1 = \frac{\vartheta_{1,1}}{a} - \frac{\vartheta_2}{\rho_1}, \qquad \kappa_2 = \frac{\vartheta_{2,2}}{b} + \frac{\vartheta_1}{\rho_2},$$

$$2\tau = \frac{a}{b}\left(\frac{\vartheta_1}{a}\right)_{,2} + \frac{b}{a}\left(\frac{\vartheta_2}{b}\right)_{,1} + \frac{\gamma - 2\omega}{2R_1} + \frac{\gamma + 2\omega}{2R_2}.$$

(1.41)

3. Hypotheses of the theory. Shell deformation

Strain and stress at any point $M(\xi, \eta, \zeta)$ of the shell are determined in terms of the deformation of the reference surface at $M(\xi, \eta, 0)$. This is achieved through the basic hypotheses of the thin-shell theory. The hypotheses can be identified as a generalization of the assumptions made in the Strength of Materials for beams.

3.1. Kirchhoff hypotheses

A basis for the subsequent purely mathematical development of thin-shell theory is provided by two assumptions (one geometric, the other static) and a criterion for the simplification of the relations in the theory, following from these assumptions.

Hypothesis of straight normals: In the analysis of deformation, particles comprising a straight line normal to the reference surface may be considered as remaining on such a normal after the deformation and the change of distance between these particles may be disregarded.

This hypothesis can be expressed by the following formula defining a radius vector R^* of a point in the deformed shell as a sum of the radius vector r^* of the deformed reference surface and a vector n^* normal to it

$$R^*(\xi, \eta, \zeta) = r^*(\xi, \eta) + n^*(\xi, \eta)\zeta . \tag{1.42}$$

In the elasticity relations and *in the expressions for elastic energy*, the stress components acting on the sections parallel to the reference surface (σ_3, σ_{31}, σ_{32} and consequently σ_{13}, σ_{23}) can be neglected.

This *hypothesis of approximately plane stressed state* leads to the following simplifications of the well-known formulas of the theory of elasticity for the deformation energy per unit volume of a body and for Hooke's law, respectively,

$$\frac{1}{2}(\sigma_1 e_1 + \sigma_2 e_2 + \sigma_{12} e_{12} + \sigma_3 e_3 + \sigma_{13} e_{13} + \sigma_{23} e_{23})$$

$$\approx \frac{1}{2}(\sigma_1 e_1 + \sigma_2 e_2 + \sigma_{12} e_{12}) , \tag{1.43}$$

$$e_1 E_1 = \sigma_1 - \nu_{12}\sigma_2 - \nu_{13}\sigma_3 \approx \sigma_1 - \nu_{12}\sigma_2 ,$$

$$e_2 E_2 = \sigma_2 - \nu_{21}\sigma_1 - \nu_{23}\sigma_3 \approx \sigma_2 - \nu_{21}\sigma_1, \qquad e_{12} G = \sigma_{12} ,$$

where σ_m and σ_{mn} are the normal and tangential stress components acting on sections $\xi, \eta, \zeta = $ const., corresponding to $m, n = 1, 2, 3$; e_m and e_{mn} are the extension and shear-strain components in the same basis; E_m, G and ν_{mn} are moduli of elasticity and Poisson's coefficient.

The hypotheses are credited to G.N. Kirchhoff and A.E.H. Love.

It has been conclusively established that the error, introduced by these hypotheses, has an order of magnitude of at most[4] the largest of the values

---

[4] The error may be larger in a narrow zone near the shell edge (cf. [155, 178]).

$$\left|\frac{h}{R_1}\right|, \quad \left|\frac{h}{R_2}\right|, \quad \left|\frac{h}{R_{12}}\right|, \quad \frac{h^2}{L_1^2}, \quad \frac{h^2}{L_2^2}. \tag{1.44}$$

The error decreases with the wall thickness of the shell relative to the radii of curvature and twist and to the *intervals of variation* (L_i) of the stressed state. The values of L_i are determined in terms of the ratios between a function $F(\xi, \eta)$, characterizing the shell deformation, and its derivatives [5]

$$\left|\frac{\partial F}{a\,\partial\xi}\right| \sim \frac{|F|}{L_1}, \quad \left|\frac{\partial F}{b\,\partial\eta}\right| \sim \frac{|F|}{L_2}. \tag{1.45}$$

The meaning of the intervals of variation, defined in (1.45), may be illustrated by approximating $F(\xi, \eta)$ in a neighbourhood of a point $m(\xi_0, \eta_0)$ by a function $C\sin(a\xi/L_1)\sin(b\eta/L_2)$.

We shall also speak of the corresponding *wavelengths* of deformation $2\pi L_1$ and $2\pi L_2$.

The intensity of variation of the stressed state naturally depends on the distribution of load. Thus, thin-shell theory cannot accurately determine the deformation caused by *local* or otherwise too strongly varying loading. The formal, mathematical accuracy can be useful only within the bounds of the accuracy of the basic hypotheses.

This leads to the following *criterion of simplification*: in thin-shell-theory relations terms of the order of magnitude of the estimates (1.44) can be omitted. While this may always be done in the expressions for the energy of the system, care is needed in the simplification of relations, where the larger terms may cancel each other.

The foregoing formulation of the basic hypotheses is the simplest one. It is similar to the traditional formulation: "Straight lines normal to the undeformed middle surface remain straight and normal to the deformed middle surface and do not change length. The normal stress acting on surfaces parallel to the middle surface may be neglected in comparison with the other stresses." These assumptions are often criticized for being contradictory. They indeed set both the transverse stress σ_3 and the extension e_3 equal to zero. The inconsistency lies in the introduction of additional equations $\sigma_3 = 0$ and $e_3 = 0$. It does not appear when the assumptions are confined (as in the foregoing) to *neglecting* specific terms in specific relations. Alternatively, consistency is achieved by

---

[5] The sign $\sim$ indicates quantities having the same order of magnitude. Thus $A \sim B$ means $A = O(B)$.

neglecting the deviation of the stressed state of a shell from the plane one (KOITER and SIMMONDS [155]) or by assuming the transverse deformation to be negligible due to a certain anisotropy of the material (SEIDE [173]).

3.2. Strain components

Consider a shell referred to coordinates ξ and η, orthogonal in the reference surface $\zeta = 0$. Components of small strain are expressed in terms of the position vector $R(\xi, \eta, \zeta)$ by formulas (ds_1^ζ is shown in Fig. 4)

$$e_1 = \frac{(ds_1^\zeta)^* - ds_1^\zeta}{ds_1^\zeta}, \quad (ds_1^\zeta)^* = |R^*_{,1}| \, d\xi \, ,$$

$$e_{12} = \frac{R^*_{,1} \cdot R^*_{,2}}{|R^*_{,1}| \, |R^*_{,2}|} - \frac{R_{,1} \cdot R_{,2}}{|R_{,1}| \, |R_{,2}|}. \tag{1.46}$$

The expressions of $R^*_{,i}$ in terms of the parameters of the shape and deformation of the reference surface follow from the hypothesis of straight normals (1.42) and the differentiation formula (1.30) for n^*

$$R^*_{,1} = r^*_{,1} + a\zeta \left(\frac{t'_1}{R'_1} + \frac{t'_2}{R'_{12}} \right) \quad (1\ 2\ a\ b). \tag{1.47}$$

From (1.46), (1.47) and multiplication formulas for the orthogonal vectors, one has the relations (illustrated in Fig. 4)

$$ds_1^\zeta = a\left(1 + \frac{\zeta}{R_1}\right) d\xi \, , \quad ds_2^\zeta = b\left(1 + \frac{\zeta}{R_2}\right) d\eta \, . \tag{1.48}$$

Substitution of the expressions for $R_{,i}$, $R^*_{,i}$ and $\varepsilon_1, \ldots, \tau$ from (1.26), (1.27) and (1.34) into (1.46) yields the basic formulas for the strain components

$$e_1 = \frac{\varepsilon_1 + \zeta\kappa_1}{1 + \zeta/R_1}, \quad e_2 = \frac{\varepsilon_2 + \zeta\kappa_2}{1 + \zeta/R_2},$$

$$e_{12} = e_{21} = \frac{\gamma + 2\zeta\tau + \zeta^2\tau(1/R_1 + 1/R_2)}{(1 + \zeta/R_1)(1 + \zeta/R_2)}. \tag{1.49}$$

From these formulas, terms of the order of magnitude of the strain components e_i, e_{12} and of h^2/R_i^2, compared to unity, have been omitted.

The formulas (1.49) determine all of the strain components considered for a thin shell. The strain at any point of the shell volume is described by the six functions of the surface coordinates $\varepsilon_1(\xi, \eta), \ldots, \tau(\xi, \eta)$.

Omitting small terms of the order of magnitude of the estimates (1.44), the strain components may be determined by the formulas

$$e_1 = \varepsilon_1 + \zeta\kappa_1, \qquad e_2 = \varepsilon_2 + \zeta\kappa_2, \qquad e_{12} = \gamma + 2\zeta\tau. \qquad (1.50)$$

However, to avoid errors caused by terms cancelling each other, the full formulas (1.49) will be preferred at all intermediate stages of derivations.

3.3. Note on definitions of κ_1, κ_2, τ

As is clear from their definitions, the parameters κ_i and τ describe the change of the normal curvature and twist. But the values (1.26) of the parameters are *not* identical to actual changes of the curvatures and twist, i.e. with

$$\frac{1}{R_1^*} - \frac{1}{R_1} = \tilde{\kappa}_1, \qquad \frac{1}{R_2^*} - \frac{1}{R_2} = \tilde{\kappa}_2, \qquad \frac{1}{R_{12}^*} - \frac{1}{R_{12}} = \tilde{\tau}.$$

The difference is quantitatively small. Replacing κ_i, τ with $\tilde{\kappa}_i$, $\tilde{\tau}$ in the formulas for the strain components, (1.50) would introduce errors of the order of magnitude of, at most, h/R_i and h/R_{12}, which are insignificant in the thin-shell theory. But this does not justify the substitution of κ_i, τ for $\tilde{\kappa}_i$, $\tilde{\tau}$ in other relations of thin-shell theory (cf. [40, p. 26]). There are cases when the replacement of κ_i, τ with $\tilde{\kappa}_i$, $\tilde{\tau}$ in equations of equilibrium of an element of the shell leads to appearance of additional terms, which though of secondary importance, are difficult to estimate [106].

For the compatibility equations, the consequences of assuming $\kappa_i = \tilde{\kappa}_i$ and $\tau = \tilde{\tau}$ are even more serious, leading to significant errors. (This is shown in different ways by KOITER [101] and AXELRAD [106], cf. [40, p. 26].)

Formulas (1.50) supply estimates of the order of magnitude of the six strain resultants $\varepsilon_1, \ldots, \tau$. As the reference surface resides within the shell volume $|\zeta| < h$ (Fig. 4). Hence for $\kappa_i h$ and τh, as well as for ε_i and γ, formulas (1.50) indicate the order of magnitude of the strain components e_i and e_{12}. When, as in this book, Hooke's law is assumed, the strain components are negligible compared to unity. Consequently,

$$|\varepsilon_i| , |\gamma| , |\kappa_i h| , |\tau h| \ll 1 . \tag{1.51}$$

4. Equilibrium equations

Equilibrium of any part of the shell is, of course, independent of the hypotheses of the theory. But, the assumption of the undeformability of linear elements normal to the reference surface influences the composition of the set of equilibrium equations needed. It is sufficient to consider the equilibrium of elements of the shell, which contain the linear elements, assumed undeformable, throughout their full length h.

4.1. Vector equations of equilibrium

Consider an element of shell volume bounded by four surfaces of the type $\xi, \eta = \text{const.}$: $\xi = \xi_c$, $\xi = \xi_c + d\xi$; $\eta = \eta_c$, $\eta = \eta_c + d\eta$ (Figs. 4 and 8). The element extends through the entire thickness h of the shell.

The stresses acting on the element side $\xi = \text{const.}$ shall be represented in the equilibrium equations by their resultant force and moment, respectively designated $T_1 \, ds_2$ and $M_1 \, ds_2$. On the side $\eta = \text{const.}$, the

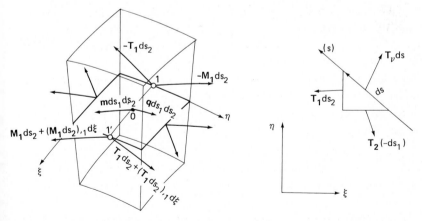

FIG. 8. Element of shell volume with stress forces and moments.

resultants are $T_2 \, ds_1$ and $M_2 \, ds_1$. This defines T_i and M_i as *resultants* related to the unit length of the section of shell, measured along a coordinate line in the undeformed reference surface.

Apart from the internal forces of the shell, the element may also be subjected to external surface and volume forces. The resultant force and moment of the external forces per unit area of the undeformed reference surface are designated by q and m.

Obviously, the forces $T_j \, ds_i$ and moments $M_j \, ds_i$ act on the opposite sides of the element in opposite directions, as shown in Fig. 8. As the element is infinitely thin, the forces are nearly the same, being action and reaction on two sides of a cross-section.

On the other hand the forces and moments $T_i \, ds_j$ and $M_i \, ds_j$ are functions of the coordinates ξ and η and their values on opposite sides of the element differ as indicated in Fig. 8. Sums of the forces and moments (with respect to a point 0 inside the element) acting on element sides $\xi = $ const. are

$$- T_1 \, ds_2 + T_1 \, ds_2 + (T_1 \, ds_2)_{,1} \, d\xi$$

$$- M_1 \, ds_2 + M_1 \, ds_2 + (M_1 \, ds_2)_{,1} \, d\xi - \overrightarrow{01} \times T_1 \, ds_2$$

$$+ \overrightarrow{01}' \times (T_1 \, ds_2 + (T_1 \, ds_2)_{,1} \, d\xi) \, .$$

As seen in the Fig. 8, $\overrightarrow{01}' - \overrightarrow{01} = t_1^* \, ds_1^* = t_1^* a^* \, d\xi$ and $\overrightarrow{01}'$ is of the same order of magnitude as $t_1^* \, ds_1^*$. Hence the term $\overrightarrow{01}' \times (T_1 \, ds_2)_{,1} \, d\xi$ is of the order of $d\xi \, d\eta \, d\xi$, compared to other terms of the order of $d\xi \, d\eta$ and must be dropped. The external loading is statically equivalent to a force $q \, ds_1 \, ds_2$ and a moment $m \, ds_1 \, ds_2$. (For oblique coordinates it would be $ds_1 ds_2 \, |t_1 \times t_2|$ instead of $ds_1 \, ds_2$.)

Taking account of forces and moments of forces acting on the sides $\xi, \eta = $ const., the force and moment equilibrium equations are

$$(T_1 \, ds_2)_{,1} \, d\xi + (T_2 \, ds_1)_{,2} \, d\eta + q \, ds_1 \, ds_2 = 0 \, ,$$

$$(M_1 \, ds_2)_{,1} \, d\xi + (M_2 \, ds_1)_{,2} \, d\eta + t_1^* \, ds_1^* \times T_1 \, ds_2 \qquad (1.52)$$

$$+ t_2^* \, ds_2^* \times T_2 \, ds_1 + m \, ds_1 \, ds_2 = 0 \, .$$

It remains to divide both sides of each equation by $d\xi\, d\eta$ (recalling from (1.2) that $ds_1 = a\, d\xi$ and $ds_2 = b\, d\eta$). The vector-form equilibrium equations for orthogonal coordinates are

$$(bT_1)_{,1} + (aT_2)_{,2} + qab = 0 ,$$

$$(bM_1)_{,1} + (aM_2)_{,2} + a^*bt_1^* \times T_1 + ab^*t_2^* \times T_2 + mab = 0 . \tag{1.53}$$

4.2. Scalar equations of equilibrium

The stress and load resultants must be represented by components relative to a chosen basis. The components in the directions t_1', t_2' and n^*—nearly tangent to the coordinate lines and normal to the deformed reference surface—are given by the formulas

$$\begin{bmatrix} T_1 \\ T_2 \\ q \end{bmatrix} = \begin{bmatrix} T_1 & S_1 & Q_1 \\ S_2 & T_2 & Q_2 \\ q_1 & q_2 & q \end{bmatrix}\begin{bmatrix} t_1' \\ t_2' \\ n^* \end{bmatrix} ,$$

$$\begin{bmatrix} M_1 \\ M_2 \\ m \end{bmatrix} = \begin{bmatrix} -H_1 & M_1 \\ -M_2 & H_2 \\ -m_2 & m_1 \end{bmatrix}\begin{bmatrix} t_1' \\ t_2' \end{bmatrix} . \tag{1.54}$$

The physical meaning of the components is illustrated in Fig. 9, where the components are shown in their positive directions.

The T_1 and T_2 are *tangential* (to the ξ- and η-lines) forces; S_1, S_2 and Q_1, Q_2 are *shear* and *lateral* forces, respectively; M_1, M_2 and H_1, H_2 denote *bending* and *torsional* moments, respectively.

Components of the stress-resultant moments M_i and of moment loading m in the direction n^* (i.e. moments acting in the tangential plane) are equal to zero. This may be inferred, directly from Fig. 8, since h is infinitely *large* compared to ds_i. Thus,

$$n^* \cdot M_1\, ds_2 \sim \sigma_1 h(ds_2)^2 , \quad n^* \cdot m\, ds_1\, ds_2 \sim p_i h\, ds_1\, ds_2\, ds_j \ (i, j = 1, 2) ,$$

while

$$M_1\, ds_2 \sim \sigma_1 h^2\, ds_2 , \qquad m_1\, ds_1\, ds_2 \sim p_i h^2\, ds_1\, ds_2 .$$

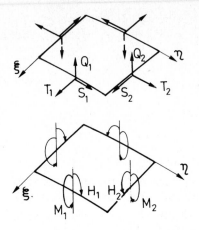

FIG. 9. Positive directions of stress resultants.

Here $t_i' p_i$ denotes external load per unit volume of the shell.

With the values of T_i and M_i known, it is easy to determine stress resultants T_ν and M_ν in *any* section of the shell. Considering the equilibrium of an element of the shell containing a triangular segment of the reference surface with dimensions $-ds_1$, ds_2 and ds (Fig. 8) yields

$$T_\nu\, ds = T_1\, ds_2 - T_2\, ds_1\,, \qquad M_\nu\, ds = M_1\, ds_2 - M_2\, ds_1\,. \qquad (1.55)$$

The determination of the components of M_ν and T_ν, defined in a manner similar to (1.54), is now straightforward.

Substituting the decompositions (1.54) into equations (1.53), differentiating with the help of (1.30) and equating the three components of both sides of each equation results in the six scalar equilibrium equations

$$\frac{(bT_1)_{,1}}{ab} + \frac{(aS_2)_{,2}}{ab} + \frac{Q_1}{R_1'} - \frac{S_1}{\rho_1'} - \frac{T_2}{\rho_2'} + \frac{Q_2}{R_{21}'} + q_1 = 0\,,$$

$$\frac{(bS_1)_{,1}}{ab} + \frac{(aT_2)_{,2}}{ab} + \frac{Q_2}{R_2'} + \frac{S_2}{\rho_2'} + \frac{T_1}{\rho_1'} + \frac{Q_1}{R_{12}'} + q_2 = 0\,, \qquad (1.56a)$$

$$\frac{(bQ_1)_{,1}}{ab} + \frac{(aQ_2)_{,2}}{ab} - \frac{T_1}{R_1'} - \frac{T_2}{R_2'} - \frac{S_1}{R_{12}'} - \frac{S_2}{R_{21}'} + q = 0\,,$$

$$\frac{(bH_1)_{,1}}{ab} + \frac{(aM_2)_{,2}}{ab} + \frac{M_1}{\rho_1'} + \frac{H_2}{\rho_2'} + m_2 = Q_2 \, ,$$

$$\frac{(bM_1)_{,1}}{ab} + \frac{(aH_2)_{,2}}{ab} - \frac{M_2}{\rho_2'} - \frac{H_1}{\rho_1'} + m_1 = Q_1 \, , \qquad (1.56b)$$

$$S_1 - \frac{H_2}{R_2'} - \frac{M_1}{R_{12}'} = S_2 - \frac{H_1}{R_1'} - \frac{M_2}{R_{21}'} = S \, .$$

Since only the small-strain components are to be considered, the nonlinear terms with factors ε_i and γ are omitted. But those of the terms stemming from the factors t_i^* in (1.53) are not as obviously negligible. When accounted for, they lead to replacement in the last three equations of (1.56) of Q_i and S_i by expressions

$$Q_2 + \frac{\gamma}{2} Q_1 \, , \qquad Q_1 + \frac{\gamma}{2} Q_2 \, , \qquad S_1 - \frac{\gamma}{2} T_1 \, , \qquad S_2 - \frac{\gamma}{2} T_2 \, .$$

The *four* stress resultants S_1, S_2, H_1 and H_2 represent the tangential-stress component $\sigma_{12} = \sigma_{21}$, whereas the normal component σ_1 or σ_2 is represented by only *two* resultants T_1, M_1 or T_2, M_2. The tangential-stress component $\sigma_{12} = \sigma_{21}$ can also be represented in terms of *two* parameters: the S, introduced in the sixth of equations (1.56), and

$$H = \tfrac{1}{2}(H_1 + H_2) \, . \qquad (1.57)$$

Thus, after substituting expressions for Q_1 and Q_2 following from the fourth and the fifth of equations (1.56), the first three of the equations contain (consistently with the theory's accuracy) only the six stress parameters

$$T_1 \, , \quad T_2 \, , \quad S \, , \quad M_1 \, , \quad M_2 \, , \quad H \, .$$

The first three equations of (1.56) retain their form except that S_1 and S_2 are each replaced by S and the two expressions $Q_i(H_1, H_2, M_1, M_2)$ are replaced by $Q_i(H, H, M_1, M_2)$.

In the linearized form, the sixth of equations (1.56) gives

$$S_1 = S + H_2/R_2 \, , \qquad S_2 = S + H_1/R_1 \, . \qquad (1.58)$$

For small displacements ($R_i' = R_i$, $\rho_i' = \rho_i$) and the coordinates ξ and η

chosen along the lines of curvature ($1/R_{12} = 0$), the mentioned three equations of equilibrium may be written in the form

$$\frac{(bT_1)_{,1}}{ab} + \frac{(a^2 S)_{,2}}{a^2 b} + \frac{R_1}{a^2 b}\left(\frac{a^2 H}{R_1^2}\right)_{,2} - \frac{b_{,1} T_2}{ab} + \frac{N_1}{R_1} + q_1 = 0,$$

$$\frac{(b^2 S)_{,1}}{ab^2} + \frac{(aT_2)_{,2}}{ab} + \frac{R_2}{ab^2}\left(\frac{b^2 H}{R_2^2}\right)_{,1} - \frac{a_{,2} T_1}{ab} + \frac{N_2}{R_2} + q_2 = 0,$$

$$\frac{(bN_1)_{,1}}{ab} + \frac{(aN_2)_{,2}}{ab} - \frac{T_1}{R_1} - \frac{T_2}{R_2} + q = 0; \qquad (1.59)$$

$$abN_1(M_1, M_2, H) = (bM_1)_{,1} - b_{,1} M_2 + (a^2 H)_{,2}/a,$$

$$abN_2(M_2, M_1, H) = (aM_2)_{,2} - a_{,2} M_1 + (b^2 H)_{,1}/b.$$

The preceding analysis of equilibrium and of deformation of the surface is exact. It yields twelve equations (1.56) and (1.34) in nineteen unknowns (T_i, S_i, Q_i, M_i, H_i, κ_i, τ_i, λ_i, ε_i, γ). Alternatively, after eliminating Q_i and λ_i, and expressing S_i, H_i and τ_i in terms of S, H, τ and γ, we have three equations of equilibrium and three equations of compatibility in twelve unknowns. The theory has to be complemented by equations connecting the strain measures to the stress parameters. This is impossible without considering the shell as a three-dimensional material body. It is in describing the three-dimensional problem in terms of the two-dimensional theory that approximations are inevitable. For the strain, the approximation was introduced in present chapter, §3 with the aid of the hypothesis of straight normals; whereas for the stress, it will be done in §§5, 6 on the basis of the hypothesis of an approximately plane stress state.

5. Elastic energy

The elastic-energy functional is determined on the basis of both of the hypotheses of the thin-shell theory.

5.1. Hooke's law

For a certain type of orthotropy and with the thin-shell hypotheses, the relations expressing Hooke's law have the simplified form (1.43). Solved with respect to the stress components these relations are

$$\sigma_1 = b_1 e_1 + b_\nu e_2 , \qquad \sigma_2 = b_2 e_2 + b_\nu e_1 , \qquad \sigma_{12} = G e_{12} . \qquad (1.60)$$

For isotropic material, the relations take on their simplest form

$$\sigma_1 = \frac{E}{1 - \nu^2}(e_1 + \nu e_2) , \quad \sigma_2 = \frac{E}{1 - \nu^2}(e_2 + \nu e_1) , \quad \sigma_{12} = G e_{12} ; \qquad (1.61)$$

i.e.

$$b_1 = b_2 = \frac{E}{1 - \nu^2} , \qquad b_\nu = \nu b_i .$$

5.2. Elastic energy of isotropic homogeneous shells

The elastic energy of a thin shell is determined by integrating the energy density (1.43), corresponding to the Kirchhoff hypothesis, over the volume (V) of the shell

$$U^V = \tfrac{1}{2} \int_V (\sigma_1 e_1 + \sigma_2 e_2 + \sigma_{12} e_{12}) \, dV . \qquad (1.62)$$

In orthogonal coordinates, the differential of the volume dV is equal to product of differentials of the lengths of the three coordinates. In view of (1.48),

$$dV = ds_1^\zeta \, ds_2^\zeta \, d\zeta = (1 + \zeta/R_1)(1 + \zeta/R_2)ab \, d\xi \, d\eta \, d\zeta .$$

Expression (1.62) may now be rewritten in the form

$$U^V = \int \int Uab \, d\xi \, d\eta , \qquad (1.63)$$

$$U = \tfrac{1}{2} \int_{\zeta_-}^{\zeta_+} (\sigma_1 e_1 + \sigma_2 e_2 + \sigma_{12} e_{12})\left(1 + \frac{\zeta}{R_1}\right)\left(1 + \frac{\zeta}{R_2}\right) d\zeta . \qquad (1.64)$$

For *homogeneous* shells, take $\zeta = 0$ on the *middle surface*, i.e. $\zeta_+ = h/2$ and $\zeta_- = -h/2$.

Representing the stress and strain components in terms of the strain resultants $\varepsilon_1, \ldots, \tau$ by using (1.61) and (1.49) and integrating in (1.64) through the wall thickness from $\zeta = -h/2$ to $\zeta = h/2$ results in

$$U = \tfrac{1}{2}B\left[\varepsilon_1^2 + \varepsilon_2^2 + 2\nu\varepsilon_1\varepsilon_2 + \frac{1-\nu}{2}\gamma^2 \right]$$

$$+ \tfrac{1}{2}D[\kappa_1^2 + \kappa_2^2 + 2\nu\kappa_1\kappa_2 + 2(1-\nu)\tau^2] + U_c, \tag{1.65}$$

$$U_c = D[(\varepsilon_1\kappa_1 - \varepsilon_2\kappa_2)(1/R_2 - 1/R_1) - \tfrac{1}{2}\gamma\tau(1-\nu)(1/R_1 + 1/R_2)];$$

$$B = \frac{Eh}{1-\nu^2}, \qquad D = \frac{Eh^3}{12(1-\nu^2)}. \tag{1.66}$$

Here the terms of order h^2/R_i^2 have been omitted.

This expression of the elastic energy per unit area of the reference surface consists of extensional-shearing term, bending-twisting term and a mixed term U_c. Substantial simplifications of the elasticity equations of thin shells become possible, when the mixed term can be omitted. This term is actually small. For isotropic homogeneous shells, it is of the order of magnitude of h/R_i, compared with other two terms of U.

To prove this estimate, we use the well-known inequality $A^2 + B^2 \geqslant 2AB$. First, replacing $2\varepsilon_1\varepsilon_2$ and $2\kappa_1\kappa_2$ in (1.65) by $-\varepsilon_1^2 - \varepsilon_2^2$ and $-\kappa_1^2 - \kappa_2^2$ transforms, the main terms of U into a sum of squares at the cost of somewhat diminishing them. The same inequality now yields an estimate for the terms of U_c,

$$\frac{|U_c|}{U} \leqslant \max\left\{ \frac{1}{1-\nu}\frac{1}{\sqrt{12}}\left|\frac{h}{R_1} - \frac{h}{R_2}\right|, \frac{1}{\sqrt{48}}\left|\frac{h}{R_1} + \frac{h}{R_2}\right| \right\}. \tag{1.67}$$

Consequently, the term U_c in (1.65) for the elastic energy may be disregarded. This introduces no additional error in the thin-shell theory (present chapter, §3).

5.3. Nonhomogeneous orthotropic shells

Consider the more general case of elastic properties of the shell material, represented by (1.60) where the coefficients b_1, b_2 and b_ν may be *variable* throughout the volume of the shell.

Using (1.60) for stress and (1.49) for strain and integrating over ζ in the energy expression (1.64), we find

$$U = \tfrac{1}{2}[B_1\varepsilon_1^2 + B_2\varepsilon_2^2 + 2B_\nu\varepsilon_1\varepsilon_2 + B_G\gamma^2$$

$$+ D_1\kappa_1^2 + D_2\kappa_2^2 + 2D_\nu\kappa_1\kappa_2 + D_G4\tau^2] + U_c , \qquad (1.68)$$

$$U_c = C_1\varepsilon_1\kappa_1 + C_2\varepsilon_2\kappa_2 + C_\nu(\varepsilon_2\kappa_1 + \varepsilon_1\kappa_2) + 2C_G\gamma\tau .$$

The elastic coefficients are

$$\begin{bmatrix} B_1 & B_2 & B_\nu \\ C_1 & C_2 & C_\nu \\ D_1 & D_2 & D_\nu \end{bmatrix} = \int\limits_{\zeta_-}^{\zeta_+} \begin{bmatrix} 1 \\ \zeta \\ \zeta^2 \end{bmatrix} \left[b_1\frac{\alpha_2}{\alpha_1} \quad b_2\frac{\alpha_1}{\alpha_2} \quad b_\nu \right] \mathrm{d}\zeta , \quad \alpha_i = 1 + \frac{\zeta}{R_i},$$

$$(1.69)$$

$$[B_G \quad C_G \quad D_G] = \int\limits_{\zeta_-}^{\zeta_+} \left[1 \quad \zeta + \frac{\zeta^2}{2R_1} + \frac{\zeta^2}{2R_2} \quad \zeta^2 + \frac{\zeta^3}{R_1} + \frac{\zeta^3}{R_2} \right] \frac{G}{\alpha_1\alpha_2} \mathrm{d}\zeta .$$

In contrast to homogeneous shells discussed in the preceding section, the reference surface $\zeta = 0$ may now have a more favourable position than the one coincident with the middle surface. It may simplify the expression for the elastic energy eliminating the mixed, extensional-flexural, term U_c, and thus reducing the energy expression to the form it has for homogeneous shells. If for some particular shell, this aim is unattainable, effort should be made to eliminate that term within U_c which happens to be particularly inconvenient in the solution of a specific problem.

The mixed term U_c can be eliminated for *any isotropic* shell, inhomogeneous in the sense that Young's modulus E may vary with ζ, provided the Poisson coefficient, ν, is (at least approximately) constant throughout the wall thickness. This case is characterized by

$$b_1 = b_2 = E(\xi, \eta, \zeta)/(1 - \nu^2) , \qquad b_\nu = \nu b_1 , \qquad G = \tfrac{1}{2}(1 - \nu)b_1 .$$

Thereby the term U_c is made negligible by fixing the reference surface, that is the side-surface coordinate ζ_- or ζ_+, according to the condition

$$\int\limits_{\zeta_-}^{\zeta_+} \frac{E\zeta}{1 - \nu^2} \mathrm{d}\zeta = 0 \quad (\zeta_+ - \zeta_- = h) . \qquad (1.70)$$

6. Constitutive equations

The constitutive equations—elasticity relations between the stress and strain resultants—and the boundary conditions for thin shells (present chapter, §7) will be derived by using the principle of virtual work.

6.1. Virtual work

In accordance with the principle of virtual work, the work done by all the external and internal forces in a virtual displacement from an equilibrium state is equal to zero

$$\delta A - \delta U^{V} = 0 . \tag{1.71}$$

Here δA is the work done by external forces; $-\delta U^{V}$ is the negative of the variation of the elastic energy, equal to the work done by the internal forces.

The work of the external forces p and f, distributed over the volume V and over the surface Π of the body (shell), for a virtual displacement δU is given by

$$\delta A = \int_{V} p \cdot \delta U \, dV + \int_{\Pi} f \cdot \delta U \, d\Pi . \tag{1.72}$$

According to Kirchhoff's first hypothesis, a virtual displacement δU can be expressed in terms of the linear and angular virtual displacements $(\delta u, \delta \vartheta)$ of the reference surface. Thus, for an infinitesimal virtual angular displacement $(\delta \vartheta)$, we have

$$\delta U = \delta u + \delta \vartheta \times n^{*} \zeta .$$

Introducing this expression into (1.72) integrating over the wall-thickness and using the notations q and m for the force and moment resultants of the external load per unit area of the reference surface yields

$$\delta A = \int \int (q \cdot \delta u + m \cdot \delta \vartheta) ab \, d\xi \, d\eta + \oint (T^{B} \cdot \delta u + M^{B} \cdot \delta \vartheta) \, ds . \tag{1.73}$$

The reader can easily reconstruct the formulas, determining q and m in terms of the volume- and surface-load intensities p and f. However, no use will be made of such formulas in the following.

The second term in (1.73) is the virtual work of the forces $f \, d\Pi$ applied on the surface of the shell edge. The symbol $\oint (\cdots) \, ds$ denotes an integral along an edge-contour line s—the boundary of the reference surface. The variables T^B and M^B introduced in (1.73) are the force and moment resultants of the external forces $(f \, d\Pi)$ per unit length of the edge contour s. (The formulas determining the T^B and M^B in terms of $f \, d\Pi$ can be written in a straightforward way.)

Consider the transformation of the external-forces-work formula (1.73) to express the work in terms of the stress and strain resultants T_i and ε_i. The replacement of m and q by their expressions in terms of the resultants T_i and M_i from the equilibrium equations (1.53) yields

$$\delta A = - \int \int \{ [(bT_1)_{,1} + (aT_2)_{,2}] \cdot \delta u + [(bM_1)_{,1} + (aM_2)_{,2} + a^* t_1^* \times bT_1$$

$$+ b^* t_2^* \times aT_2] \cdot \delta \vartheta \} \, d\xi \, d\eta + \oint (T^B \cdot \delta u + M^B \cdot \delta \vartheta) \, ds .$$

An integration by parts eliminates the four derivatives. For instance,

$$- \int \int (bT_1)_{,1} \cdot \delta u \, d\xi \, d\eta = - \oint bT_1 \cdot \delta u \, d\eta + \int \int T_1 \cdot (\delta u)_{,1} \, d\xi \, b \, d\eta .$$

The use of (1.55) for the stress resultants T_ν and M_ν acting in the section along a line not coinciding with ξ or η transforms the expression for the virtual work into

$$\delta A = \int \int \left\{ T_1 \cdot \left[\frac{1}{a} (\delta u)_{,1} + \frac{a^*}{a} t_1^* \times \delta \vartheta \right] + T_2 \cdot \left[\frac{(\delta u)_{,2}}{b} + \frac{b^*}{b} t_2^* \times \delta \vartheta \right] \right.$$

$$\left. + M_1 \cdot \frac{(\delta \vartheta)_{,1}}{a} + M_2 \cdot \frac{(\delta \vartheta)_{,2}}{b} \right\} ab \, d\xi \, d\eta + J_s , \qquad (1.74)$$

where

$$J_s = \oint [(T^B - T_\nu) \cdot \delta u + (M^B - M_\nu) \cdot \delta \vartheta] \, ds . \qquad (1.75)$$

Here we have used the triple vector-product relation $(t_i^* \times T_i) \cdot \delta\vartheta = -T_i \cdot (t_i^* \times \delta\vartheta)$.

The coefficients of T_i and M_i in (1.24) are easily recognized as variations of the strain resultants, corresponding to the small virtual displacements δu and $\delta\vartheta$ of the shell from its *deformed* shape. Indeed, one may derive the relations

$$\frac{(\delta u)_{,1}}{a} + \frac{a^*}{a} t_1^* \times \delta\vartheta = t_1' \, \delta\varepsilon_1 + t_2' \, \delta\frac{\gamma}{2} = \delta\varepsilon_1 \,,$$

$$\frac{(\delta\vartheta)_{,1}}{a} = -t_1' \, \delta\tau_1 + t_2' \, \delta\kappa_1 + n^* \, \delta\lambda_1$$

(1.76)

in the manner used to obtain the relations (1.38) and (1.39), with the exception that one now starts from arbitrarily *deformed* shell shape.

Introducing this and the component representation (1.54) for the stress resultants into (1.74) gives

$$\delta A = \int\!\!\int \left(T_1 \, \delta\varepsilon_1 + S_1 \, \delta\frac{\gamma}{2} + T_2 \, \delta\varepsilon_2 + S_2 \, \delta\frac{\gamma}{2} + M_1 \, \delta\kappa_1 + H_1 \, \delta\tau_1 \right.$$

$$\left. + M_2 \, \delta\kappa_1 + H_2 \, \delta\tau_2 \right) ab \, \mathrm{d}\xi \, \mathrm{d}\eta + J_s \,.$$

(1.77)

The final step is to represent the four shearing stress resultants S_i and H_i in δA in terms of S and H and to represent the $\delta\tau_1, \delta\tau_2$ in terms of $\delta\tau$. With (1.57), (1.58) and relations

$$\delta\tau_i = \delta\tau - \frac{\delta\gamma}{2R_i} \,,$$

(1.78)

corresponding to (1.35), the work of external forces in a virtual displacement is determined by the formula

$$\delta A = \int\!\!\int (T_1 \, \delta\varepsilon_1 + T_2 \, \delta\varepsilon_2 + S \, \delta\gamma + M_1 \, \delta\kappa_1 + M_2 \, \delta\kappa_2$$

$$+ 2H \, \delta\tau) ab \, \mathrm{d}\xi \, \mathrm{d}\eta + J_s \,.$$

(1.79)

The complete formula for δA following from (1.77) with the more complicated expressions for S_i and τ_i, stemming from the sixth of equations (1.34) and (1.56) (with $1/R_{12} = 0$), is

$$\delta A = \int \int \left[(T_1 + H_1 \tau_2)\, \delta\varepsilon_1 + (T_2 + H_2 \tau_1)\, \delta\varepsilon_2 + \left(S + M_1 \frac{\tau_1}{2} + M_2 \frac{\tau_2}{2} \right) \delta\gamma \right.$$

$$\left. + \left(M_1 - H_1 \frac{\gamma}{2} \right) \delta\kappa_1 + \left(M_2 - H_2 \frac{\gamma}{2} \right) \delta\kappa_2 + 2H\, \delta\tau \right] ab\, d\xi\, d\eta$$

$$+ J_s . \tag{1.80}$$

The nonlinear terms omitted in (1.79) are, in most cases, of the order of γ and $\tau_i h$, i.e. of the order of magnitude of the strain components, and must be neglected for consistency. In particular, the terms $H_j \tau_i$ may be identified as the second-order effect of *finite* torsion not to be accounted for by the simplest Hooke's law (1.43).

6.2. Elasticity relations of isotropic homogeneous shells

With the variation of the elastic energy of the shell

$$\delta U^V = \int \int \left(\frac{\partial U}{\partial \varepsilon_1} \delta\varepsilon_1 + \frac{\partial U}{\partial \varepsilon_2} \delta\varepsilon_2 + \frac{\partial U}{\partial \gamma} \delta\gamma \right.$$

$$\left. + \frac{\partial U}{\partial \kappa_1} \delta\kappa_1 + \frac{\partial U}{\partial \kappa_2} \delta\kappa_2 + \frac{\partial U}{\partial \tau} \delta\tau \right) ab\, d\xi\, d\eta$$

and with the expression (1.79) for δA, the virtual work equation (1.71) becomes

$$\int \int \left[\left(T_1 - \frac{\partial U}{\partial \varepsilon_1} \right) \delta\varepsilon_1 + \left(T_2 - \frac{\partial U}{\partial \varepsilon_2} \right) \delta\varepsilon_2 + \left(S - \frac{\partial U}{\partial \gamma} \right) \delta\gamma \right.$$

$$\left. + \left(M_1 - \frac{\partial U}{\partial \kappa_1} \right) \delta\kappa_1 + \left(M_2 - \frac{\partial U}{\partial \kappa_2} \right) \delta\kappa_2 + \left(2H - \frac{\partial U}{\partial \tau} \right) \delta\tau \right]$$

$$\times ab\, d\xi\, d\eta + J_s = 0 . \tag{1.81}$$

The variations $\delta\varepsilon_1, \ldots, \delta\tau$ are arbitrary functions of the coordinates ξ and η. Consequently, the coefficients of the variations $\delta\varepsilon_1, \ldots, \delta\tau$ must vanish. This renders formulas for the six force and moment stress resultants. With the elastic energy from (1.65) these formulas are

$$T_1 = B(\varepsilon_1 + \nu\varepsilon_2) + D\left(\frac{1}{R_2} - \frac{1}{R_1}\right)\kappa_1, \qquad T_2 = \cdots,$$

$$M_1 = D(\kappa_1 + \nu\kappa_2) + D\left(\frac{1}{R_2} - \frac{1}{R_1}\right)\varepsilon_1, \qquad M_2 = \cdots,$$

$$S = Gh\gamma - \frac{Gh^3}{12}\left(\frac{1}{R_1} + \frac{1}{R_2}\right)\tau,$$

$$H = \frac{Gh^3}{12}2\tau - \frac{Gh^3}{12}\left(\frac{1}{R_1} + \frac{1}{R_2}\right)\frac{\gamma}{2}. \tag{1.82}$$

When the mixed term U_c of the elastic energy is neglected, the relations of elasticity are considerably simpler

$$T_1 = B(\varepsilon_1 + \nu\varepsilon_2), \qquad\qquad M_1 = D(\kappa_1 + \nu\kappa_2),$$

$$T_2 = B(\varepsilon_2 + \nu\varepsilon_1), \qquad\qquad M_2 = D(\kappa_2 + \nu\kappa_1), \tag{1.83}$$

$$S = Gh\gamma = (1 - \nu)B\frac{\gamma}{2}, \qquad H = (Gh^3/12)2\tau = (1 - \nu)D\tau.$$

Another form of these important relations, solved explicitly for the strain resultants, is given by

$$\varepsilon_1 Eh = T_1 - \nu T_2, \qquad\qquad \kappa_1 Eh^3/12 = M_1 - \nu M_2,$$

$$\varepsilon_2 Eh = T_2 - \nu T_1, \qquad\qquad \kappa_2 Eh^3/12 = M_2 - \nu M_1, \tag{1.84}$$

$$\gamma Eh = (1 + \nu)2S, \qquad\qquad \tau Eh^3/12 = (1 + \nu)H.$$

6.3. Nonhomogeneous orthotropic shells

For a shell material with Hooke's law presented by (1.60) and the elastic parameters varying with the coordinates ξ, η and ζ, equation (1.81) and the elastic energy expression (1.68) yield

$$\begin{bmatrix} T_1 \\ T_2 \\ M_1 \\ M_2 \end{bmatrix} = \begin{bmatrix} B_1 & B_\nu & C_1 & C_\nu \\ B_\nu & B_2 & C_\nu & C_2 \\ C_1 & C_\nu & D_1 & D_\nu \\ C_\nu & C_2 & D_\nu & D_2 \end{bmatrix} \begin{bmatrix} \varepsilon_1 \\ \varepsilon_2 \\ \kappa_1 \\ \kappa_2 \end{bmatrix}, \qquad \begin{array}{l} S = B_G\gamma + C_G 2\tau, \\[2mm] H = C_G\gamma + D_G 2\tau. \end{array} \tag{1.85}$$

The terms of the elasticity relations containing the factors C_i, C_ν and C_G complicate the solution of shell problems considerably.

When the elastic properties are *constant* through the wall-thickness, taking the reference surface coincident with the middle surface makes the C terms of (1.85) negligibly small. The elastic relations (1.85) then become almost as simple as for an isotropic shell,

$$T_1 = B_1 \varepsilon_1 + B_\nu \varepsilon_2 , \qquad M_1 = D_1 \kappa_1 + D_\nu \kappa_2 ,$$

$$T_2 = B_2 \varepsilon_2 + B_\nu \varepsilon_1 , \qquad M_2 = D_2 \kappa_2 + D_\nu \kappa_1 , \qquad (1.86)$$

$$S = B_G \gamma , \qquad\qquad H = D_G 2\tau .$$

Solved for the strains, the elasticity relations of an orthotropic shell have the form

$$\varepsilon_1 = B'_1 T_1 - B'_\nu T_2 , \qquad \kappa_1 = D'_1 M_1 - D'_\nu M_2 ,$$

$$\varepsilon_2 = B'_2 T_2 - B'_\nu T_1 , \qquad \kappa_2 = D'_2 M_2 - D'_\nu M_1 , \qquad (1.87)$$

$$\gamma = B'_G S , \qquad\qquad 2\tau = D'_G H .$$

For many problems of shells with the elastic properties varying through the wall-thickness, it is possible to reduce the elasticity equations (1.85) to the simple form (1.86). For *isotropic* shells, the necessary minimization of the U_c term in the elastic energy expression is achieved by choosing the position of the reference surface in accordance with (1.70). The resulting elasticity equations may be written in a form differing from (1.83) of the homogeneous shell only in the appearance of different "reduced" Poisson's coefficients ν' and ν'' for membrane and for bending deformation,

$$T_1 = B(\varepsilon_1 + \nu' \varepsilon_2) , \qquad T_2 = B(\varepsilon_2 + \nu' \varepsilon_1) , \qquad S = \frac{1 - \nu'}{2} B\gamma ,$$

$$M_1 = D(\kappa_1 + \nu'' \kappa_2) , \qquad M_2 = D(\kappa_2 + \nu'' \kappa_1) , \qquad H = (1 - \nu'')D\tau ,$$

$$[B \quad \nu'B \quad D \quad \nu''D] = \int_{\zeta_-}^{\zeta_+} [1 \quad \nu \quad \zeta^2 \quad \nu\zeta^2] \frac{E}{1 - \nu^2} \, d\zeta . \qquad (1.88)$$

The error of omitting the C terms in the elasticity equations is of the order of magnitude of U_c/U. When Poisson's ratio is approximately constant through the wall thickness or E and v vary symmetrically with respect to the middle surface, the equations of the type (1.88) can be as accurate as the full equations (1.85). Thus, solutions obtained for homogeneous isotropic shells can be extended to nonhomogeneous ones. As a rule, variations in v present no difficulty. Even materials whose other properties differ widely may have nearly equal values of v.

Temperature variations influence v comparatively mildly. And, above all, the value of v is not generally known with the certainty comparable to that of the data on E. The possible deviations of the actual v-values from nominal ones are usually of the same order of magnitude as variations of the nominal v-value through the shell. Thus, the retention of appropriately minimized C terms in the elasticity equations provides a dubious gain in accuracy.

Naturally, the reduction of the stress resultants to a reference surface, which is not a middle surface, does not change the form of any of the relations of the present chapter, §§1–4. The same applies to the strain resultants $\varepsilon_1, \ldots, \tau$ describing deformation of a surface in the same way, whether it is a middle surface or not.

6.4. Stresses

The strain or stress resultants determine the stress as well as the strain components at any point of a shell according to the hypotheses of thin-shell theory and to Hooke's law.

For an isotropic material, with (1.50) and (1.61),

$$\sigma_1 = \frac{E}{1-v^2}[\varepsilon_1 + v\varepsilon_2 + \zeta(\kappa_1 + v\kappa_2)], \qquad \sigma_{12} = \sigma_{21} = G(\gamma + 2\zeta\tau),$$

$$\sigma_2 = \frac{E}{1-v^2}[\varepsilon_2 + v\varepsilon_1 + \zeta(\kappa_2 + v\kappa_1)], \tag{1.89}$$

In terms of the stress resultants, the components of stress are represented for homogeneous shells with the help of (1.84) and (1.89) as

$$\sigma_1 = \frac{T_1}{h} + \zeta\frac{M_1}{h^3/12}, \quad \sigma_2 = \frac{T_2}{h} + \zeta\frac{M_2}{h^3/12}, \quad \sigma_{12} = \sigma_{21} = \frac{S}{h} + \zeta\frac{H}{h^3/12}.$$

$$\tag{1.90}$$

Obviously, these formulas give the (maximum) values of stress at the points $\zeta = \pm h/2$.

The formulas for stress components in nonhomogeneous and orthotropic shells are similar to those presented and follow directly from (1.50), (1.60) and (1.87).

6.5. Simplification of elasticity relations for H_1, H_2, S_1, S_2

In the thin-shell equations, the tangential stresses $\sigma_{12} = \sigma_{21}$ are represented by stress resultants H and S. (In the present chapter, §7, this is done for the boundary conditions also.) It is, however, useful to have elasticity relations for the *four* resultants H_1, H_2, S_1 and S_2 themselves. (For one, H_i and S_i have direct physical interpretation as components of T_i and M_i, and appear in formulas for a change of coordinates.)

The elasticity equations for H_i and S_i follow directly from the equations presented for homogeneous shells, since in this case $H_1 \cong H_2$.

Indeed, consider the resultant torsional moment $H_2 \, ds_1$ of the stress acting on a side (which is flat for $1/R_{12} = 0$) of an element of the shell (Figs. 4 and 9). Expressing σ_{21} in terms of γ and τ by using (1.49) and (1.61) yields

$$H_2 \, ds_1 = \int_{-h/2}^{h/2} \sigma_{21} \, ds_1^{\zeta} \zeta \, d\zeta$$

$$= G \int_{-h/2}^{h/2} \left[\gamma + \zeta 2\tau + \zeta^2 \tau \left(\frac{1}{R_1} + \frac{1}{R_2} \right) \right] \frac{ds_1}{1 + \zeta/R_2} \zeta \, d\zeta \; .$$

Integration results in the formula for H_2. When terms of the order of magnitude of $h^2/(20R_iR_j)$ are neglected, the formula for H_2—and a similar one for H_1—become

$$H_1 = \frac{Gh^3}{12} \left(2\tau - \frac{\gamma}{R_1} \right), \qquad H_2 = \frac{Gh^3}{12} \left(2\tau - \frac{\gamma}{R_2} \right). \qquad (1.91)$$

Expressing $H_1 - H_2$ in terms of $S = Gh\gamma$, (1.83) results in the equation [6]

$$H_1 - H_2 = S \frac{h}{12} \left(\frac{h}{R_2} - \frac{h}{R_1} \right). \qquad (1.92)$$

[6] A.L. GOL'DENVEIZER, *Prikl. Math. i Mekhanika* (PMM) **4** (1940) 35–42.

Inserting $H_i = H + (H_i - H_j)/2$ into the equations of equilibrium (1.56) reveals that with a relative error of the order of magnitude of $h^2/12R_iR_j$ or $h^2/R_i\rho_j$, it can be assumed that $H_1 = H_2 = H$. With this, the sixth equation of equilibrium gives expressions of S_i in terms of S, H and M_i. Thus, the elasticity relations (1.83) are extended to include the formulas for H_i and S_i. For the important case $1/R_{12} = 0$, these formulas may be written as

$$H_1 = H_2 = H = (1 - \nu)D\tau ,$$

$$S_1 = Gh\left(\gamma + \frac{\tau h^2}{6R_2}\right), \qquad S_2 = Gh\left(\gamma + \frac{\tau h^2}{6R_1}\right), \tag{1.93}$$

where small nonlinear terms have been neglected.

These equations supplement the elasticity equations (1.83) at the same level of accuracy, consistent with the basic hypotheses. Inspection of (1.56), (1.34) and (1.93) relating S_i and S and τ_i and τ indicates a possibility of further simplification. The terms determining the difference between S_i and S or τ_i and τ have factors h/R_i. For thin enough shells this difference seems to be small. Hence, it appears possible to use the equations of elasticity with $S_i = S$ and $\tau_i = \tau$:

$$T_1 = B(\varepsilon_1 + \nu\varepsilon_2) , \qquad\qquad M_1 = D(\kappa_1 + \nu\kappa_2) ,$$

$$T_2 = B(\varepsilon_2 + \nu\varepsilon_1) , \qquad\qquad M_2 = D(\kappa_2 + \nu\kappa_1) ,$$

$$S_1 = S_2 = \frac{1 - \nu}{2}B\gamma \equiv Gh\gamma , \quad H_1 = H_2 = (1 - \nu)D\tau \equiv \frac{Gh^3}{12}2\tau , \tag{1.94}$$

$$\tau = \tau_1 = \tau_2 .$$

These equations are known as Love's "simplest approximation". They are as simple as the corresponding equations for plates. For *most* shell *problems*, they are practically equivalent to (1.83). This is substantiated when $\gamma = S/Gh$ and τ are replaced in the expression (1.65) for the elastic energy by $S_i/Gh = \gamma + \tau h^2/6R_i$ and $\tau_i = \tau - \gamma/2R_i$ (in accordance with (1.93) and (1.35)). The additional terms of U generated by $S_i - S$ and $\tau_i - \tau$ are easily estimated as was done for U_c in the

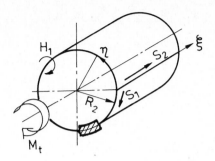

FIG. 10. Open circular cylinder under Saint-Venant torsion.

present chapter, §5. They turn out to be of the same, negligible, order of h/R_{min}. This explains why, despite the obvious discrepancy of the relations $S_1 = S_2$ and $\tau_1 = \tau_2$ with equilibrium and compatibility equations and corresponding deviations from the reciprocity theorem and variational principles, Love's simplest approximation can be so widely and effectively used.

For over 75 years there was "no single instance known, when this led to errors above those of thin-shell hypothesis" (NOVOZHILOV [40]). Clearly, the error may be large when the value of H_i/R_i or $\gamma/2R_i$ constituting $S_i - S$ or $\tau_i - \tau$ is of the order of S or τ and when the deformation consists mainly of torsion or shear of the shell wall.

The example was provided by REISSNER [91]—an open circular cylinder under Saint-Venant torsion (Fig. 10). In this case, the *simplest* approximation renders $S_1 = S_2 = 0$ instead of $S_1 - H_2/R_2 = S = S_2 = 0$. The resultant $S_1 = H_2/R_2$, which is left out, accounts for half of the torsional moment. The error doubles the maximum stress. (This example is further considered in the present chapter, §7.1.)

A complementary example is provided by the torsion of a *closed* cylinder. The well-known result—cross-sections undeformed ($\kappa_2, \tau_2 = 0$), generators twisted into spiral form ($\tau_1 = \gamma/R_2$)—indicates erroneousness of the relation $\tau_1 = \tau_2$.

Both examples are atypical of thin-shell situations; they represent the exceptional circumstances in which the "simplest" relations become inaccurate.

7. Boundary conditions. Temperature effects

Conditions on the shell edges may involve the forces or the geometry of deformation or they may be of a mixed nature. At a boundary between two parts of a shell, adequate conditions of continuity—of both forces and deformation—must be considered. Boundary conditions may also include temperature effects.

7.1. Edge forces and displacements

Conditions on the shell edges follow from the virtual work equation. As the surface integral in the virtual work equation (1.81) is equal to zero, the second term of the equation—the boundary-contour integral J_s—must also vanish

$$J_s = \oint_\nu [(T^B - T_\nu) \cdot \delta u + (M^B - M_\nu) \cdot \delta \vartheta] \, ds = 0. \qquad (1.95)$$

As a rule, the boundary contour consists of parts running along one of the coordinate lines ξ and η. Consider an edge part with boundary contour coinciding with an η-line (Fig. 11). The general case of arbitrary oriented boundary contour can be analysed similarly, starting directly from (1.95).

For the boundary to be discussed, $ds = b \, d\eta$, $T_\nu = T_1$ and $M_\nu = M_1$, so that (1.95) has the form

$$\oint [(T^B - T_1) \cdot \delta u + (M^B - M_1) \cdot \delta \vartheta] b \, d\eta = 0. \qquad (1.96)$$

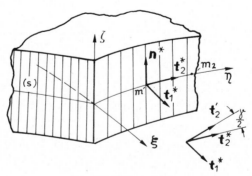

FIG. 11. Edge surface.

As a consequence of the first Kirchhoff hypothesis (present chapter, §3), the angular displacement $\delta\vartheta$ is dependent on the linear displacement of the reference surface δu. (For actual small displacements, this relation is presented by the last three formulas in (1.40).)

Define the components of virtual displacements from an arbitrarily deformed shape of the surface in the "rotated" directions t_i' and n,

$$\delta u = t_1' \, \delta u + t_2' \, \delta v + n^* \, \delta w \, , \qquad \delta\vartheta = -t_1' \, \delta\vartheta_2 + t_2' \, \delta\vartheta_1 + n^* \, \delta\omega \, .$$

$$(1.97)$$

With these decompositions of δu and $\delta\vartheta$, the equation $\delta\varepsilon_2 \cdot n^* = 0$ (where $\delta\varepsilon_2$ is expressed similarly to (1.76) and the derivative $(\delta u)_{,2}$ is determined by formulas (1.46)) gives an expression for $\delta\vartheta_2$ in terms of δu, δv and δw,

$$\delta\vartheta_2 \frac{b^*}{b} = -\frac{(\delta w)_{,2}}{b} + \frac{\delta v}{R_2'} + \frac{\delta u}{R_{21}'}, \qquad \frac{b^*}{b} = 1 + \varepsilon_2 \approx 1 \, . \qquad (1.98)$$

Equation (1.96), after introduction of the expressions for δu and $\delta\vartheta$ according to (1.97) and (1.98) and the integration by parts,

$$-\oint (H_1^B - H_1)(\delta w)_{,2} \, \mathrm{d}\eta = -(H_1^B - H_1) \, \delta w \big|_0^{II} + \oint \delta w (H_1^B - H_1)_{,2} \, \mathrm{d}\eta$$

$$(1.99)$$

may be put in the form

$$\oint \left\{ \left(T_1^B - T_1 + \frac{H_1^B - H_1}{R_{21}'} \right) \delta u + \left(S_1^B - S_1 + \frac{H_1^B - H_1}{R_2'} \right) \delta v \right.$$

$$\left. + \left[Q_1^B - Q_1 + \frac{1}{b}(H_1^B - H_1)_{,2} \right] \delta w + (M_1^B - M_1) \, \delta\vartheta_1 \right\} b \, \mathrm{d}\eta = 0 \, .$$

$$(1.100)$$

The first term of (1.99) vanishes. It represents the difference between the values of $(H^B - H_1) \, \delta w$ at the beginning and at the end (denoted 0 and II) of the boundary contour. But since the contour is in fact closed, its endpoints coincide. The same situation exists for a boundary contour of a more general form, consisting of parts running along different coordinate lines or not following any coordinate lines at all.

The equation similar to (1.100) can be obtained from (1.100) by interchanging the indices 1, 2 with 2, 1 and replacing u, v and $b\, d\eta$ with v, u and $a\, d\xi$, respectively.

Equation (1.100) represents a full set of four boundary conditions. Consider the boundary conditions for two basic cases:

(a) The edge of the shell is fully restrained and gets fixed linear and angular displacements u^B and $\boldsymbol{\vartheta}^B$. This is expressed by four conditions defining the components of displacement of the boundary contour and the angle of rotation of the boundary surface around the contour

$$u = u^B, \qquad v = v^B, \qquad w = w^B, \qquad \vartheta_1 = \vartheta_1^B.$$

These are the four displacement components *corresponding to the four virtual displacements in* (1.100). Naturally, since the four displacements are fixed, the corresponding virtual displacements are impossible:

$$\delta u = 0, \qquad \delta v = 0, \qquad \delta w = 0, \qquad \delta\vartheta_1 = 0,$$

and (1.100) is thus satisfied.

(b) The edge is free. The virtual displacements δu, δv, δw and $\delta\vartheta_1$ of a free edge are arbitrary functions of the coordinate of a boundary-contour point. In (1.100) the variations are *arbitrary* functions of η.

This implies that each of the coefficients of δu, δv, δw or $\delta\vartheta$ is equal to zero. Thus, on a free edge the following four stress resultants must be equal to specified resultants of the edge loading

$$T_{(1)} = T_1 + \frac{H_1}{R_{21}}, \quad S_{(1)} = S_1 + \frac{H_1}{R_2}, \quad Q_{(1)} = Q_1 + \frac{1}{b}H_{1,2}, \quad M_{(1)} = M_1.$$

$$(1.101)$$

By taking R_{21}, R_2 instead of R_{21}', R_2', the nonlinear small terms $H_1\tau_2$ and $H_1\kappa_2$ have been omitted here.

The edge conditions can be expressed in terms of the resultants S, H, instead of S_i, H_i. Thus, for $1/R_{21} = 0$, the four resultants may be written with (1.57) and (1.58) in the form

$$T_{(1)} = T_1, \qquad S_{(1)} = S + 2H/R_2, \qquad Q_{(1)} = N_1 + H_{,2}/b,$$

$$M_{(1)} = M_1.$$

$$(1.102)$$

The *five* stress resultants of the section ξ = const. are represented on the edge by the *four* parameters (1.101). The reduction is a consequence of the Kirchhoff hypothesis, linking the angular displacement $\delta\vartheta_2$ to the linear displacements v and w.

Example: To illustrate the application of the boundary conditions and the possible inaccuracy of the "simplest-approximation" relations $S_1 = S_2$, consider the Saint-Venant problem of pure *torsion* of an *open* cylinder (Fig. 10).

On the free longitudinal edges $\eta = \eta_1, \eta_2$, we have the four conditions $T_2, S_2, Q_{(2)}, M_2 = 0$. Applying the semi-inverse method we satisfy the four conditions on each of the edges and all the equations of equilibrium (1.59) (where for the cylinder $a = b = R_2 = $ const.; $R_1, \rho_1 = 0$) by the solution $M_1 = M_2 = T_1 = T_2 = S = 0$, $H = $ const. The relations (1.58) then render $S_2 = 0$ and $S_1 = H_2/R_2$. From (1.92) and $S = 0$ follows $H_2 = H_1 = H$. The corresponding uniform strain state satisfies the compatibility equations identically. Finally, the static equivalence between the forces in any cross-section and the applied moment M_t (Fig. 10) renders the relation between H and M_t,

$$M_t = \int_{\eta_1}^{\eta_2} (H_1 + S_1 R_2)R_2 \, d\eta = \left(H + \frac{H}{R_2}R_2\right)s = 2Hs , \qquad (*)$$

where $s = R_2(\eta_2 - \eta_1)$ is the length of the profile. With H from (1.83), we come to the well-known result $M_t = \frac{1}{3}Gh^3 s\tau$.

On the other hand, the approximate relations (1.94) $S_1 = S_2$ would lead from the correct relation $S_2 = 0$ to the erroneous $S_1 = 0$. The static equivalence (*) with $S_1 = 0$ would then give only *one half* of the correct value of the moment

$$M_t = \tfrac{1}{6}Gh^3 s\tau .$$

7.2. Edge strain conditions

Any use of displacements (u, ϑ) in the field equations of shell theory has been in the foregoing avoided. To make the theory virtually *intrinsic*, and thus make the problems involving finite rotations more tractable (cf. p. 246), we state now in terms of the strains $\varepsilon_1, \ldots, \tau$ the kinematical edge conditions. (The displacement u, ϑ, when involved, e.g., in determination of loading and its components, can be calculated in terms of strains. This fits well into iterative incremental solutions of the nonlinear problems including calculation of the current shapes.)

Consider a boundary surface $\xi = $ const. shown in Fig. 11. This surface consists of straight ζ-lines, normal to the reference surface. In any deformation these lines remain both straight and normal to the reference surface (the Kirchhoff hypothesis). The boundary surface can deform only in a manner similar to a beam with infinitely narrow cross-sections, i.e. it can be extended, change its curvature and twist. Extension of the boundary contour which coincides with the η-line is defined by ε_2. The bending and twisting of the surface $\xi = $ const. will be defined by a vector (cf. (1.23) and (1.25))

$$b\kappa_{(2)} = k^*_{(2)} - k_{2R} , \qquad \kappa_{(2)} = -t'_1\kappa_{(2)} + t'_2\tau_{(2)} + n^*\lambda_{(2)} . \qquad (1.103)$$

The parameter k_{2R} is that defined in present chapter, §2.1. Namely, $k_{2R}\, d\eta$ is an angle between the triad t'_1, t'_2 and n^* at two points of the reference surface: at $m(\xi, \eta)$ and at $m_2(\xi, \eta + d\eta)$. When the points m and m_2 belong to the boundary contour, the vectors t'_2 and n' are tangent to the locally undeformed boundary surface (as are n^* and t^*_2 in Fig. 11). The vector k_{2R} describes the curvature and twist of the *boundary* surface, in its *undeformed* state.

To determine the flexure and twist of the boundary surface by (1.103), it remains to define the $k^*_{(2)}$ to be a parameter representing the shape of the surface *after* deformation, as k_{2R} does for the undeformed shape. Thus, $k^*_{(2)}\, d\eta$ is an angle between vector pair t^*_2 and n^* tangent to the boundary surface at the point m and the corresponding pair at the point m_2. We need an expression of $k^*_{(2)}$ in terms of the curvature parameter k^*_2. An angle between the pair t'_2 and n^* at m and at m_2 is (present chapter, §§1, 2) $k^*_2\, d\eta$. To obtain the $k^*_{(2)}\, d\eta$ we must add to $k^*_2\, d\eta$ the decrement $-(n^*\gamma/2)_{,2}\, d\eta$ of the angle between t'_2 and t^*_2 (Fig. 11). We have thus

$$k^*_{(2)} = k^*_2 - \tfrac{1}{2}(n^*\gamma)_{,2} .$$

Equation (1.103) together with the relation (1.23) and formula (1.30) result now in

$$\kappa_{(2)} = \kappa_2 - \tfrac{1}{2}(n^*\gamma)_{,2}/b = -t_1'(\kappa_2 + \gamma/2R_{21}') + \cdots.$$

The three components of $\kappa_{(2)}$ together with ε_2 constitute a set of four parameters necessary and sufficient to determine any deformation of the boundary surface of a shell ([176, p. 49]; cf. [232, p. 337])

$$\kappa_{(2)} = \kappa_2 + \frac{\gamma}{2R_{21}'}\,, \qquad \tau_{(2)} = \tau_2 - \frac{\gamma}{2R_2'}\,, \qquad \lambda_{(2)} = \lambda_2 - \gamma_{,2}/2b\,, \qquad \varepsilon_2\,.$$

(1.104)

The small nonlinear terms $\tau_2\gamma$ and $\kappa_2\gamma$ will be dropped setting $R_{21}' = R_{21}$, $R_2' = R_2$ in $\kappa_{(2)}$, $\tau_{(2)}$.

With the aid of (1.35), the parameter $\tau_{(2)}$ in (1.104) may be expressed in terms of τ

$$\kappa_{(2)} = \kappa_2 + \frac{\gamma}{2R_{21}}\,, \qquad \tau_{(2)} = \tau - \frac{\gamma}{R_2}\,, \qquad \lambda_{(2)} = \lambda_2 - \gamma_{,2}/2b\,, \qquad \varepsilon_2\,.$$

(1.105a)

Naturally, the preceding discussion of the conditions on the edge running along the η-line (on the edge surface $\xi = $ const.) may be extended to the edge following the other coordinate line or indeed any other curve.

Thus, deformation of a boundary surface $\eta = $ const. is determined by four strain parameters defined by formulas similar to (1.105a),

$$\kappa_{(1)} = \kappa_1 + \frac{\gamma}{2R_{12}}\,, \qquad \tau_{(1)} = \tau_1 - \frac{\gamma}{2R_1} = \tau - \frac{\gamma}{R_1}\,,$$

$$\lambda_{(1)} = \lambda_1 + \frac{\gamma_{,1}}{2a}\,, \qquad \varepsilon_1\,.$$

(1.105b)

The boundary conditions in terms of the strain parameter $\kappa_{(i)}$ and ε_j do not determine the rigid-body displacement and rotation of an edge of a shell. These remain to be specified.

The conditions formulated assume the distribution of edge tractions to correspond to that of stresses within the shell. (A concept in fact similar to that in the Saint-Venant problems.) However, the actual distribution is rarely known with any precision.

7.3. Continuity conditions

Consider the conditions on a line separating two zones of a reference surface, which differ in the form of solution. The difference may stem from the shell geometry, loading or mathematical expediency. On such a boundary line the static equivalence of the stress resultants (forces and moments) and the congruence of the two adjoining parts of the shell have to be observed. This means that there are eight boundary conditions which must be met: four conditions for the stress resultants and four for the strain resultants. The conditions concern the parameters of stress and strain, identified in the two preceding sections as representative of the boundary conditions.

For the boundary line running, for instance, along an η-line and for $1/R_{21} = 0$, the continuity conditions may consist of the equality of the parameters listed in (1.102) and (1.105a). On the two sides of the boundary line, the following quantities must be equal:

$$\kappa_2 , \qquad \tau - \gamma/R_2 , \qquad \lambda_2 - \gamma_{,2}/2b , \qquad \varepsilon_2 ;$$
$$T_1 , \qquad S + 2H/R_2 , \qquad N_1 + H_{,2}/b , \qquad M_1 . \tag{1.106}$$

Alternatively, the four geometric conditions may be expressed by equating the displacements u, v, w and ϑ_1, on both sides of the boundary line.

It is easy to write the parameters corresponding to those of (1.106) for a boundary running along another coordinate line or even along a line coinciding with none of the coordinate lines.

The stated conditions are applicable with obvious restrictions when the two parts of a shell meet on the boundary line at a finite angle (measured between tangent planes of the two reference surfaces). In such cases the components of the stress and strain resultants on the two sides of the boundary line have different directions and hence may be different in value.

Coordinate lines may be closed curves, as for instance in tubes, toroidal shells and shells of revolution, discussed in this book. The stress and strain resultants as well as the displacements must then be periodic functions of these coordinates.

7.4. Thermal-expansion problem

The equations and the boundary conditions discussed in the foregoing may be augmented to take into account the effects of temperature variation. The deformation of a shell caused by a given distribution of the thermal expansion of the material may be determined as if it were caused by an easily calculable "thermal loading". Consider this "loading" for an *isotropic* shell. The more general problem—orthotropic shells—requires only a straightforward modification of the "thermal loading" formulas (it can be found in [79]).

Consider a shell having a temperature extension βt, known as a function of the coordinates ξ, η and ζ at any point of shell volume. The problem is to find the stress and strain caused by a change of temperature from an unstressed reference state $t = 0$.

To take the thermal expansion into account the corresponding part of the strain components must be explicitly set apart. Hooke's law determines stress components for the purely *elastic* share of the strain components (hypothesis of J.M.C. Duhamel and F. Neumann). After a substitution of the extension components e_1 and e_2 by the corresponding components of *elastic* deformation $e_i - \beta t$, the elasticity relations (1.61) become

$$\sigma_1 = \frac{E}{1 - \nu^2}(e_1 + \nu e_2) - \frac{E\beta t}{1 - \nu}, \qquad \sigma_2 = \frac{E}{1 - \nu^2}(e_2 + \nu e_1) - \frac{E\beta t}{1 - \nu},$$

$$\sigma_{12} = \sigma_{21} = G e_{12} . \tag{1.107}$$

It may be shown by inserting (1.107) into (1.64) (or by means of the static-equivalence relations) that, with (1.107), the elasticity relations (present chapter, §6) acquire additional thermal terms in the formulas for T_i and M_i. For any thermoelastically isotropic shell, homogeneous or not, these equations become

$$T_1 = T_1^e - T^t, \qquad T_2 = T_2^e - T^t, \qquad \begin{bmatrix} T^t \\ M^t \end{bmatrix} = \int_{\zeta_-}^{\zeta_+} \frac{E\beta t}{1 - \nu}\begin{bmatrix} 1 \\ \zeta \end{bmatrix} d\zeta,$$

$$M_1 = M_1^e - M^t, \qquad M_2 = M_2^e - M^t,$$

$$S_i = S_i^e, \qquad H_i = H_i^e \qquad (S = S^e, H = H^e) . \tag{1.108}$$

Here T_i^e, M_i^e, S_i^e and H_i^e are the values of the stress resultants which are determined by the *actual* strain *as if* it were purely *elastic*—as if there were no thermal expansion and the deformation were caused by loading alone.

Inserting the expressions (1.108) into the equilibrium equations (1.56) reveals that a shell has stress resultants equal to the T_i^e and M_i^e, and displacements equal to those caused by the thermal expansion βt, when loaded by distributed forces and moments equal to

$$q_1^t = -T_{,1}^t/a\,, \qquad q_2^t = -T_{,2}^t/b\,, \qquad q^t = (1/R_1' + 1/R_2')T^t\,,$$
$$m_1^t = -M_{,1}^t/a\,, \qquad m_2^t = -M_{,2}^t/b\,, \tag{1.109}$$

and by an appropriate thermal loading on the edges.

The substitution of expressions (1.108) of the stress resultants in terms of the "elastic parts" T_i^e and M_i^e, and of the parameters T^t and M^t into (1.101) results in the boundary conditions on a free edge,

$$T_1^e + \frac{H_1}{R_{21}'} = T^t\,, \qquad S_1 = \frac{H_1}{R_2'} = 0\,,$$

$$Q_1 + \frac{H_{1,2}}{b} = 0\,, \qquad M_1^e = M^t\,. \tag{1.110}$$

This gives the *edge* thermal "loading" consisting of a force resultant T^t, stretching the reference surface, and a moment resultant M^t, bending the surface.

Consider an elementary but useful application of the thermal load determined by (1.109) and (1.110), namely, the problem when the thermal expansion βt is a function of the coordinate ζ only, i.e. varying only through the shell-thickness. In this case, formulas (1.108) give T^t, M^t = const. and hence $q_i^t = 0$ and $m_i^t = 0$. Thus, the thermal loading consists only of a normal distributed load q^t and the resultants T^t and M^t on the edges.

It follows from the equilibrium equations and the elasticity relations (1.88) that the corresponding deformation of a shell consists of two parts:

(a) Elongation, uniform all over the shell and in all directions, determined by the boundary force T^t and by the normal load q^t

$$T_i^e = T^t\,, \qquad \varepsilon_1 = \varepsilon_2 = \frac{1}{B(1+\nu')}T^t\,. \tag{1.111}$$

(b) Deformation caused by the action of constraints and by the moment M^t distributed along the shell edges.

8. Static-geometric analogy. Novozhilov's equations

The linear shell equations display a duality extending to the boundary conditions. The analogy allows a reduction of the system of shell equations to three equations with three complex unknowns. The simplicity of the reduced system facilitates its analysis and the identification of negligible terms.

8.1. Static-geometric duality

The similarity between the equilibrium equations (1.56)–(1.59) and the compatibility equations (1.34)–(1.36) is difficult to overlook. This similarity reveals a far-reaching static-geometric analogy. The linear compatibility equations can be obtained from the linear equations of equilibrium (without their load terms) by a substitution of the strain resultants for the stress resultants as indicated in the following matrix scheme

$$
\begin{bmatrix}
T_1 & T_2 & M_1 & M_2 \\
T_1 & S_2 & -H_1 & -M_2 \\
S_1 & T_2 & M_1 & H_2 \\
Q_1 & Q_2 & 0 & 0
\end{bmatrix}
\leftrightarrows
\begin{bmatrix}
\kappa_2 & -\kappa_1 & \varepsilon_2 & -\varepsilon_1 \\
-\kappa_2 & \tau_1 & \tfrac{1}{2}\gamma & -\varepsilon_1 \\
\tau_2 & -\kappa_1 & \varepsilon_2 & -\tfrac{1}{2}\gamma \\
\lambda_2 & -\lambda_1 & 0 & 0
\end{bmatrix},
\quad
\begin{bmatrix}
S \\ H \\ N_1 \\ N_2
\end{bmatrix}
\leftrightarrows
\begin{bmatrix}
\tau \\ -\tfrac{1}{2}\gamma \\ \lambda_2 \\ -\lambda_1
\end{bmatrix}.
$$

$$(1.112)$$

This analogy between the equilibrium and compatibility equations was discovered, together with the latter equations, by GOL'DENVEIZER [20]. The duality indicated in (1.112) may also be observed between the two sets of parameters of *edge* stress and deformation, listed in (1.101)–(1.106).

The elasticity relations display a static-geometric duality of a somewhat more general form. Formulas (1.83) determining the stress resultants are transformed into (1.84) determining the strain resultants by the substitution of the resultants as indicated in (1.112) and by an interchange of the elasticity (stiffness) parameters,

$$[B \; \nu] \rightleftarrows \left[\frac{-1}{D(1-\nu^2)} \; -\nu \right], \qquad [D \; \nu] \rightleftarrows \left[-\frac{1}{Eh} \; -\nu \right]. \qquad (1.113)$$

The duality extends to *all* equations and boundary conditions of *linear* thin-shell theory. It is widely used, making derivation and checking of any equation easier. Any relation between the quantities listed in (1.112) and (1.113) has its dual, obtainable by a substitution of the resultants as indicated in (1.112) and (1.113).

One manifestation of the duality has particularly fruitful uses in the shell theory. The expressions for the strain in terms of displacement functions satisfy the compatibility equations identically. This implies for the stress dual relations, which satisfy the equilibrium equations identically. Thus, the general expressions of strain resultants (in terms of the displacement components u, v and w) (1.40) and (1.41) spawn dual formulas for $T_1, \ldots, H$ in terms of *stress functions*. (These formulas were stated independently by LUR'E [21] and GOL'DENVEIZER [20].) Further information concerning the use of stress functions may be found in the monographs of GOL'DENVEIZER [48] and NOVOZHILOV [40].

Another important consequence of the duality is discussed in the next two sections.

8.2. Complex form of the shell equations

Consider transformation of the linear equations of equilibrium (1.59) and of compatibility (1.36) into a single system involving six complex variables. Define new dependent variables as the following combinations of the dual stress and strain resultants:

$$\begin{aligned}
\tilde{T}_1 &= T_1 - iEhh'\kappa_2, & \tilde{M}_1 &= M_1 + iEhh'\varepsilon_2, \\
\tilde{T}_2 &= T_2 - iEhh'\kappa_1, & \tilde{M}_2 &= M_2 + iEhh'\varepsilon_1, \qquad (1.114) \\
\tilde{S} &= S + iEhh'\tau, & \tilde{H} &= H - iEhh'\gamma/2\,;
\end{aligned}$$
$$iEhh' = \sqrt{-EhD} \quad (i = \sqrt{-1}\,), \quad h' = h/\sqrt{12(1-\nu^2)}\,.$$

Multiply each of the compatibility equations (1.36) (without their nonlinear terms) by the factor $iEhh'$ and add to each of them the dual equilibrium equation of (1.59). The process just sketched leads to three equations for the six complex stress resultants defined in (1.114).

With the operator notation $N_i(\ ,\ ,\ )$ introduced in (1.59), the three equations may be written in the form

$$N_1(\tilde{T}_1,\ \tilde{T}_2,\ \tilde{S}) + \frac{1}{R_1}N_1(\tilde{M}_1,\ \tilde{M}_2,\ \tilde{H}) + \frac{R_1}{a^2b}\left(\frac{a^2\tilde{H}}{R_1^2}\right)_{,2} = -q_1 \quad (1\ 2\ a\ b),$$

(1.115)

$$\frac{1}{ab}\frac{\partial}{\partial\xi}bN_1(\tilde{M}_1,\ \tilde{M}_2,\ \tilde{H}) + \frac{1}{ab}\frac{\partial}{\partial\eta}aN_2(\tilde{M}_2,\ \tilde{M}_1,\ \tilde{H}) - \frac{\tilde{T}_1}{R_1} - \frac{\tilde{T}_2}{R_2} = -q\ .$$

(1.116)

The notation $(1\ 2\ a\ b)$ implies one more equation differing from the one written by replacing the indices and parameters $1, 2, a, b$ by $2, 1, b, a$, respectively.

The complex variables $\tilde{M}_1$, $\tilde{M}_2$ and $\tilde{H}$ can be expressed in terms of the $\tilde{T}_1$, $\tilde{T}_2$ and $\tilde{S}$, and their complex conjugates $\bar{T}_1$, $\bar{T}_2$ and $\bar{S}$ by using elasticity relations (1.83) and (1.84). This gives

$$\tilde{M}_1 = ih'(\tilde{T}_2 - \nu\tilde{T}_1)\ , \qquad \tilde{M}_2 = ih'(\tilde{T}_1 - \nu\tilde{T}_2)\ ,$$

(1.117)

$$\tilde{H} = -ih'(\tilde{S} + \nu\bar{S})\ .$$

These relations and two identities, following from the definitions (1.59) of the operators $N_i(\ ,\ ,\ )$:

$$N_1(\tilde{T}_1,\ \tilde{T}_2,\ \tilde{S}) + N_1(\tilde{T}_2,\ \tilde{T}_1,\ -\tilde{S}) \equiv \frac{1}{a}(\tilde{T}_1 + \tilde{T}_2)_{,1} \quad (1\ 2\ a\ b),$$

give the expressions

$$N_1(\tilde{M}_1,\ \tilde{M}_2,\ \tilde{H}) = ih'\left\{\frac{1}{a}(\tilde{T}_1 + \tilde{T}_2)_{,1} - N_1(\tilde{T}_1,\ \tilde{T}_2,\ \tilde{S})\right.$$

$$\left. -\nu N_1(\bar{T}_1,\ \bar{T}_2,\ \bar{S})\right\} \quad (1\ 2\ a\ b)\ .$$
(1.118)

Clearly, to make (1.115), (1.116) and (1.118) to an effective resolving system, the complex conjugates $\bar{T}_1$, $\bar{T}_2$ and $\bar{S}$ must be eliminated.

8.3. Novozhilov's equations

Consider a simplification of the system (1.115) and (1.116), where the complex moments $\tilde{M}_i$ and $\tilde{H}$ have been eliminated by using (1.118) and (1.117). We neglect, after substitution of $\tilde{M}_i$ and $\tilde{H}$, in (1.115) those terms with ih'/R_i, relative to similar terms without this factor,

$$N_1(\tilde{T}_1, \tilde{T}_2, \tilde{S}) - i\frac{h'}{R_1}[N_1(\tilde{T}_1, \tilde{T}_2, \tilde{S}) + \nu N_1(\tilde{T}_1, \tilde{T}_2, \tilde{S})] \approx N_1(\tilde{T}_1, \tilde{T}_2, \tilde{S}),$$

$$(1.119)$$

$$(a^2\tilde{S})_{,2} - ih'R_1\left[\frac{a^2}{R_1^2}(\tilde{S} + \nu\tilde{S})\right]_{,2} \approx (a^2\tilde{S})_{,2}.$$

The last equality means neglecting the third term of (1.115) in comparison with the $\tilde{S}$ term in $N_1(\tilde{T}_1, \tilde{T}_2, \tilde{S})$.

Equations (1.115) then become

$$N_1(\tilde{T}_1, \tilde{T}_2, \tilde{S}) = -i\frac{h'}{R_1}\frac{(\tilde{T}_1 + \tilde{T}_2)_{,1}}{a} - q_1 \quad (1\ 2\ a\ b).$$

$$(1.120)$$

By using (1.118) and (1.120) the $N_1(\tilde{M}_1, \ldots)$ and $N_2(\tilde{M}_2, \ldots)$ terms of (1.116) may be expressed in terms of $\tilde{T}_1 + \tilde{T}_2$,

$$N_1(\tilde{M}_1, \tilde{M}_2, \tilde{H}) = ih'\left\{\frac{1}{a}(\tilde{T}_1 + \tilde{T}_2)_{,1}\right.$$

$$\left. + ih'\frac{1}{R_1 a}(\tilde{T}_1 + \tilde{T}_2 + \nu\tilde{T}_1 + \nu\tilde{T}_2)_{,1} + q_1 + \nu q_1\right\}.$$

Neglect here terms with small factor h'/R_1 by assuming

$$\tilde{T}_1 + \tilde{T}_2 + i\frac{h'}{R_k}(\tilde{T}_1 + \tilde{T}_2 + \nu\tilde{T}_1 + \nu\tilde{T}_2) = \tilde{T}_1 + \tilde{T}_2 \quad (k = 1, 2). \quad (1.121)$$

Equations (1.120) together with the simplified equation (1.116) constitute Novozhilov's system. After some elementary transformation, the equations become

$$(b\tilde{T}_1)_{,1} - \tilde{T}_2 b_{,1} + (a^2\tilde{S})_{,2}/a + i(h'/R_1)b(\tilde{T}_1 + \tilde{T}_2)_{,1} = -abq_1,$$

$$(1.122)$$

$$(a\tilde{T}_2)_{,2} - \tilde{T}_1 a_{,2} + (b^2\tilde{S})_{,1}/b + i(h'/R_2)a(\tilde{T}_1 + \tilde{T}_2)_{,2} = -abq_2,$$

$$\frac{\tilde{T}_1}{R_1} + \frac{\tilde{T}_2}{R_2} - ih'\nabla^2(\tilde{T}_1 + \tilde{T}_2) = q + ih'\frac{1+\nu}{ab}[(bq_1)_{,1} + (aq_2)_{,2}] \, ;$$

$$\nabla^2 = \frac{1}{ab}\left(\frac{\partial}{\partial \xi} \frac{b}{a} \frac{\partial}{\partial \xi} + \frac{\partial}{\partial \eta} \frac{a}{b} \frac{\partial}{\partial \eta} \right). \tag{1.123}$$

An inspection of the omissions made in deducing Novozhilov's system reveals estimates for the accuracy of the equations. The possible error is determined by the assumptions (1.119) and (1.121), which are roughly equivalent to

$$\tilde{S} - i\frac{h}{3R_j}(\tilde{S} + \nu\tilde{S}) \approx \tilde{S} \, , \qquad \tilde{T}_1 + i\frac{h}{3R_1}\tilde{T}_1 \approx \tilde{T}_1 \quad (1\ 2). \tag{1.124}$$

Separation of the real and imaginary parts of (1.124) gives *sufficient* conditions of accuracy of the complex equations in the form

$$6\frac{|R_1|}{h} \gg \left| \frac{6H/h^2}{S/h} \right| \gg \frac{h}{2|R_1|},$$

$$6\frac{|R_1|}{h} \gg \left| \frac{6(M_2 - \nu M_1)/h^2}{T_1/h} \right| \gg \frac{h}{2|R_1|} \quad (1\ 2). \tag{1.125}$$

Thus, equations (1.122) correctly describe the stress states of a *mixed character*—when the stresses are due not only to bending and twisting deformation of the shell-wall but also to its extension and shear. It is noteworthy that conditions (1.125) compare bending in the ξ-direction with tension in η-direction and vice versa, *not* the two kinds of shell strain *in* any *one* direction. (This has its origin in the relations (1.112) of the static-geometric duality.)

But, as mentioned, the conditions (1.125) are *not* indispensable conditions. The bounds of applicability of the Novozhilov equations are considerably wider.

On the other hand, a complex transformation is possible only for the linear shell equations, and the Novozhilov equations describe only the *linear* approximation. Extension to the nonlinearity is discussed in [208].

9. The Donnell equations and the membrane model of a shell

Most results of practical value were achieved in thin-shell theory with the help of its ramifications simplified at the expense of limiting the corresponding scope of application. Consider briefly the two most important branches, covering two basically opposite types of problems, and the use of the first of these in setting up the bifurcation-stability problem.

9.1. Equations of strongly varying deformation

The general shell equations can be substantially simplified for the stressed states satisfying the following restrictions:

(a) Any of the stress and strain resultants $T_1, \ldots, \tau$, collectively referred to as F, must vary with the coordinates ξ and η so strongly that (for $1/R_{12} = 0$)

$$\left| \frac{\partial^2 F}{a^2 \, \partial \xi^2} \right| \gg \left| \frac{F}{R_1 R_2} \right|, \quad \left| \frac{\partial^2 F}{b^2 \, \partial \eta^2} \right| \gg \left| \frac{F}{R_1 R_2} \right| \quad \text{or} \quad L_1^2, L_2^2 \ll |R_1 R_2|,$$

(1.126)

where L_1 and L_2 denote the intervals of variation of the stress and strain defined in (1.45). The conditions (1.126) mean that the stressed state must vary much more markedly than the unit normal vector $\mathbf{n}$. This interpretation becomes clear when we recall the differentiation formulas (1.8) for $\mathbf{n}$.

(b) The deformation in the direction of each of the coordinates must be of *mixed* nature. That is, the resultants of *membrane* stress and strain on the one side, and bending and torsional resultants on the other, must not differ too much in their orders of magnitude. This can be specified by relations between the corresponding parts of the strain and stress,

$$[|\kappa_1| \, |\kappa_2| \, |\tau| \, |T_1| \, |T_2| \, |S|] \frac{h}{2} \sim [|\varepsilon_1| \, |\varepsilon_2| \, |\gamma| \; 6|M_1| \; 6|M_2| \; 6|H|] \varphi,$$

(1.127)

$$\frac{h}{|R_k|} \ll \varphi \ll \frac{|R_k|}{h} \quad (k = 1, 2).$$

Under the stated conditions the terms Q_1/R_1', Q_2/R_2' and $\lambda_1/R_2, \lambda_2/R_1$ in the first two equations of equilibrium (1.56) and compatibility (1.36), respectively, may be neglected. This may be checked easily when the resultants Q_i and λ_i are expressed in terms of M_i, H and ε_i, γ with the help of the fourth and the fifth of equations (1.56) and (1.36).

The conditions (1.127) and the formula (1.92) imply $H_1 = H_2 = H$ (as in (1.94)).

(c) It is assumed finally that for the simplified theory being considered in all but the third of equilibrium and compatibility equations (1.56) and (1.36), the nonlinear terms may be dropped (making $\rho_i' = \rho_i$, $1/R_{ij}' = 0$). Ignored terms include $\lambda_j T_j$, $\lambda_j S_j$, $\lambda_j H_j$, $\lambda_j M_j$, $\lambda_i \kappa_j$, $\lambda_j \varepsilon_i$, $\lambda_j \gamma$ and $Q_j/R_{ji}' = \tau_j Q_j$. These terms are as a rule small. Indeed, according to (1.36) the values of $a\lambda_1$ and $b\lambda_2$ have the order of magnitude of the strain components. As to the terms $Q_j/R_{ji}' = Q_i \tau_i$, they are normally smaller than those terms, Q_j/R_j', already neglected according to the preceding assumption.

Thus, to the accuracy of conditions (1.126) and (1.127), all equations of equilibrium and compatibility (1.56) and (1.36), except the third pair, can (for $1/R_{12} = 0$ and $q_1 = q_2 = 0$) be satisfied by the expressions

$$T_1 = d_2 \Psi , \qquad T_2 = d_1 \Psi , \qquad S_1 = S_2 = S = -d_{12} \Psi ,$$

$$\kappa_1 = -d_1 W , \qquad \kappa_2 = -d_2 W , \qquad \tau_1 = \tau_2 = \tau = -d_{12} W ; \tag{1.128}$$

$$d_2 \Psi = \frac{1}{b}\left(\frac{\Psi_{,2}}{b}\right)_{,2} + \frac{b_{,1}}{ba^2}\Psi_{,1}$$

$$2d_{12} \Psi = \frac{b}{a}\left(\frac{\Psi_{,2}}{b^2}\right)_{,1} + \frac{a}{b}\left(\frac{\Psi_{,1}}{a^2}\right)_{,2} \qquad (1\,2\,a\,b). \tag{1.129}$$

Since the elasticity equations give expressions for M_i and H in terms of κ_i and τ and expressions for ε_i and γ in terms of T_i and S, equations (1.129) and (1.128) express all the stress and strain resultants in terms of the two functions Ψ and W. In this way, the third of the equations of equilibrium (1.56) and of compatibility (1.36) give as a system for W and Ψ, the pair

$$D\nabla^4 W + \left(\frac{1}{R_1} + \kappa_1\right)T_1 + \left(\frac{1}{R_2} + \kappa_2\right)T_2 + 2S\tau = q ,$$

$$\frac{1}{Eh}\nabla^4 \Psi + \frac{\kappa_1}{R_2} + \frac{\kappa_2}{R_1} + \kappa_1 \kappa_2 - \tau^2 = 0 \quad (\nabla^2 = d_1 + d_2) . \tag{1.130}$$

This system is to be considered in conjunction with the boundary conditions (four on an edge) of the general shell theory (present chapter, §7).

Equations (1.130) and (1.128) display the static-geometric duality indicated in (1.112) and (1.113).

9.2. Bifurcation-stability equations

Consider the stability of equilibrium of a shell in some deformed state. Let the bifurcated deviation from the "0" state be represented by infinitely small increments W^* and Ψ^* resulting in stress and strain functions after bifurcation $W + W^*$ and $\Psi + \Psi^*$. The variables $W + W^*$ and $\Psi + \Psi^*$ describe a *possible* equilibrum state of the shell and thus satisfy (1.129) and (1.130) just as this system of equations was satisfied by W and Ψ before bifurcation. This yields as equations for the bifurcation increments W^* and Ψ^*,

$$D\nabla^4 W^* + \frac{1}{R_1'} d_2 \Psi^* + \frac{1}{R_2'} d_1 \Psi^* - T_1 d_1 W^* - T_2 d_2 W^*$$

$$- 2S d_{12} W^* - 2\tau d_{12} \Psi^* = 0, \qquad \left(\frac{1}{R_i'} = \frac{1}{R_i} + \kappa_i \right) \quad (1.131)$$

$$\frac{1}{Eh} \nabla^4 \Psi^* - \frac{1}{R_2'} d_1 W^* - \frac{1}{R_1'} d_2 W^* + 2\tau d_{12} W^* = 0 .$$

Since the bifurcation *increments* are *infinitesimally small*, all of the terms that are nonlinear in the increments W^* and Ψ^* are omitted. The stability equations (1.131) are linear. The loading determining the values of stress resultants and shell-form parameters T_i, S, $1/R_i'$ and τ, which allow a nonzero solution of the stability equations for homogeneous boundary conditions, is the *critical* loading.

For an important class of problems the stability equations (1.131) become much simpler. When the buckling mode varies so strongly that for some appropriate coordinates ξ and η, the zone of the initial buckling is *shallow* enough to assume $d_i(AW^*) = AW^*_{,ii}$ for $A = a$ or b and $i = 1$ or 2, equations (1.131) can be presented in the form

$$\left[(\partial_1^2 + \partial_2^2)^2 - \frac{2}{T_*}(T_1 \partial_1^2 + T_2 \partial_2^2 + 2S \partial_1 \partial_2) \right] W^* + QF = 0 ,$$

$$QW^* - (\partial_1^2 + \partial_2^2)^2 F = 0, \quad Q = k_2 \partial_1^2 + k_1 \partial_2^2 - 2\tau R \partial_1 \partial_2 , \quad (1.132)$$

where it is denoted

$$\partial_1 = \frac{\partial}{\partial x}, \quad \partial_2 = \frac{\partial}{\partial y}, \quad x = a\frac{\xi - \xi_0}{c}, \quad y = b\frac{\eta - \eta_0}{c},$$

$$c = \left(\frac{hR}{\sqrt{12(1 - \nu^2)}}\right)^{1/2}, \tag{1.133}$$

$$T_* = Eh^2 \frac{1}{R\sqrt{3(1 - \nu^2)}}, \quad k_1 = \frac{R}{R_1'}, \quad k_2 = \frac{R}{R_2'}, \quad F = 2\Psi^*/(T_*R).$$

The dimensional parameter R can be chosen arbitrarily. In what follows (Chapter 3, §9), we set $R = R_2'(\xi_0, \eta_0)$ with ξ_0 and η_0 being coordinates of a reference point of the buckling zone.

If also the prebuckling curvatures and twist vary so much less strongly than the buckling mode, that for $A = 1/R_i'$ and τ we have $(AW^*)_{,j} = AW^*_{,j}$ and $(AF)_{,j} = AF_{,j}$ $(j = 1, 2)$, one of the variables can be eliminated from (1.132). This leads to a single stability equation

$$DW^* = 0,$$

$$D = (\partial_1^2 + \partial_2^2)^4 - \frac{2}{T_*}(T_1\partial_1^2 + T_2\partial_2^2 + 2S\partial_1\partial_2)(\partial_1^2 + \partial_2^2)^2 + Q^2. \tag{1.134}$$

The assumption made in eliminating the stress function F from the equations (1.132) is widely used for *shallow shells* (cf., for instance, WLASSOW [36]).

9.3. On the membrane model of a shell

A thin shell can carry a load effectively only when its form and edge conditions assure a working state without too high a wall-bending stress. The situation is similar to that for a slender arch. Even small bending moments may cause large stress. Thus, a *membrane* stress state—one free of bending—is aimed for in load-carrying shells.

(On the contrary, with respect to the flexibility of the shell, the membrane state of stress is definitely unfavourable. It is easily estimated that for a bending deformation, the elastic displacements in a shell may reach an order of magnitude of L/h greater than that of the membrane deformation, with L as an overall characteristic dimension of the shell.)

A brief note on the membrane shell theory is presented here, just what is to be used in the following chapters. (The membrane theory is comprehensively discussed in the monographs of FLÜGGE [152], SEIDE [173], GOL'DENVEIZER [48, 178], WLASSOW [36] and others.)

The lateral stress-resultants Q_i are expressed in terms of the moment stress-resultants as follows from the fourth and fifth of the equilibrium equations (1.56). When the moment stress-resultants are negligible, the Q_i terms of the first three of the equilibrium equations are also negligible. The three equations are thus reduced to the following system:

$$\frac{(bT_1)_{,1}}{ab} + \frac{(aS)_{,2}}{ab} - \frac{S}{\rho_1'} - \frac{T_2}{\rho_2'} + q_1 = 0 ,$$

$$\frac{(bS)_{,2}}{ab} + \frac{(aT_2)_{,2}}{ab} + \frac{S}{\rho_2'} + \frac{T_1}{\rho_1'} + q_2 = 0 ; \tag{1.135}$$

$$\frac{T_1}{R_1'} + \frac{T_2}{R_2'} + \frac{S}{R_{12}'} + \frac{S}{R_{21}'} = q \quad (S = S_1 = S_2) . \tag{1.136}$$

Other equations and boundary conditions of membrane theory are implied by the assumption $M_i = 0$ and $H_i = 0$ and the static-geometric duality.

The scope of applicability of the membrane theory may be surmised on the basis of an analysis of the equations of equilibrium. A clear indication of the type of problems adequately described by this theory follows from the expressions of Q_1 and Q_2 in terms of the M_i and H_i, rendered by the equations of equilibrium (1.56). Equations (1.56) indicate the bending terms dropped in the equations (1.135) and (1.136) to be the smaller, the slower the moments M_i and H_i vary. The terms with $(bQ_1)_{,1}$ and $(aQ_2)_{,2}$, neglected in the membrane equation (1.136), depend on the *second* derivatives of the moments.

Hence the membrane equations can describe *stress states slowly varying* with ξ and η. Such states occur in shells with slowly varying curvatures due to a smoothly distributed load and appropriate edge conditions.

Commentary

To §§1–6: The first general tensor-formulation of the shell theory (for the equilibrium, also the vector formulation) has been given by LUR'E [21]. This work clearly marked the way to the intrinsic nonlinear theory.

To §§1, 2: The vector-form description of the curvature of a surface may be traced back to the "method of moving axes" used in the book of LOVE [28] (1944 and earlier editions). Later this method was abandoned in favour of the description of geometry in terms of second-order tensors: In the author's opinion, the vector approach is somewhat more graphic, presenting as it does the mathematical tool and the geometric interpretation simultaneously.

The Gauss–Codazzi vector equation (1.10) is given for orthogonal coordinates in the book of LAGALLY [33]. Extension (to *oblique* coordinates ξ and η) necessary to describe the *deformed* surface, is achieved by introducing the auxiliary vector t of (1.4).

The strain vectors ε_i and κ_i were introduced by GÜNTHER [80] together with vector compatibility relations. But the analysis was a strictly linear one.

The nonlinear shell theory in the tensor form including the finite rotation and stress-function vectors is due to SIMMONDS and DANIELSON [50].

REISSNER [163, 171] developed the vector-form theory for the moment-stress elasticity.

The *vector* description of the surface shape and its deformation presented in §2 is proposed in [176, 197] as an extension of the mentioned investigations. In particular, the scalar strain resultants and the scalar compatibility equations (1.34) basically coincide with those given by REISSNER [163].

The strains of the well-known second-order tensor formulations differ from those in §2 by terms of the order of h/R_i (which is negligible).

The nonlinear compatibility equations were first derived by K.Z. Galimov in 1953. The equations may be found in the first book on nonlinear shell theory, that of MUSHTARI and GALIMOV [59]. (These equations are commented upon in §3.3.)

LIBAI and SIMMONDS [206] gave new derivation of equations (1.33). They also pointed out that the integrability condition for the finite-rotation vector, obtained by PIETRASKIEWICZ [188], constitutes an equation identical to the first of (1.33).

To §3.1: An indication that the errors of the Kirchhoff–Love hypothesis are of the order of magnitude of h/R_i as in (1.44) was made already by BASSET [3], immediately after the classic article of LOVE [2]. An elegant presentation of the main equations of the thin-shell theory, fully consistent with the accuracy of the basic hypotheses, has been given in the REISSNER [23, 24] work. A review of the investigation of the basic problems of the shell theory, during the first hundred years of its existence, can be found in the work of KOITER and SIMMONDS [155] and KOITER [124].

To §4.2: The reduction of the four stress resultants S_i and H_i to the two S and H was first proposed by TREFFTZ [17]. LUR'E [37] gave a physical interpretation of the circumstances making the reduction possible. A comprehensive discussion of this problem may be found in the work of KOITER [101], SANDERS [92], DANIELSON and SIMMONDS [122]; the relevant investigations have been reviewed by KOITER and SIMMONDS [155].

To §5.2: The estimate (1.67) for the mixed membrane-bending term of the elastic energy and its minimization by way of choosing the reference surface according to (1.70) appears to be proposed in [61, 176]. (Cf. NOVOZHILOV [40].) A fully general analysis leading to the estimate of the mixed term is due to KOITER [77].

To §6.2: The equations of the form (1.83) and (1.84), first of all those for H, S, γ and τ, are associated in the Russian literature with the names of L.I. Balabukh and V.V. Novozhilov (cf. the note to §4.2).

To §6.3: The discussion is based on the articles [61, 79]; cf. [181, 230]. Stresses in two-layer plates are considered in [233] and in AXELRAD, E.L., *Trudy LIAP* **24** (1957) 41–96 (in Russian). Comprehensive treatment of layered shells is given by LIBRESCU [169].

To §6.5: The purpose of these deliberations is to illustrate that, with the exception of some rather extreme cases, the "Love's simplest approximation" formulas (1.94) are as exact as can be in the thin-shell theory. This is so for all problems discussed in this book. Moreover, the simplest relations (1.94) provide the basis of the two main branches of the shell theory—the Donnell theory (§9.1) and the membrane theory (§9.3)—as well as of the flexible-shell theory (Chapter 2, §§4, 5).

To §7.4: The present discussion leans on the article [61].

To §8: The first case of a compatibility equation and of the static-geometric analogy is in fact due to REISSNER [5]. To this work of H. Reissner (on spherical shells) may also be traced the origin of the complex transformation of shell equations and of their asymptotic integration. These methods, being developed by V.V. Novozhilov, A.L. Gol'denveizer and many other investigators, play a prominent role in the thin-shell theory.

The derivation of the Novozhilov equations given in §8 is somewhat similar to the one given by CHERNYKH [86]. An alternative (to the Novozhilov's) complex transformation has been proposed by SANDERS [127].

An example of a stressed state, which can be described by V.V. Novozhilov's equations only with substantial error, is provided by the work of COHEN [75].

To §9.1: The equations of the type (1.129) and (1.130) were first proposed for the cylinder shell by DONNELL [13]. Simultaneously (and to solve the same buckling problem), the equilvalent equations were proposed by MUSHTARI [14]. The starting point was provided by the expressions of κ_i and τ in terms of the displacement component w, which follow from (1.41) and (1.40) with (1.34) when the u and v terms are dropped. The equations were developed further by WLASSOW [36] in 1944. A substantial improvement of the equations is due to KOITER [101]: the equations were made intrinsic and, in particular, independent of the *linear* relations between κ_i and τ and displacements. This work provides also the bibliography of the theory.

To §§1–9: The tensor component formulation of the shell theory is known since 1940 (LUR'E [21]). This form became popular in theoretical studies after BUDIANSKY and SANDERS [88, 92] (1963) convincingly argued for its esthetic supremacy in the general theory. Consider translation of the basic relations presented above in vector form into the tensor-component statement. This will throw some additional light on the physical meaning of certain tensor components.

For oblique coordinates X^α ($\alpha = 1, 2$) we define instead of (1.1) the basis $a_\alpha = r_{,\alpha}$, a^α, n and we introduce an (unconventional) normal-curvature vector $b_\alpha = -n_{,\alpha}$. This leads to the fundamental tensors: $a_\alpha a^\alpha = a_{\alpha\beta}a^\alpha a^\beta$, $b_\alpha a^\alpha = b_{\alpha\beta}a^\alpha a^\beta$. The curvature vector of (1.5), (1.8) obtains the expression $k_\alpha = -n \times b_\alpha + n/R_{\alpha\beta}$. Employing further the superscripts * ' for the deformed-state characteristics and the rotated basis $a_{\alpha R}$ we express the strain vectors (1.23) in terms of components denoted $E_{\alpha\beta}$, $\bar{\rho}_{\alpha\beta}$:

$$\varepsilon_\alpha a_\alpha \equiv E_\alpha = E_{\alpha\beta}a^\beta \,, \qquad a_\alpha |a_\alpha| \,, \qquad a^\beta = |a^\beta|$$

$$k_\alpha a_\alpha = -n \times \rho_\alpha + \lambda_\alpha n \,, \qquad \rho_\alpha \equiv \bar{\rho}_{\alpha\beta}a^\beta = b_\alpha^* - b_{\alpha\beta}a_R^\beta \,,$$

$$b_\alpha^* = b_{\alpha\beta}^* a^{*\beta} = b_{\alpha\beta}' a_R^\beta \,.$$

These definitions directly lead to expressions of the classical strain-tensor components [155] $2\gamma_{\alpha\beta} = a_{\alpha\beta}^* - a_{\alpha\beta}$, $'\rho_{\alpha\beta} = b_{\alpha\beta}^* - b_{\alpha\beta}$:

$$2\gamma_{\alpha\beta} = E_{\alpha\beta} + E_{\beta\alpha} + \varepsilon_\alpha \cdot \varepsilon_\beta a_\alpha a_\beta \,, \tag{a}$$

$$'\rho_{\alpha\beta} = \bar{\rho}_{\alpha\beta} + b_{\alpha .}^{'\lambda}\gamma_{\lambda\beta} \,, \qquad b_{\alpha .}^{'\lambda} = b_\alpha^* \cdot a_R^\lambda \quad (b_\alpha^* \cdot a^{*\lambda} = b_{\alpha .}^{*\lambda}) \,. \tag{b}$$

Appropriate choice of the rotated basis (indicated in §2.1) assures $E_{\alpha\beta} = E_{\beta\alpha}$ (cf. the polar decomposition in [133, p. 461]). Thus for small strain (a) means $\gamma_{\alpha\beta} = E_{\alpha\beta}$. The relation (b) reveals the physical meaning of the "modified tensor of changes of curvature" $\rho_{\alpha\beta}$, constructed in [88, 92] for the "best" linear theory and in [155, (4.13)] for the intrinsic nonlinear theory in the form

$$\rho_{\alpha\beta} = {}^{\backprime}\rho_{\alpha\beta} - \tfrac{1}{2} \left(b_{\alpha}^{\kappa} \gamma_{\kappa\beta} + b_{\beta}^{\kappa} \gamma_{\kappa\alpha} \right). \tag{c}$$

For small strain there is in (c) $b_{\alpha}^{\kappa} \gamma_{\kappa\beta} \cong b_{\alpha}^{\prime\kappa} \gamma_{\kappa\beta}$. Thus, the "modified changes of curvature" $\rho_{\alpha\beta}$ embody the components $\bar{\rho}_{\alpha\beta}$ of the vectors $\boldsymbol{\rho}_{\alpha}$: $\rho_{\alpha\beta} = (\bar{\rho}_{\alpha\beta} + \bar{\rho}_{\beta\alpha})/2$.

The vector equations (1.33), (1.53) a.o. can be translated into tensor-component equations with the aid of tensor analysis. It is sufficient to note that $a_{R}^{\alpha} E_{\alpha}$, $a_{R}^{\alpha} k_{\alpha} a_{\alpha}$, $a_{\alpha R} N^{\alpha} a^{\alpha}$, $a_{\alpha R} M^{\alpha} a^{\alpha}$ constitute dyad tensors. (For oblique coordinates X^{α} the factors ab in (1.53) are replaced by $|r_{,1} \times r_{,2}|$.)

CHAPTER 2

FLEXIBLE-SHELL THEORY

CHAPTER 2

FLEXIBLE-SHELL THEORY

The general equations of thin shells are here specialized and simplified for the class of flexible shells. Limits of applicability of the flexible-shell theory and special cases of its equations are examined.

But first (§§1–3) we review the basic problems of the flexible shells, formulate that feature of the stress state, necessary for flexibility, and discuss the semi-inverse solutions of the basic problems.

1. Basic flexible-shell problems

The main types of flexible shells are represented in Fig. 12. Characteristic of these applications of shells is the realization of a significant elastic displacement of one edge of the shell relative to another edge. The load is usually applied to the movable edge and/or is represented by a uniform normal pressure. Such features of loading are common with beams. In many cases the similarity may be observed also in the shell shape (Fig. 12).

The main problems of flexible shells include those analogous to the bending and torsion of a beam.

Solution of the problems can be facilitated by the semi-inverse method of Saint-Venant, well developed for elastic beams. This is used by setting the boundary conditions in the spirit of the principle of Saint-Venant—prescribing only the resultants of load (or of displacement) on an edge, *not* its *distribution* along the boundary contour. In this manner, the problem of a flexible shell loaded on its edges is reduced to a one-dimensional problem. One variable, ξ, is eliminated. Problems of this type are represented by the six schemes in Fig. 13.

Clearly, the simplified boundary conditions are sufficient only for a limited number of real problems. The rest remain two-dimensional. Their solution will be facilitated by specializing the general shell

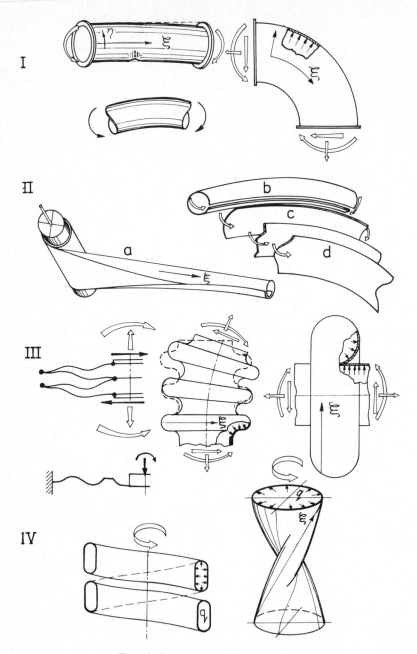

FIG. 12. Four classes of flexible shells.

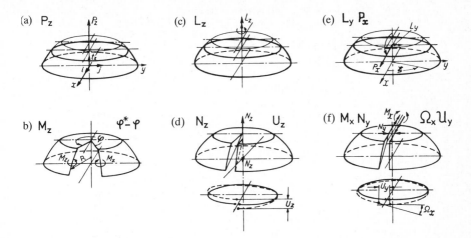

FIG. 13. The Saint-Venant problems.

equations and boundary conditions—by using a *flexible-shell model* and the corresponding theory. For the more general situations, the six Saint-Venant's problems provide limit-case models and reference solutions. It is with this end in view they will be considered.

1.1. *Rotationally symmetric problems*

Rotationally symmetric problems—with stress and strain resultants independent of the coordinate ξ (Fig. 5, p. 11)—are represented in Fig. 13 by schemes (a)–(d).

Deformation of types (a) and (b) leaves the shell form axisymmetric. It is the symmetric *deformation of a full shell of revolution* and the *pure bending of a curved beam*. In both of these problems the meridional planes $\xi = $ const. are free of shear stress. This implies that all shear-stress and shear-strain resultants, with exception of Q_2 (Fig. 14), namely

$$\gamma, \quad \tau_1, \quad \tau_2, \quad \tau, \quad H_1, \quad H_2, \quad H, \quad S_1, \quad S_2, \quad S, \quad Q_1, \quad (2.1)$$

are equal to zero. Under these conditions, three of the equations of equilibrium (1.56) and three of the compatibility equations (1.36) are satisfied identically. The remaining six equations contain unknown functions of η only,

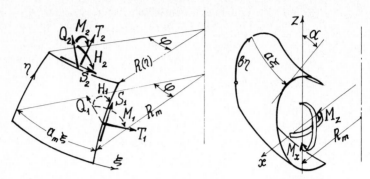

FIG. 14. Geometry and stress resultants of flexible shells.

$$\varepsilon_1, \quad \varepsilon_2, \quad \kappa_1, \quad \kappa_2, \quad \lambda_1, \quad M_1, \quad M_2, \quad T_1, \quad T_2, \quad Q_2. \quad (2.2)$$

In these problems the distributed load must act in the meridian plane ($q_1, m_1 = 0$) and be constant along the ξ-lines. The bending load moment M_z acts in case (b) on the ends of a curved beam, i.e. on edges ξ = const. of a sector of a shell of revolution. It must be assumed distributed along the edges in the form of resultants M_1^B and T_1^B equal to the stress resultants $M_1(\eta)$ and $T_1(\eta)$, respectively, of *any* section ξ = const.

The third and fourth axisymmetric Saint-Venant's problems ((c) and (d) in Fig. 13) are (c) *torsion of a shell of revolution* by forces applied at the edges (equivalent to a moment L_z) and by distributed load $q_1(\eta)$ and $m_1(\eta)$; (d) *torsion of a curved thin-walled beam*, viz. sector of a shell of revolution, by forces distributed on the edges ξ = const. in the same way as stresses in any section ξ = const.

The problems (c) and (d) involve the resultants listed in (2.1), the resultants (2.2) being equal to zero. For the case (d) this remains true only for the linear approximation. Only as far as the torsion changes the axisymmetric character of the shell-form insignificantly, the case may be considered as a problem involving pure shear strain.

Problems (b) and (d) may also be regarded as Volterra's distortion problems. If a shell occupying an angle less than 2π (as in Fig. 13(b)) is elastically deformed by matching the edges ξ = const., the stress state ensuing after connecting the edges corresponds to that of problem (b). Similarly, the stress state of problem (d) may be caused by a distortion, which consists of an axial displacement U_z of the sides of a meridional-plane cut. This is illustrated for one of the ξ-lines in Fig. 13(d).

1.2. Lateral and space bending

Problems (e) and (f) (Fig. 13), to be discussed now, are nonaxisymmetric. The stress and the strain vary in *both* of the coordinates ξ and η. However, in the *linear* Saint-Venant's problem the variables can be separated by simply representing the solution in the form of series in $\cos j\xi$ and $\sin j\xi$.

Any load can be presented as a sum of two parts, varying symmetrically and skew-symmetrically with respect to the plane $\xi = 0$. Consider the case of a symmetric load. (The complementary case is completely similar in treatment.) The load components and the corresponding stress and strain resultants can be represented by a Fourier series of the form

$$
\begin{bmatrix} T_1 & T_2 & M_1 & M_2 & Q_2 & q_2 & q \\ \kappa_2 & \kappa_1 & \varepsilon_2 & \varepsilon_1 & \lambda_1 & m_2 & 0 \end{bmatrix}
$$

$$
= \sum_j \begin{bmatrix} T_1^j & T_2^j & M_1^j & M_2^j & Q_2^j & q_2^j & q^j \\ \kappa_2^j & \kappa_1^j & \varepsilon_2^j & \varepsilon_1^j & \lambda_1^j & m_2^j & 0 \end{bmatrix} \cos j\xi , \qquad (2.3)
$$

$$
\begin{bmatrix} S_i & H_i & Q_1 & q_1 \\ \tau_i & \gamma/2 & \lambda_2 & m_1 \end{bmatrix} = \begin{bmatrix} S_i^0 & H_i^0 & Q_1^0 & q_1^0 \\ \tau_i^0 & \frac{1}{2}\gamma^0 & \lambda_2^0 & m_1^0 \end{bmatrix}
$$

$$
+ \sum_j \begin{bmatrix} S_i^j & H_i^j & Q_1^j & q_1^j \\ q_i^j & \frac{1}{2}\gamma^j & \lambda_2^j & m_1^j \end{bmatrix} \sin j\xi \quad (i = 1, 2; \ j = 0, 1, 2, \ldots) .
$$
$$
(2.4)
$$

Substituting the expansions into the field equations and boundary conditions and equating the Fourier coefficients of both sides of each equation eliminates the variable ξ. Within the linear problem this results in a separate system of equations and boundary conditions for the nth amplitudes $T_1^n, \ldots, \lambda_2^n$ ($n = 0, 1, \ldots$).

For $n = 0$ the system falls apart further into *two* independent systems: one system determines $T_1^0, \ldots, \lambda_1^0$; the other, the remaining resultants $S_1^0, \ldots, \lambda_2^0$. This corresponds to the four axisymmetric problems discussed in the preceding section.

The lateral and space bending (problems (e) and (f) in Fig. 13) are described by the *first harmonics* of the series (2.3) and (2.4):

$$T_1^1(\eta) \cos \xi, \ldots, \lambda_2^1(\eta) \sin \xi \, .$$

This deformation is caused by a *"wind-type" loading*—one distributed along the edges η = const. and on the surface of shell as cos ξ or sin ξ. On the edges η = const., the wind-type loading is statically equivalent to a lateral force P_x and to a moment L_y, shown in diagram (e) of Fig. 13.

A curved beam (Fig. 13(f)) has a deformation varying as cos ξ, sin ξ when loaded on the edges ξ = const. by forces with resultants N_y and M_x shown in Fig. 13(f).

The deformation of a curved shell-beam may also be thought of as a result of Volterra's *distortion*, i.e. of the displacement U_y and the rotation Ω_x of an edge ξ = const. with respect to the other edge (Fig. 13(f)).

Thus, the six Saint-Venant's problems ((a)–(f) in Fig. 13) correspond to the harmonics $j = 0, 1$ in the Fourier expansion of the resultants. These cases are singled out by the fact that forces in the section η = const. have a resultant force $P(\eta)$ and a moment $L(\eta)$, whereas the stresses, corresponding to the harmonics $j \geq 2$ of the solution, are self-equilibrated. The resultant force P and moment L are determined from statics, in terms of load of the axisymmetric and wind types ($j = 0, 1$). The expressions of components of P and L in terms of the stress-resultant amplitudes T_i^0, T_i^1, . . . yield *integrals of the equations of equilibrium*.

Expressions for the distortion parameters Ω and U in terms of the strain-resultant amplitudes κ_i^0, κ_i^1, . . . provide static-geometric dual *integrals of the compatibility equations*. The algebraic equations provided by the integrals result in substantial simplification and lower the order of the system of differential equations of the Saint-Venant problems.

Thus, the solution of problems (a), (b), (e) and (f) is determined by systems of two second-order differential equations. The systems are derived in the present chapter, §2 (for problems (a) and (b)) and §3 (for problems (e) and (f)).

The Saint-Venant's torsion problems (c) and (d) allow a closed-form solution which is presented in Chapter 3, §5.

1.3. Flexible-shell stress state. Basic hypothesis

A glance at the schemes presented in Fig. 12 conveys that in any of the cases the stress state varies much more markedly along the η-lines than in the other coordinate, ξ. Further analysis substantiates this conclusion and renders a mathematical measure for the difference in the variation intensities.

Despite the manifold diversity of flexible shells, seen in Fig. 12, their flexibility is connected in all cases with the same basic feature of the stress state. The deformation varies in the direction of one of the surface coordinates η so much more strongly than in the other coordinate ξ that for any of the functions $F(\xi, \eta)$, describing the important stress and strain resultants, the second derivative $\partial^2 F/a^2 \partial\xi^2$ is negligible compared to $\partial^2 F/b^2 \partial\eta^2$. This property of the stress state will be used as a *basic hypothesis* of the theory of flexible shells. The general thin-shell theory will be simplified by the following mathematical expression[1] of the hypothesis (Fig. 14)

$$\left| \frac{\partial^2 F}{a^2 \partial\xi^2} \right| \sim \left| \frac{\partial^2 F}{b^2 \partial\eta^2} \right| \varepsilon , \qquad \varepsilon = \left(\frac{b/n}{a/j} \right)^2 \ll 1 , \qquad (2.5)$$

with the values of a/j, b/n defined as the intervals of variation of the stress state

$$\left| \frac{\partial F}{a \partial\xi} \right| \sim \frac{|F|}{a/j} , \qquad \left| \frac{F}{b \partial\eta} \right| \sim \frac{|F|}{b/n} . \qquad (2.6)$$

In what follows, shells of types I–III in Fig. 12 are considered. The local shape of these shells is identical to that of the shells of revolution, and the coordinates ξ and η will be mostly chosen as in the Figs. 5 and 14. This means that $a = R(\eta)$ while $b = $ const. Using intervals of variation of the stress state R/j and b/n (we have, clearly, $j \geqslant 1$), the flexible-shell hypothesis may thus be expressed with the condition

$$\left| \frac{\partial^2 F}{R^2 \partial\xi^2} \right| \sim \left| \frac{\partial^2 F}{b^2 \partial\eta^2} \right| \varepsilon , \qquad \varepsilon = \left(\frac{bj}{nR} \right)^2 \ll 1 . \qquad (2.7)$$

---

[1] The $\sim$ sign is explained on p. 21.

The possibility of the stress or strain function $F(\xi, \eta)$ being a *vector* is allowed for here.

The hypothesis leads to a certain simplified model of a flexible shell. It will be shown (present chapter, §4) that the stress state of a flexible shell is semi-momentless (or semi-membrane)—the moments M_1 and H_1 in the sections $\xi = $ const. and the lateral force Q_1 (Fig. 14) can be disregarded in the equations of equilibrium. Analogous simplifications are introduced in the compatibility equations. Here terms including ε_2, γ and λ_2 are dropped.

In short, flexible shells have a semi-momentless deformation. Substantial wall-bending occurs only in the direction of *one* of the surface coordinates. Variation of the stress state with this coordinate is stronger than with the other.

Conversely, inspection of the known cases suggests that any substantially large elastic displacements of shells should exhibit the basic features of semi-membrane deformation just stated (e.g. [221, 222, 228]).

The domain of the flexible-shell theory within the general framework of the theory of thin shells can be delimited by comparison with the two branches of the thin-shell theory, responsible for most of its applications. The membrane theory (Chapter 1, §9.3) describes stress states varying weakly with respect to *both* coordinates. The Donnell–Mushtari–Koiter theory (Chapter 1, §9.1) encompasses those stress states varying strongly with *both* coordinates. As indicated above, the semi-membrane theory of flexible shells covers the deformation which varies with respect to *one* of the surface coordinates *much more strongly* than with respect to the other.

2. Reissner equations

For the two basic axisymmetric Saint-Venant problems a system of two second-order nonlinear equations, the boundary conditions and formulas for stress, strain and displacements are derived.

2.1. Compatibility and equilibrium equations and their integrals

The deformation to be considered here preserves the rotational symmetry of the shell form. The geometry of deformed shell is thus described by formulas differing from (1.15) and (1.16) in that the *values* of variables are referred to the *deformed* shell—as indicated by an asterisk $(\alpha^*, R_i^*, R^*, \rho_i^*)$.

With the angle between two meridional planes denoted by φ, we have the relation (Figs. 5 and 13)

$$1 + \varepsilon_1 = \frac{ds_1^*}{ds_1} = \frac{R^*\varphi^*}{R\varphi} \quad \text{or} \quad \frac{1+\varepsilon_1}{R^*} = \frac{k}{R}, \quad k = \frac{\varphi^*}{\varphi}. \tag{2.8}$$

Now (1.15), (1.16), (1.26) and (1.29) give

$$\left[\frac{1}{R_1'} \quad \frac{1}{\rho_1'} \quad \frac{1}{R_2'}\right] = \left[\frac{kc^*}{R} \quad \frac{ks^*}{R} \quad \frac{\dot{\alpha}^*}{b}\right], \qquad \left[\begin{array}{c} c^* \\ s^* \end{array}\right] = \left[\begin{array}{c} \cos\alpha^* \\ \sin\alpha^* \end{array}\right],$$

$$[\kappa_1 \quad \lambda_1 \quad \kappa_2] = \left[\frac{kc^* - c}{R} \quad \frac{ks^* - s}{R} \quad \frac{\dot{\vartheta}}{b}\right], \tag{2.9}$$

where

$$\vartheta(\eta) = \alpha^*(\eta) - \alpha(\eta), \qquad (\ )^{\cdot} = \frac{d(\ )}{d\eta}. \tag{2.10}$$

The angle of rotation ϑ of a tangent to a meridian in the plane $\xi = \text{const.}$ is identical to the ϑ_2 introduced in (1.37) for small displacements. The subscript "2" may now be omitted and the value of ϑ is unrestricted.

The analysis of deformation culminates in a compatibility equation following from the differentiated relation (2.8). With $\dot{R}^* = -(1 + \varepsilon_2)bs^*$ and $\dot{R} = -bs$ we have

$$\frac{1}{b}(\varepsilon_1 R)^{\cdot} + (1 + \varepsilon_2)ks^* - s = 0. \tag{2.11}$$

This is an *integral of* the nonlinear *compatibility equations* (1.36). It can, of course, be obtained by direct integration, as is in fact done for a more general case of deformation in the present chapter, §4. Explicitly, (2.84) is reduced to (2.11) after inserting the expression for λ_1 from the fourth of equations (1.36), $1/\rho_1'$ from (2.9), $a = R$ and dropping the nonsymmetric terms (γ, p_i).

As mentioned in the present chapter, §1.1, for the axisymmetric problems being discussed three of the equilibrium equations (1.56) are satisfied identically. The nonzero stress resultants appear only in the remaining three of the equations of equilibrium, which become

$$(T_2 R)^{\cdot} + Q_2 R\dot\alpha^* + T_1 kbs^* + q_2 bR = 0 \,,$$

$$-T_2 R\dot\alpha^* + (Q_2 R)^{\cdot} - T_1 kbc^* + qbR = 0 \,; \tag{2.12}$$

$$\frac{1}{b}(M_2 R)^{\cdot} + M_1 ks^* - Q_2 R + m_2 R = 0 \,. \tag{2.13}$$

Equations (2.12) may be integrated. Multiplying the first of the equations by c^* or by s^* and adding the second equation multiplied by s^* or $-c^*$ gives (with $\dot s^* = c^* \dot\alpha^*$ and $\dot c^* = -s^* \dot\alpha^*$)

$$[R(T_2 c^* + Q_2 s^*)]^{\cdot} = -Rb(q_2 c^* + qs^*) \,,$$

$$[R(T_2 s^* - Q_2 c^*)]^{\cdot} = Rb(-q_2 s^* + qc^*) - T_1 kb \,. \tag{2.14}$$

Upon integration, these equations lead to expressions for T_2 and Q_2. Introducing a new unknown—stress function $V(\eta)$—and using the notation $P_z(\eta)$ for the axial stress resultant (Fig. 13) in the section $\eta = \text{const.}$ of breadth φR, we have

$$T_1 b = V^{\cdot} \,, \qquad T_2 R = -Vks^* + F_1 \,, \qquad Q_2 R = Vkc^* + F \,;$$

$$\begin{bmatrix} F_1 \\ F \end{bmatrix} = \begin{bmatrix} s^* \\ -c^* \end{bmatrix} \int R(-q_2 s^* + qc^*) b \, \mathrm{d}\eta + \begin{bmatrix} c^* \\ s^* \end{bmatrix} \frac{P_z}{\varphi} \,, \tag{2.15}$$

$$\frac{1}{\varphi} P_z = \frac{1}{\varphi} P_z(\eta_0) - \int_{\eta_0}^{\eta} R(q_2 c^* + qs^*) b \, \mathrm{d}\eta \,.$$

2.2. Reduced system

The two equations (2.11) and (2.13) can be transformed into a system involving only the two unknowns ϑ and V. The strain resultants ε_i are expressed in terms of the T_i and the M_i in terms of the κ_i by using the elasticity relations. It remains to introduce (2.9) and (2.15).

For an orthotropic and nonhomogeneous shell—with the elasticity relations (1.86) and (1.87)—the system is

$$(\dot{V}RB_1')^{\cdot}\frac{1}{b^2} - V\left[(s^*k)^2\frac{B_2'}{R} - (s^*B_\nu')^{\cdot}\frac{k}{b}\right] + s^*k - s$$

$$= (F_1B_\nu')^{\cdot}\frac{1}{b} - \frac{s^*}{R}kF_1B_2',$$

$$(\dot{\vartheta}RD_2)^{\cdot}\frac{1}{b^2} + s^*k\left[(kc^* - c)\frac{D_1}{R} + \dot{\vartheta}\frac{1}{b}D_\nu\right]$$

$$+ \frac{1}{b}(kc^*D_\nu - cD_\nu)^{\cdot} - kc^*V = F - m_2R.$$

(2.16)

The system determining the ϑ and V may be derived in a similar manner for the more general elastic relations (1.85). However, in what follows the intention is to look into the basic features of the flexible shells. This is more effectively done for the simplest elastic properties—for isotropic, homogeneous shells with the elasticity relations (1.83) and (1.84) $(B_i' = 1/Eh, \ldots)$. Further, it is expedient to bring out the dependent function ϑ in all terms of the equations, where it is involved implicitly in c^* and s^*. To this end, introduce the expressions

$$s^* = s + \vartheta c', \quad c'(\alpha, \vartheta) = c\frac{\sin \vartheta}{\vartheta} + s\frac{\cos \vartheta - 1}{\vartheta} = c - \tfrac{1}{2}s\vartheta - \cdots,$$

$$c^* = c - \vartheta s', \quad s'(\alpha, \vartheta) = -c\frac{\cos \vartheta - 1}{\vartheta} + s\frac{\sin \vartheta}{\vartheta} = s + \tfrac{1}{2}c\vartheta - \cdots;$$

(2.17)

transforming the equations to a form differing from the linear approximation only in the values of coefficients attached to the unknowns. In nondimensional form the system of equations becomes

$$\left(\psi\frac{r}{t}\right)^{\cdot} - \psi r\left[\frac{1}{t}\beta^2 s^{*2} - \beta\nu\left(\frac{s^*}{t}\right)^{\cdot}\right] + \mu^*c'\vartheta$$

$$= -ms - h^\circ\frac{\beta}{t}s^*F_1^\circ + h^\circ\nu\left(\frac{F_1^\circ}{t}\right)^{\cdot},$$

$$(\dot{\vartheta}rt^3)^{\cdot} - \vartheta r[\beta^2 s^*s't^3 + \beta\nu(s't^3)^{\cdot}] + \dot{\vartheta}r\nu\beta(s^* - s')t^3 - \mu^*c^*\psi$$

$$= F^\circ - mh^\circ[s^*c\beta t^3 + \nu(ct^3)^{\cdot}] - m_2 b^2 r/D_m.$$

(2.18)

Here we have introduced the dimensionless stress function

$$\psi = V\sqrt{12(1 - \nu^2)}/Eh_m^2 \tag{2.19}$$

and dimensionless parameters of curvature, bending deformation and wall-thickness

$$\mu = \sqrt{12(1-\nu^2)}\frac{b^2}{R_m h_m}, \quad \mu^* = k\mu ,$$

$$m = \mu^* - \mu = \mu\frac{\varphi^* - \varphi}{\varphi}, \quad h^\circ = \frac{h_m}{b\sqrt{12(1-\nu^2)}}. \tag{2.20}$$

Other symbols used in (2.18) are

$$r = \frac{R}{R_m}, \quad t = \frac{h}{h_m}, \quad [F_1^\circ \; F^\circ] = \frac{b^2}{R_m D_m}[F_1 \; F], \quad \beta = \frac{kb}{R},$$

$$D_m = Eh_m^3/[12(1-\nu^2)]. \tag{2.21}$$

The constants R_m and h_m are equal to chosen mean values of $R(\eta)$ and $h(\eta)$.

For the case of a constant normal pressure q and $q_1 = q_2 = 0$, the load functions are (in P° we set $\varphi = 2\pi$ as only $\varphi^*/\varphi = k$, not φ or φ^*, appear in the analysis)

$$F^\circ = P^\circ s^* + q^\circ p , \quad F_1^\circ = P^\circ c^* + q^\circ p_1 ,$$

$$\begin{bmatrix} p \\ p_1 \end{bmatrix} = \begin{bmatrix} -s^* \\ -c^* \end{bmatrix}\int_{\eta_0}^{\eta} rs^*\, d\eta + \begin{bmatrix} -c^* \\ s^* \end{bmatrix}\int_{\eta_1}^{\eta} rc^*\, d\eta ; \tag{2.22}$$

$$P^\circ = P_z(\eta_0)\frac{b^2}{2\pi R_m D_m}, \quad q^\circ = q\frac{b^3}{D_m}.$$

The values of η_0 and η_1 may be chosen to simplify the functions ψ, p and p_1. The stress function ψ, besides being an integral of T_1, is simply related to the radial stress resultant T_r,

$$T_r = Q_2 c^* - T_2 s^* = \frac{D_m}{b^2}\mu^*\frac{\psi}{r} - \frac{b}{r}\int_{\eta_1}^{\eta}(qc^* - q_2 s^*)r\, d\eta . \tag{2.23}$$

All stress and strain resultants are explicitly expressed in terms of the functions ψ and ϑ. The following formulas are useful in this respect:

$$T_1 \frac{b^2}{D_m} = \frac{\dot{\psi}}{h^\circ}, \qquad T_2 \frac{b^2}{D_m} = -\psi\mu^* \frac{s^*}{r} + \frac{F_1^\circ}{r},$$

$$\kappa_2 b = \dot{\vartheta}, \qquad \kappa_1 \frac{b}{h^\circ} = -\vartheta\mu^* \frac{s'}{r} + m\frac{c}{r}. \tag{2.24}$$

Similar expressions may also be written for nonhomogeneous ortho-tropic shells.

The deformed shape of the shell is fully determined by c^*, s^* and ε_2 in terms of ϑ and ψ. Indeed, with (2.20) and $[R^* \; z^*]^{\displaystyle\cdot} = b(1 + \varepsilon_2)[-s^* \; c^*]$ we have (cf. Fig. 5) for the deformed-meridian coordinates and the displacements

$$[R^* \; z^*] = b \int (1 + \varepsilon_2)[-s^* \; c^*]\,\mathrm{d}\eta \; ;$$

$$u_r = R^* - R = -b \int (\vartheta c' + \varepsilon_2 s^*)\,\mathrm{d}\eta \; , \tag{2.25}$$

$$u_z = z^* - z = -b \int (\vartheta s' - \varepsilon_2 c^*)\,\mathrm{d}\eta \; .$$

For a shell closed along the ξ-line (parallel circle), one obviously has

$$u_r = R\varepsilon_1 \; . \tag{2.26}$$

2.3. Simplified system for flexible shells

Equations (2.16) and (2.18) are valid for *any* rotationally symmetric deformation. For *flexible shells* the equations may be considerably simplified. The terms of equations (2.18) with factors β^2 and $\nu\beta$ may be omitted.

Indeed, compared to the first terms of the corresponding equation the β^2 and $\nu\beta$ terms have orders of magnitude indicated by

$$\left(\frac{b}{nR^*}\right)^2, \quad \nu \frac{R^*}{R_2^*}c^*\left(\frac{b}{nR^*}\right)^2, \quad \nu\left(\frac{b}{nR^*}\right)\frac{\vartheta c}{2}, \quad |\dot{\vartheta}| \sim n|\vartheta|, \quad |\dot{\psi}| \sim n|\psi|. \tag{2.27}$$

The value $2\pi b/n$, with n as introduced in (2.27), is a wavelength of deformation in the η-direction. The value of $2\pi R$ is the wavelength of variation of the (vector) parameters of the deformation ϑt_1 and ψt_1 in the circumferential direction. In the axisymmetrical case being discussed only the *direction* of the vectors ϑt_1 and ψt_1 varies with ξ. For *flexible shells* the variation in the direction of η is so much more intensive than in the coordinate ξ, that the ratio $b^2/(nR)^2$ of the squares of the two wavelengths of variation is negligible compared to unity. (The characteristic feature of the type of stress state, which assures the flexibility.) The ratio $b^2/(nR)^2 \ll 1$ is substantiated in what follows for the problems considered.

There is yet another way of estimating the β^2 and $\nu\beta$ terms of the equations.

Consider a linear approximation of the equations (2.18). For $h =$ const. the equations are static-geometric duals. This allows their conversion to complex form.

Setting $s^* = s' = s$ and $c^* = c' = c$ in (2.18) (linearization), multiplying the first of the equations by $i = \sqrt{-1}$ and adding it to the second one results in a single equation for the complex variable $\bar{\vartheta}$,

$$(r\dot{\bar{\vartheta}})^{\cdot} - \bar{\vartheta}r\beta^2 s^2 - \bar{\vartheta}\nu r\beta \dot{s} + i\mu c\bar{\vartheta}$$

$$= F° - ih°(s\beta F_1° - \nu \dot{F}_1°) - ims - mh°(sc\beta + \nu \dot{c}), \qquad (2.28)$$

$$\bar{\vartheta} = \vartheta + i\psi, \qquad \bar{\vartheta} = \vartheta - i\psi, \qquad i = \sqrt{-1}.$$

The load functions $F°$ and $F_1°$, as defined in (2.22), are of the same order of magnitude. Thus, the terms with $h°F_1°$ are of the order of $ih/3R$ times the $F°$ term.

The two β terms on the left-hand side of the equation have the estimates

$$\frac{\bar{\vartheta}r\beta^2 s^2}{i\mu c\bar{\vartheta}} \sim i\frac{k^2 b/R}{\cos\alpha}\frac{h}{3R}, \qquad \frac{\bar{\vartheta}\nu r\beta \dot{s}}{i\mu c\bar{\vartheta}} \sim \frac{\bar{\vartheta}k}{i\bar{\vartheta}}\frac{\nu h}{3R_2}.$$

Terms of relative order of h/R_j are negligible. However, the terms being compared here are complex and the relevant small ratio is not $h/3R$ but $ih/3R$. Omitting these terms we neglect $\vartheta h/3R$ compared to ψ and

$\psi h/3R$ compared to ϑ (cf. Chapter 1, §8.3). This is correct if the orders of magnitude of the two resolving functions are not too different. This relation is implied by the static-geometric duality: ϑ and ψ are represented in (2.18) in a similar way. In all the flexible shell problems there actually is $|\psi| \sim |\vartheta|$.

Thus, (2.28) takes the form

$$(r\dot{\bar{\vartheta}})^{\cdot} + i\mu c\bar{\vartheta} = F^{\circ} - ims . \tag{2.29}$$

A case, for which this simplification is *not* possible, is presented by a plate or shallow shell, when $\cos \alpha^* \ll 1$. Here, the second term in (2.28) and sometimes the F_1 terms must be retained. The terms involving F_1 account for the load parallel to the middle plane of a plate. The F_1 terms can also be significant when the load function F approaches zero. A case of this sort—circular toroidal shell under normal pressure—is discussed in Chapter 5, §5.

Without those terms which are small for flexible shells (those dropped in (2.29)), equations (2.18) become

$$\left(\psi\frac{r}{t}\right)^{\cdot} + \mu^* c'\vartheta = -ms \quad (c'\vartheta \equiv s^* - s) ,$$
$$(\dot{\vartheta} rt^3)^{\cdot} - \mu^* c^* \psi = P^{\circ} s^* + q^{\circ} p . \tag{2.30}$$

These equations can be extended to nonhomogeneous orthotropic shells. An analysis similar to the foregoing evolves (2.16) to the form (2.30) with

$$\frac{1}{t} = \frac{B_1'}{B_{1m}'}, \quad t^3 = \frac{D_2}{D_{2m}}, \quad \psi = \frac{V}{\sqrt{D_{2m}/B_{1m}'}}, \quad \mu = \frac{b}{R_m h^{\circ}},$$
$$F^{\circ} = Fb^2/(R_m D_{2m}), \quad h^{\circ} = \sqrt{B_{1m}' D_{2m}}/b . \tag{2.31}$$

Here, D_{2m} and B_{1m}' are chosen mean values of $D_2(\eta)$ and $B_1'(\eta)$.

The system (2.30) is of fourth order. This means that only *two* boundary conditions can be met on any edge $\eta = $ const. One of the *four* conditions on an edge of a thin shell (discussed in Chapter 1, §7) is fulfilled identically. It is the condition involving S_i and H_i or τ_j and γ.

These resultants vanish for any axisymmetric deformation. Another condition is, in fact, substituted by the determination of one of the constants, P_z and $\varphi^* - \varphi$, introduced in the integrals of the equilibrium and of the compatibility equations. This is done in terms of one of the following conditions: (a) the flexure parameter m (representative of $\varphi^* - \varphi$) is determined in terms of the bending moment M_z in the section $\xi = $ const.; (b) the axial force P_z is determined in terms of displacement Δ_z of one edge of the shell ($\eta = \eta_1$) with respect to the other edge ($\eta = \eta_2$):

$$\int_{\eta_1}^{\eta_2} [T_1(R^* - R_m^*) + M_1 c^*] b \, \mathrm{d}\eta = M_z ,$$

$$u_z(\eta_2) - u_z(\eta_1) = \Delta_z . \tag{2.32}$$

When the shell is closed along the parallels ($\varphi = 2\pi$), the first of equations (2.32) is obviously irrelevant.

Equations (2.32) are static-geometric duals of each other in their linear approximation. This becomes evident when the u_z is written in terms of κ_2 and ε_2 or T_1, M_1 and u_z in terms of ϑ and ψ.

The z-components of the stress and strain resultants in a section $\eta = $ const. are expressed directly in terms of P_z and $k - \cos \vartheta$,

$$T_2 c^* + Q_2 s^* = \frac{1}{2\pi R} P_z , \qquad \kappa_1 c^* + \lambda_1 s^* = \frac{1}{R}(k - \cos \vartheta) .$$

The two conditions on an edge $\eta = $ const. may involve besides M_2 and ε_1 the *radial* components of the stress and strain resultants T_2 and κ_1,

$$Q_2 c^* - T_2 s^* , \qquad \lambda_1 c^* - \kappa_1 s^* = -(\sin \vartheta)/R .$$

For a shell with *closed* meridians (η-lines) the boundary conditions are replaced by continuity conditions for stress, strain and displacement (Chapter 1, §7).

2.4. Middle-range-thickness shells

The thinness condition $h/R_j \ll 1$, necessary for the adequacy of the thin-shell theory, is not fulfilled for two types of shells discussed in the following sections. In Bourdon tubes and in bellows, the radius of curvature R_2 reaches values of the same order as the wall-thickness h. Also, the minimum values of R_2 occur in just that zone of shell where the stresses are maximal.[2] To determine the stress in such situations we need a theory that is valid for shells with a wall-thickness in the range

$$h^2/R_j^2 \ll 1 . \qquad (2.33)$$

Consider briefly a derivation of the formulas for the strain and the stress components and of the system of equations for the shells of intermediate-range thickness (2.33). We follow essentially the work of REISSNER [47] (the main difference being in the factors by σ_2 and σ_{23} in (2.36)).

The hypotheses of thin-shell theory must now be replaced by less restrictive ones. Assume an expression for displacement allowing for the extension of a normal linear element and for its inclination with respect to the middle surface, determined by the angle γ_2 representing the *transverse shear* (cf. (1.42))

$$U = u + \zeta \boldsymbol{\vartheta} \times n + \left(e'\zeta + \frac{1}{h}\zeta^2 e'' \right) n , \quad \boldsymbol{\vartheta} = -(\vartheta^0 + \gamma_2)t_1 . \qquad (2.34)$$

The vector $\boldsymbol{\vartheta}$ $(= -\vartheta t_1)$ is the angle of small rotation of the linear element $n\zeta$ *normal* to the middle surface before deformation. The angle $\vartheta^0 t_1$ of rotation of the *tangent* plane is different (as opposed to the thin-shell theory where it is the same by hypothesis).

The components of the rotationally symmetric strain, obtained by substitution of (2.34) for U into (1.38) generalized for a layer $\zeta \neq 0$, are

$$\left(1 + \frac{\zeta}{R_2}\right)e_2 = \varepsilon_2 + \zeta\kappa_2 + \frac{\zeta}{R_2}e' + \frac{\zeta^2}{hR_2}e'' \quad (2\ 1) ,$$

$$e_3 = \frac{\partial}{\partial\zeta}(U \cdot n) = e' + \frac{2\zeta}{h}e'' , \qquad (2.35)$$

$$\left(1 + \frac{\zeta}{R_2}\right)e_{23} = \gamma_2 + \zeta\frac{\dot{e}'}{b} + \frac{\zeta^2}{h}\frac{\dot{e}''}{b}, \quad (\ )^{\cdot} = \frac{\mathrm{d}(\ )}{\mathrm{d}\eta} .$$

_____

[2] The zone does not occur in the vicinity of shell edges. Therefore, the difficult problem of boundary conditions need not be discussed.

All the equilibrium and compatibility equations presented for thin shells remain valid. However, we need two additional equilibrium equations, corresponding to the new degrees of freedom—the newly introduced displacement parameters e' and e''. The elasticity equations must be generalized to take the transverse extension and the transverse shear into account. This could be done with the aid of the principle of virtual work without any further assumptions (as in Chapter 1, §6). A simpler (and more symmetric with respect to stress and strain) approach to deriving the equations is to *assume* expressions for the *stress* distribution, corresponding to the strain expressions (2.35),

$$\left(1 + \frac{\zeta}{R_2}\right)\sigma_2 = \sigma_2' + \frac{2\zeta}{h}\sigma_2'' \quad (2 \ 1),$$

$$\left(1 + \frac{\zeta}{R_2}\right)\sigma_{23} = \sigma_{23}'\left(1 - \frac{4\zeta^2}{h^2}\right), \tag{2.36}$$

$$\left(1 + \frac{\zeta}{R_1}\right)\left(1 + \frac{\zeta}{R_2}\right)\sigma_3 = \left(\sigma_3' + \frac{2\zeta}{h}\sigma_3''\right)\left(1 - \frac{4\zeta^2}{h^2}\right).$$

The stress parameters σ_2', σ_2'' and σ_{23}' introduced here are related to the stress resultants by the conditions of static equivalence, which can be constructed directly. Based on the diagram in Fig. 4 for $1/R_{12} = 1/R_{21} = 0$ and on the definitions of the stress resultants (Chapter 1, §4), one has

$$\int_{-h/2}^{h/2} [\sigma_2 \ \sigma_2\zeta \ \sigma_{23}]\left(1 + \frac{\zeta}{R_1}\right) d\zeta = [T_2 \ M_2 \ Q_2] \quad (2 \ 1). \tag{2.37}$$

The two additional equilibrium equations and the seven elasticity equations expressing ε_i, κ_i, γ_2, e' and e'' in terms of T_i, Q_2, M_i, σ_3' and σ_3'' follow from the REISSNER [49] variational theorem

$$\delta\left\{\int\int_V \left[\sum \sigma_{ij}e_{ij} - \Psi\right] dV - \int_{\Pi_1} f \cdot U \, d\Pi - \int_{\Pi_2} (U - U_B) \cdot f^R \, d\Pi\right\} = 0. \tag{2.38}$$

Here f and f^R are the intensities of the distributed load on the parts Π_1 and Π_2 of the surface of the shell, where the forces are active (prescribed) and reactive, respectively; U_B denotes the prescribed displacement of Π_2. The function $\Psi(\sigma_{ij})$ represents Hooke's law by formulas

$$e_{ij} = \partial\Psi/\partial\sigma_{ij}.$$

Equating the cofactors of variations of parameters σ_3', σ_3'', ε_i, κ_i, e', e'' and γ_2 on the left-hand side of (2.38) to zero, we obtain the two additional equilibrium equations and the elasticity relations just mentioned. With these equations and (2.24) and (2.33), equations (2.11) and (2.13) yield a system for the two unknowns ϑ and ψ.

The new system differs from the thin-shell equations (2.18) in the meaning of ϑ and in the appearance of an additional term in the first equation. The additional term represents the transverse shear γ_2. Its order of magnitude, compared to the leading term $\ddot{\psi}$ of the equation, is $h^2/(6R_i^2)$. In accordance with condition (2.33) this term must be omitted. Thus, equations (2.18), or their version (2.30), also describe the deformation of shells of the *intermediate-range thickness*. The remaining difference to the thin-shell equations lies in the meaning of ϑ and is of no practical importance. The influence of the transverse extension and the transverse shear on the values of the stress and strain resultants as well as on the displacements appears to be negligible also for the shells of an intermediate-range thickness.

On the contrary, the corrections are considerable in the formulas for the *stresses*. According to (2.36) the maximum through the wall-thickness stresses for the shells with $|h/R_2| < 0.4$ and $|h/R_1| \ll 1$ (the usual dimensions for flexible shells) are

$$\sigma_1 = \frac{T_1}{h} \pm \frac{Eh/2}{1-\nu^2}\left(\kappa_1 + \nu\frac{\dot{\vartheta}}{b}k_R\right), \quad k_R = \frac{1 \pm h/(6R_2)}{1 \pm h/(2R_2)},$$

$$\sigma_2 = \frac{T_2}{h}\left(1 + \frac{1+1.2\nu}{2}\frac{h}{R_2}\right) \pm \frac{Eh/2}{1-\nu^2}\left(\frac{\dot{\vartheta}}{b}k_R + \nu\kappa_1\right). \tag{2.39}$$

3. Schwerin–Chernina equations

For the basic nonaxisymmetric Saint-Venant's linear problems (Fig. 13), namely, for the lateral bending and torsion of shells of revolution and curved beam shells, a system of two second-order equations of the

Reissner–Meissner type and the appropriate boundary conditions are discussed.[3]

3.1. Integrals of the equilibrium and compatibility equations

Consider a shell, corresponding to the diagrams (e) and (f) of Fig. 13, subject to a loading which varies as cos ξ or sin ξ. With appropriately distributed forces on the edges η = const. (Saint-Venant's problem), the stress and the strain resultants will contain in the linear approximation only sin ξ and cos ξ harmonics in (2.3) and (2.4). The equilibrium and compatibility equations, as well as boundary conditions and elasticity relations, can be reduced to a form determining the *amplitudes* of the resultants $T_1^1(\eta), \ldots, \gamma^1(\eta)$. In particular, the linearized fourth equations of equilibrium and of compatibility (1.56) and (1.36) give, after substitution of (2.3) and (2.4),

$$\frac{1}{b}(R\varepsilon_1^1)^{\cdot} - \frac{\gamma^1}{2} + s\varepsilon_2^1 + R\lambda_1^1 = 0 ,$$

$$\frac{1}{b}(RM_2^1)^{\cdot} + H_1^1 + sM_1^1 - RQ_2^1 = 0 . \tag{2.40}$$

Equilibrium of the part of the shell between the edge $\eta = \eta_1$ and some section η = const. determines the resultant force $P(\eta)$ and moment $L(\eta)$ of the stresses acting over the section in terms of the edge-load resultants $P(\eta_1)$ and $L(\eta_1)$ and the distributed load intensities q and m. Upon integration with respect to ξ, the static-equivalence relations,

$$\int_0^{2\pi} T_2 R \, d\xi = P , \qquad \int_0^{2\pi} [M_2 + R \times T_2] R \, d\xi = L \quad (R = ix + jy) , \tag{2.41}$$

become equations for the components of T_2 and M_2, providing integrals of the equilibrium equations. To obtain explicit expressions for these integrals, the stress resultants in the vector equations (2.41) must be presented in terms of the components $T_2, S_2, \ldots$ and of the unit vectors of the reference system x, y, z.

---

[3] The equations were proposed by SCHWERIN [7] for spherical shells and by CHERNINA [66] for any shell of revolution. The extension to the curved-beam problem is due to WAN [135, 136].

Therefor we obtain from Fig. 5

$$\begin{bmatrix} t_1 \\ t_2 \\ n \end{bmatrix} = \begin{bmatrix} -\sin \xi & \cos \xi & 0 \\ -s \cdot \cos \xi & -s \cdot \sin \xi & c \\ c \cdot \cos \xi & c \cdot \sin \xi & s \end{bmatrix} \begin{bmatrix} i \\ j \\ t_z \end{bmatrix}. \tag{2.42}$$

With the expansions (1.54), where t_i' and n' are taken respectively equal to t_i and n (linear approximation), expressions $T_1 = T_1^1(\eta) \cos \xi, \ldots$, and (2.42), equations (2.41) give[4]

$$-T_2^1 s + Q_2^1 c - S_2^1 = \frac{1}{\pi R} P_x ,$$

$$-H_2^1 s - M_2^1 - T_2^1 Rc - Q_2^1 Rs = \frac{1}{\pi R} L_y . \tag{2.43}$$

Consider now the stress resultants in the section $\xi = \text{const.}$ The moment $M_z(\xi, \eta)$ of stresses acting in the part of this section between $\eta = \eta_1$ and a current point $m(\xi, \eta)$ is expressed in terms of a new dependent function $V(\eta)$ introduced by

$$\int_{\eta_1}^{\eta} (T_1^1 R + M_1^1 c) \cos \xi b \, d\eta = M_z(\xi, \eta) = VR \cos \xi$$

or

$$M_z(0, \eta) = VR . \tag{2.44}$$

Differentiation gives

$$T_1^1 R + M_1^1 c = \frac{1}{b} (RV)^{\cdot} . \tag{2.45}$$

We now form the equation of equilibrium of moments about the axis z (Fig. 13) for a part of the shell enclosed between surfaces $\eta = \eta_1$ and some $\eta = \text{const.}$ and the meridional planes $\xi = 0, \pi$. The equation is

---

[4] Expressions for P_x and L_y in terms of distributed load are written out in Chapter 3, §4.

$$M_z(\pi, \eta) - M_z(0, \eta) + \int_0^\pi \sin \xi \, d\xi (S_2^1 R^2 + cH_2^1 R + fR)_{\eta_1}^\eta = 0 ,$$

$$f = \frac{b}{R} \int_{\eta_1}^\eta (Rq_1^1 + m_1^1 c)R \, d\eta .$$

(2.46)

The solution of algebraic equations (2.43), (2.45) and (2.46) yields expressions for the stress resultants in terms of the function V and of the load functions f_1 and f_2. They are

$$RT_1^1 = \frac{1}{b}(RV)^{\cdot} - cM_1^1 ,$$

$$RT_2^1 = -sV - cM_2^1 - f_1 ,$$

$$RS^1 = RS_2^1 - cH_1^1 = V - 2cH^1 - f , \qquad \begin{bmatrix} f_1 \\ f_2 \end{bmatrix} = \begin{bmatrix} s & c \\ -c & s \end{bmatrix} \begin{bmatrix} (P_x/\pi) - f \\ L_y/\pi R \end{bmatrix} .$$

$$RQ_2^1 = cV - H_2^1 - sM_2^1 - f_2 ,$$

(2.47)

The analysis of deformation will give relations that are dual to (2.47).

Integration of equations (1.38) and (1.39) provides relations between the strain resultants ε_1 and κ_1 and the new distortion parameters Ω and U,

$$-\int_0^{2\pi} \kappa_1 R \, d\xi = \Omega , \quad \int_0^{2\pi} (\varepsilon_1 + R \times \kappa_1)R \, d\xi = [R \times \vartheta + u]_0^{2\pi} = U .$$

(2.48)

A further relation exists between the resultants ε_2 and κ_2 and displacement of a line $\eta = $ const. relative to the edge $\eta = \eta_1$,

$$\int_{\eta_1}^\eta [\varepsilon_2 + (R + zt_z) \times \kappa_2]b \, d\eta = [(R + zt_z) \times \vartheta + u]_{\eta_1}^\eta .$$

(2.49)

With the aid of (2.42), the projection of this equation onto the z-axis can be written in the form dual to (2.44)

$$\int_{\eta_1}^\eta (-\kappa_2^1 R + \varepsilon_2^1 c) \cos \xi b \, d\eta = [-R\vartheta_2 + u_z]_{\eta_1}^\eta = -\vartheta R \cos \xi .$$

(2.50)

The new dependent function $\vartheta(\eta)$ introduced here has a specific geometric meaning: $\vartheta(\eta)$ is the angle of rotation of a tangent plane relative to a plane passing through the line $\eta = $ const. and connected to it during any deformation of the shell.

Equations (2.48)–(2.50) are dual to the relations (2.41) and (2.44) in the sense of static-geometric duality (1.112). They follow from (2.41) and (2.44) if the parameters there are replaced in accordance with (1.112) and the additional duality relations

$$[P \ P_x \ L \ L_y \ V] \rightleftarrows [-\Omega \ -\Omega_x \ -U \ -U_y \ -\vartheta] \qquad (2.51)$$

are used.

The force and moment resultants P and L are constants in the absence of distributed load (when $q, m = 0$). Consequently (see Chapter 1, §8), their static-geometric duals Ω and U are constants, too.

The integrals (2.48) and (2.50) of the compatibility equations and an equation dual to (2.46) lead to expressions for the strain resultants, dual to (2.47). Transforming (2.47) according to the static-geometric duality relations (1.112) and (2.51) yields

$$R\kappa_2^1 = (R\vartheta)^{\cdot}/b + c\varepsilon_2^1,$$

$$R\kappa_1^1 = -s\vartheta + c\varepsilon_1^1 - g_1,$$

$$R\tau^1 = -\vartheta + c\gamma^1,$$

$$\begin{bmatrix} g_1 \\ g_2 \end{bmatrix} = \begin{bmatrix} s & c \\ -c & s \end{bmatrix} \begin{bmatrix} \Omega_x/\pi \\ U_y/\pi R \end{bmatrix}. \qquad (2.52)$$

$$R\lambda_1^1 = c\vartheta - \tfrac{1}{2}\gamma^1 + s\varepsilon_1^1 - g_2,$$

The equilibrium and compatibility equations of the problem are thus complete.

3.2. Reduced system

Equations (2.40) become a system of two equations in two unknowns ϑ and V, when all of the stress and strain resultants have been expressed in terms of ϑ and V. These expressions follow from formulas (2.47) and (2.52) and from the elasticity relations of Chapter 1, §6. However, they contain a number of terms of order that must be omitted in the theory of thin shells. There are two kinds of small terms. The first are terms of the order of magnitude (compared to the main term of the corresponding expression) of

$$D_i B_j' / (R_k R) \quad (i, j = 1, 2, \nu, G; \; R_k = R_1, R_2) \,. \tag{2.53}$$

Here D_i and B_j' are the elastic constants of an orthotropic nonhomogeneous shell (Chapter 1, §6). For the usual case of an isotropic homogeneous material we have, from (1.83), (1.84) and (1.66),

$$D_i B_j' \sim \frac{h^2}{12}, \tag{2.54}$$

and the values of terms, indicated in (2.53), are negligible—far below the level of error of thin-shell theory, which is h/R_k. Clearly, this remains true for a wide range of orthotropic and nonhomogeneous shell materials.

Another group of small terms of expressions $M_1^1(\vartheta, V), \dots,$ $\gamma^1(\vartheta, V)$, being discussed, comprise those terms, which can be estimated *after* the substitution into (2.40). In the reduced system of equations they represent terms, which are of the order of magnitude of

$$\frac{D_i B_j'}{R R_k}, \quad \frac{(D_i B_j' V)^{\displaystyle\cdot}}{R b V}, \quad \frac{(D_i B_j' \dot{V})^{\displaystyle\cdot}}{b^2 V}, \quad \frac{(D_i B_j' \vartheta)^{\displaystyle\cdot}}{R b}, \quad \frac{(D_i B_j' \dot{\vartheta})^{\displaystyle\cdot}}{b^2 \vartheta},$$

$$\tag{2.55}$$

as compared to the dominant terms of the equation. For homogeneous shells $(D_i B_j' = h^2/12)$, the values are between $h^2/(10 R R_k)$ and $h^2/(10 b^2 / n^2)$, with b/n, determined as in (2.27), being an interval of variation of the stress state in η. According to Chapter 1, §3.1 the corresponding terms must be omitted. The conclusion remains valid for a range of nonhomogeneous orthotropic shells—as long as the values of $D_i B_j'$ are of the same order of magnitude as $h^2/12$.

Neglecting the terms discussed, the expressions of stress and strain resultants following from (2.47), (2.52), (1.86) and (1.87) are

$$M_2^1 = D_2 \frac{(R\vartheta)^{\bullet}}{Rb} - D_\nu \frac{s\vartheta + g_1}{R},$$

$$M_1^1 = D_\nu \frac{(R\vartheta)^{\bullet}}{Rb} - D_1 \frac{s\vartheta + g_1}{R},$$

$$H^1 = -D_G 2 \frac{\vartheta}{R},$$

$$\varepsilon_1^1 = B_1' \frac{(RV)^{\bullet}}{Rb} + B_\nu' \frac{sV + f_1}{R},$$

$$\varepsilon_2^1 = -B_\nu' \frac{(RV)^{\bullet}}{Rb} - B_2' \frac{sV + f_1}{R},$$

$$\gamma^1 = B_G' \frac{V - f}{R}.$$

(2.56)

It is easy to see that simplifications leading to these formulas are equivalent to using, instead of (2.47) and (2.52), of the simplified relations

$$RT_1^1 = \frac{1}{b}(RV)^{\bullet}, \qquad RT_2^1 = -sV - f_1, \qquad RS^1 = V - f,$$

$$R\kappa_2^1 = \frac{1}{b}(R\vartheta)^{\bullet}, \qquad R\kappa_1^1 = -s\vartheta - g_1, \qquad R\tau^1 = -\vartheta.$$

(2.57)

With the expressions (2.47) and (2.52) for Q_2^1 and λ_1^1 and with (2.56) and (2.57), the equations (2.40) become a system of the Schwerin–Chernina type for orthotropic nonhomogeneous shells,

$$(\dot{V}RB_1')^{\bullet} \frac{1}{b^2} - V\left[(B_1' + B_2' - 2B_\nu')\frac{s^2}{R} + B_G' \frac{1}{R} + (sB_1' - sB_\nu')^{\bullet} \frac{1}{b}\right] + c\vartheta$$

$$= -g_2 + f_1 B_2' \frac{s}{R} - \left(\frac{f_1}{R} B_\nu'\right)^{\bullet} \frac{R}{b} - \frac{f}{R} B_G',$$

$$(\dot{\vartheta}RD_2)^{\bullet} \frac{1}{b^2} - \vartheta\left[(D_2 + D_1 + 2D_\nu)\frac{s^2}{R} + D_G \frac{4}{R} + (sD_2 + sD_\nu)^{\bullet} \frac{1}{b}\right] - cV$$

$$= -f_2 + g_1 D_1 \frac{s}{R} + \left(\frac{g_1}{R} D_\nu\right)^{\bullet} \frac{R}{b} - Rm_2^1.$$

(2.58)

It is convenient to use these equations in a nondimensional form.

With the notations introduced in (2.19)–(2.21) (where of course the physical meaning of the functions ϑ and V is different from that defined in (2.44) and (2.50)), equations (2.58) for isotropic homogeneous shells may be written in the form

$$
\left(\dot{\psi}\frac{r}{t}\right)^{\cdot} - \psi\lambda\left[2\lambda\frac{1+s^2+\nu c^2}{rt} + \left(\frac{1-\nu}{t}s\right)^{\cdot}\right] + \vartheta c\mu
$$

$$
= g_2^\circ - h^\circ\left[\frac{1+\nu}{rt}2\lambda f^\circ + \left(\frac{\nu f_1^\circ}{rt}\right)^{\cdot}r - \frac{\lambda s}{rt}f_1^\circ\right],
$$

$$
(\dot{\vartheta}rt^3)^{\cdot} - \vartheta\lambda\left[2\lambda t^3\frac{1+s^2-\nu c^2}{r} + (1+\nu)(st^3)^{\cdot}\right] - \psi c\mu \tag{2.59}
$$

$$
= -f_2^\circ + h^\circ\left[\left(t^3\frac{\nu}{r}g_1^\circ\right)^{\cdot}r + \frac{s}{r}\lambda t^3 g_1^\circ\right] - rm_2^\circ .
$$

$$
[f_i^\circ\ f^\circ] = [f_i\ f]b^2/(R_m D_m), \quad g_i^\circ = g_i\mu, \quad \lambda = b/R_m . \tag{2.60}
$$

The resolving system may be substantially simplified for the *flexible* shells. Thus, under conditions stated in the present chapter, §1.3 and discussed in §2 for the symmetrical case, the terms of equations (2.59) with coefficients λ and h° may be omitted.

Indeed, the second terms of equations (2.59) are of the order of magnitude of

$$
\left(\frac{b/n}{R}\right)^2\left(2 + c\frac{R}{R_2}\right) \qquad
$$

(with n determined by $|\dot{\vartheta}| \sim n|\vartheta|, |\ddot{\vartheta}| \sim n^2|\vartheta|, |\dot{\psi}| \sim n|\psi|)$,

compared to the first terms of the equations. As discussed in connection with (2.27), the value of $b^2/(nR)^2$ is negligible for flexible shells. The *interval of variation* b/n is smaller than the radius of curvature R_2 of the meridian. Thus, the second terms of equations (2.59) are negligible, compared to the first.

The small terms become easier to evaluate when (2.59) are presented in complex form, similar to (2.28). Consider now a constant-thickness shell ($t = 1$). Multiplying the first of equations (2.59) by $i = \sqrt{-1}$ and adding the second equation produces an equation for $\bar{\vartheta} = \vartheta + i\psi$ and

$$\bar{\vartheta} = \vartheta - i\psi,$$

$$(\dot{\bar{\vartheta}}r)^{\cdot} - \bar{\vartheta}\lambda\left[2\lambda\frac{1+s^2}{r} + \dot{s}\right] + \bar{\vartheta}\nu 2\lambda^2\frac{c^2}{r} + ic\mu\bar{\vartheta}$$

$$= -f_2^{\circ} + ig_2^{\circ} - rm_2^{\circ} + h^{\circ}\left[\nu\left(\frac{g_1^{\circ} - if_1^{\circ}}{r}\right)^{\cdot}r + \lambda\frac{s}{r}(g_1^{\circ} + if_1^{\circ}) + i2\lambda\frac{1+\nu}{r}f^{\circ}\right].$$

$$(2.61)$$

The terms involving f_1°, f° and g_1° are of the order of magnitude of nh/b or h/R compared to f_2° and g_2° terms. Hence, with reservations stated in the present chapter, §2.3, these terms are negligible for thin shells. The two terms in (2.61) involving $\bar{\vartheta}$ and $\bar{\vartheta}$ are of the order of magnitude of $(b/nR)^2$ or $b^2/(RR_2n^2)$ compared to the term $(\dot{\bar{\vartheta}}r)^{\cdot}$. These terms prove to be small for the *flexible* shells.

The complex equation becomes

$$(\dot{\bar{\vartheta}}r)^{\cdot} + ic\mu\bar{\vartheta} = -f_2^{\circ} + ig_2^{\circ} - rm_2^{\circ}. \qquad (2.62)$$

This equation directly confirms the possibility to omit in the system (2.59) the $\psi\lambda$ and $\vartheta\lambda$ terms. It provides also an estimate for the value of n. For the case of edge loading, (2.62) gives $\bar{\vartheta} \sim ic\mu\bar{\vartheta}$ or $|n^2| \sim \mu c$. This implies the flexible-shell relation $|b/nR|^2 \ll 1$ for the stress state. (The quantity n^2 may be complex.)

The simplifications discussed can be extended, subject to certain restrictions, to shells of variable thickness ($t \neq 1$), nonhomogeneous and orthotropic elastic properties.

The system (2.59) thus may be used in the simple form

$$\left(\psi\frac{r}{t}\right)^{\cdot} + \vartheta c\mu = c\Omega^{\circ} - \frac{s}{r}U^{\circ},$$

$$(\vartheta rt^3)^{\cdot} - \psi c\mu = cP_x^{\circ} - \frac{s}{r}L^{\circ} - cf^{\circ},$$

$$[\Omega^{\circ}\ U^{\circ}] = -\frac{\mu}{\pi}\left[\Omega_x\ \frac{U_y}{R_m}\right],$$

$$[P_x^{\circ}\ L^{\circ}] = \frac{b^2}{\pi R_m D_m}\left[P_x\ \frac{L_y}{R_m}\right].$$

$$(2.63)$$

Here and in the following the distributed *moment-load* term, m_2, is *omitted*.

Let us note an important feature of (2.63). The derivation of the equations contains, in fact, dropping of all terms with ε_2, γ, λ_2 and M_1, H, Q_1 in the compatibility and equilibrium equations, respectively. This corresponds to the flexible-shell model discussed in the present chapter, §4.2. Consistently, contributions from ε_2, γ and λ_2 are not included in $\Omega°$ and $U°$ and the stress resultants M and N (Fig. 13) determined by a solution of (2.63) do not take into account M_1, H or Q_1. (An instance of the influence of γ is provided by the solution of a tube problem in Chapter 3, §4.2.)

3.3. Boundary conditions

The system of equations derived for the problem is of fourth order. Thus, *two* boundary conditions can and must be fulfilled on each edge $\eta = $ const. Similar to the axisymmetrical case discussed in the present chapter, §2, the remaining conditions are satisfied by a solution of the equations identically.

Consider *force boundary conditions*. In principle, they define on an edge the four stress resultants listed in (1.101). In the problems being discussed here, the stresses vary along the edge as $\cos \xi$ or $\sin \xi$ and the four resultants are represented by the expressions

$$T_2^1 \cos \xi , \qquad \left(Q_2^1 + \frac{1}{R} H_2^1 \right) \cos \xi = Q_{(2)}^1 \cos \xi ,$$

$$\left(S^1 + 2 \frac{c}{R} H^1 \right) \sin \xi = S_{(2)}^1 \sin \xi , \qquad M_2^1 \cos \xi . \tag{2.64}$$

Thus, the boundary conditions include the four *amplitudes* of the resultant $T_2^1(\eta), \ldots$ But in the present solution the four are not independent. They are connected by the relations (2.43), which can be easily transformed to

$$-T_2^1 s + Q_{(2)}^1 c - S_{(2)}^1 = P_x / \pi R ,$$

$$-Q_{(2)}^1 Rs - M_2^1 - T_2^1 Rc = L_y / \pi R . \tag{2.65}$$

Equations (2.43) have been taken into account in the reduced equations

(2.58) and any solution of the equations satisfies (2.43) or (2.65) identically. Thus, only two of the resultants (2.64) are independent— two boundary conditions are necessary and sufficient. For instance, on a *free edge*, η = const., the absence of loading is expressed by *any two* of resultants (2.64) being equal to zero (as well as, naturally, zero values of the resultant force and moment for the entire edge P_x, $L_y = 0$). The remaining two of the stress parameters (2.64) are equal to zero automatically, as a consequence of relations (2.65).

The stress resultants are expressed in the boundary conditions in terms of the functions ϑ and ψ by using (2.56) and (2.57).

The *geometric boundary conditions* can be formulated in terms of the displacements or in terms of the strain parameters. In both cases they can be ultimately expressed in terms of the dependent variables ϑ and ψ. First consider the displacements.

The angle $\vartheta_2^1(\eta)$ of rotation of a tangent to the meridian in the plane of symmetry of displacements may be expressed in terms of the function ϑ with the aid of relations (2.52) and (1.41),

$$\vartheta_2^1 = \int \kappa_2^1 b \, d\eta = \int [(R\vartheta)^{\cdot} + \varepsilon_2^1 cb]\frac{1}{R} \, d\eta . \qquad (2.66)$$

The displacement $u_z = u_z^1 \cos \xi$ in the direction of the z-axis and the corresponding angle u_z^1/R of rotation of the parallel circle (ξ-line) are expressed in terms of ϑ with the aid of (2.50). Substitution of ϑ_2^1 according to (2.66) and integration by parts gives

$$\frac{u_z^1}{R} = \int (-\vartheta s + \varepsilon_2^1 c)\frac{b}{R} \, d\eta . \qquad (2.67)$$

Displacement of the meridian $\xi = 0$ in the direction of x-axis (Fig. 5) and the equal value u_r^1 of the amplitude of radial displacement are defined directly according to the scheme of Fig. 5. With $\alpha^* = \alpha + \vartheta_2^1$,

$$u_x^1 = u_r^1 = -\int [b(1 + \varepsilon_2^1)s^* - bs] \, d\eta = -\int (\vartheta_2^1 c + \varepsilon_2^1 s)b \, d\eta . \quad (2.68)$$

The constants of integration in (2.66)–(2.68) characterize the rigid-body rotations and displacements of the shell. They are equal to the rotation and the displacement of some parallel circle η = const. For a shell closed along the lines $\eta = $ const., these lines remain as undeformed circles (for the problem under consideration here).

With the static-geometric duality the foregoing discussion of static boundary conditions is carried over to the kinematic boundary conditions. The strain conditions on an edge can be expressed in terms of the four strain resultants listed in (1.105a). For the present problem these are

$$\kappa_1^1 \cos \xi \,, \qquad \lambda_{(1)}^1 \cos \xi = \left(\lambda_1^1 - \frac{1}{2R} \gamma^1 \right) \cos \xi \,,$$

$$\tau_{(1)}^1 \sin \xi = \left(\tau_1^1 - \frac{c}{2R} \gamma^1 \right) \sin \xi \,, \qquad \varepsilon_1^1 \cos \xi \,. \tag{2.69}$$

The four parameters κ_1^1, $\lambda_{(1)}^1$, $\tau_{(1)}^1$ and ε_1^1 expressed in terms of ϑ and ψ satisfy the equations dual to (2.43) (obtainable as projections of (2.48) on the x- and y-axes),

$$\tau_{(1)}^1 + \lambda_{(1)}^1 - \kappa_1^1 s = \frac{1}{\pi R} \Omega_x \,, \qquad \lambda_{(1)}^1 Rs - \varepsilon_1^1 R + \kappa_1^1 Rc = \frac{1}{\pi R} U_y \,. \tag{2.70}$$

Thus, the conditions on an edge can prescribe two of the four strain parameters (expressed in terms of ϑ and ψ) or their independent linear combinations with Ω_x and U_y. Two geometric boundary conditions are necessary and sufficient along an edge.

Consider, for instance, the conditions on an edge $\eta = \eta_1$ clamped in a rigid body. This implies zero deformation of the edge surface. In accordance with Chapter 1, §7, the extension of the boundary contour and the torsion of the edge surface must be equal to zero,

$$\eta = \eta_1 : \qquad \varepsilon_1^1 = 0 \,, \quad \tau_{(1)}^1 = \tau_1^1 - \frac{c}{2R} \gamma^1 = 0 \,.$$

The corresponding conditions for the functions ϑ and ψ follow from substitution of the expressions for ε_1^1 and $\tau_{(1)}^1$ according to (1.84), (2.19) and (2.57). The conditions are, for $\eta = \eta_1$,

$$(r\psi)^{\bullet} + \nu(\lambda\psi s + h°f_1°) = 0 \,,$$

$$-\vartheta - \frac{1+\nu}{rt} ch°(\lambda\psi - h°f°) = 0 \,. \tag{2.71}$$

The term with the small coefficient $\lambda h^\circ \sim h/(3R_m)$ may be omitted.

3.4. Parallelism of the six basic problems

Let us briefly review the solutions of the Saint-Venant's problems represented by diagrams (a), (b), (e) and (f) in Fig. 13.

The two systems of equations (2.30) and (2.63) have in the linear approximation identical left-hand sides.

Equations (2.30), in linearized form, and (2.63) can be presented as a single system describing any of the six problems. For the constant wall thickness $(t = 1)$, considered in this section, the system may be written in the form

$$\begin{aligned}(\psi_k r)^{\cdot} + \mu c \vartheta_k = A_k, \\ (\vartheta_k r)^{\cdot} - \mu c \psi_k = B_k, \end{aligned} \quad \text{or} \quad \begin{aligned} L(\psi_k, \vartheta_k) = A_k \\ L(\vartheta_k, -\psi_k) = B_k. \end{aligned} \tag{2.72}$$

The individual problems are described by solutions ϑ_k, ψ_k corresponding to the different right-hand sides of the equations.

The dual form of (2.72) and the similarity of the right-hand sides A_k and B_k as well as of the respective edge conditions make the functions ψ and ϑ for three of the problems identical to the functions ψ and ϑ of the remaining three. This parallelism is indicated in Table 1.

TABLE 1.

Loading or distortion case, notation of Fig. 13	A_k	B_k	ψ	ϑ
(a) Axial extension force P_z	0	$P^\circ s$	$P^\circ \psi_P$	$P^\circ \vartheta_P$
(b) Bending of a curved beam in plane	$-ms$	0	$-m\vartheta_P$	$m\psi_P$
(e) Bending of a shell of revolution by moment L_y	0	$-L^\circ \dfrac{s}{r}$	$L^\circ \psi_L$	$L^\circ \vartheta_L$
(e) Bending by lateral force P_x	$P_x^\circ \cdot \text{const.}$	$P_x^\circ\left(c - \dfrac{cz}{R}\right)$	$P_x^\circ \psi_{Px}$	$P_x^\circ \vartheta_{Px}$
(f) Lateral plane bending— distortion U_y	$U^\circ \dfrac{s}{r}$	0	$-U^\circ \vartheta_L$	$U^\circ \psi_L$
(f) Bending of a curved beam out of plane—distortion Ω_x	$\Omega^\circ\left(c - \dfrac{cz}{R}\right)$	$\Omega^\circ \cdot \text{const.}$	$\Omega^\circ \vartheta_{Px}$	$-\Omega^\circ \psi_{Px}$

There are problems, for instance in curved-tube analysis, where one can assume $r = R/R_m = 1$ and thus $\psi_L = \psi_P$, $\vartheta_L = \vartheta_P$. The six problems then are described by merely *two* sets of functions: ψ_P, ϑ_P and ψ_{Px}, ϑ_{Px}.

3.5. Asymptotic solution

Consider shells of constant thickness. Equations (2.29) of the axisymmetric deformation and equations (2.62) can be written in the following unified form, simplified by the assumption $r = 1$ (to be discussed later)

$$\tilde{\vartheta}'' + i\mu c\tilde{\vartheta} = g_k \quad (\tilde{\vartheta} = \vartheta + i\psi, \; 'i = \sqrt{-1}; \; {}_i k = 0, 1) ,$$
$$g_0 = s(P^\circ - im) + q^\circ p , \qquad g_1 = -s(L^\circ + iU^\circ) + c(P_x^\circ + i\Omega^\circ) . \tag{2.73}$$

For the axisymmetric case $g_k = g_0$; for the stress state varying as $\cos \xi$ and $\sin \xi$ in (2.73) $g_k = g_1$. The expressions of g_0 and g_1 are taken from (2.22) and (2.63).

We restrict the analysis to shells thin enough to have $\mu \gg 1$ and with the shape smooth enough to make $\dot{g}_k$ and $\ddot{g}_k$ of the order of magnitude of g_k or less. (These conditions are fulfilled for flexible shells in most cases.)

For the shells just defined, there follows $\mu c \gg 1$ in the zone where $\cos \alpha \sim 1$. There, (2.73) has the particular solution

$$\tilde{\vartheta} = g_k/(i\mu c) , \quad g_k = g_{kr} + ig_{ki} . \tag{2.74}$$

This function $\tilde{\vartheta}$ represents the *membrane* stress state.

Consider now the part of the shell where the opposite situation prevails—the term of the equation $i\mu \cos \alpha$, representing the membrane stress and strain, is small. It is the zone around the point at which $\cos \alpha = 0$, $\alpha = \pm\pi/2$. Introduce the coordinate η to have $\eta = \pi/2$ at the point $\alpha = \pi/2$ and represent the functions $c(\eta) = \cos \alpha$ and g_k by the first nonzero term of the respective Taylor series,

$$c(\eta) \approx 0 + \left(\eta - \frac{\pi}{2}\right)\dot{c}\left(\frac{\pi}{2}\right), \quad \dot{c}\left(\frac{\pi}{2}\right) = -b/R_2(\pi/2); \quad g_k(\eta) \approx g_k\left(\frac{\pi}{2}\right).$$

With this, (2.73) may be written—in the vicinity of the point $\eta = \pi/2$, i.e. $\alpha = \pi/2$—in the form

$$\bar{\vartheta}_0^{\cdots} - \left(\eta - \frac{\pi}{2}\right) i\mu_a^3 \bar{\vartheta}_0 = g_k\left(\frac{\pi}{2}\right), \quad \mu_a^3 = \mu\frac{b}{R_2(\pi/2)}. \quad (2.75)$$

The quantity $|\eta - \pi/2|b$ is in fact the distance from the point where $\alpha = 0$.

Introduction of new dependent and independent variables e and x_0 reduces the equation to a well-known type for $e(x_0)$,

$$\bar{\vartheta}_0 = g_k\left(\frac{\pi}{2}\right)\mu_a^{-2}e(x_0), \quad x_0 = \left(\frac{\pi}{2} - \eta\right)\mu_a; \quad \frac{d^2 e}{dx_0^2} + ix_0 e = 1. \quad (2.76)$$

The particular solution of this equation, satisfying the condition

$$x_0 e(x_0)\big|_{x_0 \gg 1} \approx -i,$$

allows a generalization of the solution $\bar{\vartheta}_0$ extending its validity on the other parts of the shell, including even the zone of membrane stress state. This is achieved by replacing in the expression $\bar{\vartheta}_0$ the values $g_k(\pi/2)$ and x_0 by the variables $g_k(\eta)$ and x,

$$\bar{\vartheta} = g_k(\eta)\frac{1}{\mu_a^2}e(x), \quad x = c(\eta)\frac{\mu_a}{b}R_2\left(\frac{\pi}{2}\right). \quad (2.77)$$

This function provides the necessary asymptotic solution. Indeed, for $c \to 0$, i.e. for $\eta \to \pi/2$, $x \to x_0$, formula (2.77) yields $\bar{\vartheta} = \bar{\vartheta}_0$, which reproduces the solution of the governing equation in the zone of wall-bending—near the point where $c = \cos \alpha = 0$. On the other hand, in the region where $c \sim 1$ (and the membrane stresses prevail), there is $x \gg 1$ and thus $e(x) \approx -i/x$. It makes the expression (2.77) for $\bar{\vartheta}$ coincident with the membrane solution (2.74).

Thus, the solution (2.77) is valid both for the area where $\mu c \gg 1$ and for $c \ll 1$. It is the more precise the larger the value of μ_a.

In what follows we use solutions of (2.73) for the boundary conditions

$$\alpha = 0: \qquad \vartheta = 0, \quad \psi = 0;$$
$$\alpha = \pi/2: \qquad \vartheta = 0, \quad \dot{\psi} = 0. \qquad (2.78)$$

The solution (2.77) satisfies these conditions when the auxiliary function $e(x) = e_r(x) + ie_i(x)$ satisfies the conditions

$$x = 0 \ (\cos \alpha = 0): \qquad e_i = 0, \quad \frac{de_r}{dx} = 0;$$

$$x \gg 1 \ (\cos \alpha \sim 1): \qquad |e_r|, \quad \left|\frac{de_i}{dx}\right|, \quad \left|\frac{de_r}{dx}\right| \ll 1, \quad e_i \approx \frac{1}{x}.$$

The graphs of the functions e_r and e_i and their derivatives are presented, following the work of CLARK and REISSNER [39], in Fig. 15.

Formula (2.77) yields for the two governing functions

$$\begin{bmatrix} \vartheta & \dot{\vartheta} \\ \psi & \dot{\psi} \end{bmatrix} = \begin{bmatrix} g_{kr} & -g_{ki} \\ g_{ki} & g_{kr} \end{bmatrix} \begin{bmatrix} \dfrac{e_r}{\mu_a^2} & -\dfrac{\sin\alpha}{\mu_a}\dfrac{de_r}{dx} \\[2ex] \dfrac{e_i}{\mu_a^2} & -\dfrac{\sin\alpha}{\mu_a}\dfrac{de_i}{dx} \end{bmatrix}. \qquad (2.79)$$

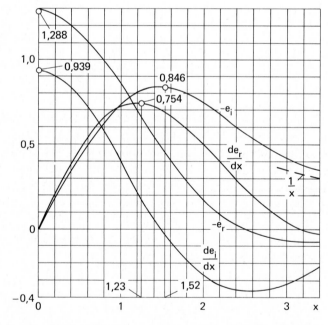

FIG. 15. Graphs of real and imaginary parts e_r and e_i of the function $e(x)$ and their derivatives.

These formulas are simplified by relations $(g_k e)^{\cdot} \approx g_k \dot{e}$ and $\dot{x} \approx -\mu_a \sin \alpha$. That is, terms that are small, according to the assumed smooth variation of g_k or consistent with the adopted accuracy of the representation of $c(\eta)$, are omitted.

The asymptotic solution (2.77) may be applied to determine the deformation for arbitrary boundary conditions. This is done with the aid of two particular solutions of the homogeneous equation corresponding to the Airy equation (2.76). The reader can find the asymptotic solution for the general boundary conditions in the monographs of CHERNYKH [86] and CHERNINA [114].

The maximum values of $|\dot{\vartheta}|$ and $|\dot{\psi}|$ and thus of the stress resultants occur for the solution (2.77) and (2.78) in the region of intensively varying deformation of the shell wall—near the points where $\cos \alpha = 0$. In this region it is consistent to calculate stresses with the T_1 and M_2 determined by (1.83), (2.24) and (2.57), where the ϑ and ψ terms are neglected compared to the $\dot{\vartheta}$ and $\dot{\psi}$ terms. This gives formulas for the stresses

$$\sigma_{T1} = \frac{T_1}{h} = Eh^\circ \dot{\psi}, \qquad \sigma_{M2} = \frac{M_2}{h^2/6} = \frac{D6}{bh^2} \dot{\vartheta}. \qquad (2.80)$$

4. Governing equations and the model of flexible shell

Starting from the hypothesis on the stress-state variation in flexible shells, a governing system of two integro-differential equations is derived. This extends the Reissner equations to nonsymmetrical non-linear deformation. For the corresponding linear Saint-Venant problems the equations are equivalent to the Schwerin–Chernina system.

Analysis of the governing equations displays a semi-momentless (semi-membrane) character of the stress state—the bending and torsion moments M_1 and H_1 in sections $\xi = $ const. may be neglected (or assumed constant in η).

Shells having in their initial state the local geometry of a shell of revolution (Fig. 14) are considered.

4.1. Integrals of the field equations

In this section all the stress and strain parameters will be expressed in terms of two functions ϑ and $V = Ehh°\psi$ introduced similarly to the axisymmetrical case. Setting $b = \text{const.}$ we write

$$\kappa_2 = \frac{1}{b}\dot{\vartheta}(\xi, \eta) , \qquad T_1 = \frac{1}{b}\dot{V}(\xi, \eta) \equiv \dot\psi Ehh° , \qquad (\ )^{\displaystyle\cdot} = \frac{\partial(\ )}{\partial\eta} .$$

$$(2.81)$$

Consider the second and third of the compatibility equations (1.36) and of the equilibrium equations (1.56) to obtain integrals analogous to those in (2.9) and (2.15).

The two compatibility equations (1.36) written in the suitable form—with all nonaxisymmetrical terms transferred to the right-hand side and denoted concisely p_2 and p_3—are

$$(\kappa_1 R)^{\displaystyle\cdot} + \lambda_1 R\dot\alpha^* + \dot\vartheta s = \tau_{,1}b - \gamma_{,1}b/2R_2 \equiv p_2 Rb ,$$

$$\kappa_1 R\dot\alpha^* - (\lambda_1 R)^{\displaystyle\cdot} + \dot\vartheta c = \tau^2 Rb - \lambda_{2,1}b \equiv p_3 Rb .$$

$$(2.82)$$

As in the axisymmetrical case, we set $a = R$ and denote (Fig. 14)

$$\alpha^* = \alpha + \vartheta , \quad \dot\alpha^* = b/R_2' , \quad c^* = \cos\alpha^* , \quad s^* = \sin\alpha^* = -\dot R^*/b .$$

$$(2.83)$$

In the more general case here under discussion, the variables are functions not only of η but also of coordinate ξ, as $\vartheta = \vartheta(\xi, \eta)$. Moreover, the functions ϑ and α^* may not have the meaning of geometric angles (in contrast to α).

Multiplying the first of equations (2.82) by s^* or c^* and adding the second of the equations multiplied by c^* and $-s^*$, respectively, leads to equations with left-hand sides forming the derivatives $(\kappa_1 s^* R - \lambda_1 c^* R + \sin\vartheta)^{\displaystyle\cdot}$ and $(\kappa_1 c^* R + \lambda_1 s^* R + \cos\vartheta)^{\displaystyle\cdot}$. Integrating, with the additive of integration included in ϑ in the first equation and denoted explicitly by $k(\xi)$ in the second, gives upon solving for κ_1 and λ_1,

$$\kappa_1 R = k(\xi)c^* - c + I(p_i) , \qquad \lambda_1 R = k(\xi)s^* - s + J(p_i) ,$$

$$\begin{bmatrix} I(p_i) \\ J(p_i) \end{bmatrix} \equiv \begin{bmatrix} s^* \\ -c^* \end{bmatrix} \int_{\eta_0}^{\eta} (p_2 s^* + p_3 c^*)Rb\,\mathrm{d}\eta + \begin{bmatrix} c^* \\ s^* \end{bmatrix} \int_{\eta_1}^{\eta} (p_2 c^* - p_3 s^*)Rb\,\mathrm{d}\eta .$$

$$(2.84)$$

The limits of integration, denoted η_0 and η_1, can be chosen arbitrarily. We derive now similar integrals of the equilibrium equations (1.56).

With expressions (2.84) substituted into $1/R_1' = c/R + \kappa_1$ and $1/\rho_1' = s/R + \lambda_1$, the second and third of equilibrium equations (1.56) may be presented in the form (anticipating later verification, we write $S_i = S$ and $\tau_i = \tau$)

$$(T_2 R)^{\cdot} + Q_2 R\dot{\alpha}^* + \dot{V}ks^* = -S_{,1}b - \lambda_2 SRb - \dot{V}J(p_i) - Q_1 Rb\tau - q_2 Rb$$

$$\equiv \hat{q}_2 Rb ,$$

$$T_2 R\dot{\alpha}^* - (Q_2 R)^{\cdot} + \dot{V}kc^* = Q_{1,1}b - 2S\tau Rb + qRb - \dot{V}I(p_i)$$

$$\equiv \hat{q}_3 Rb .$$

(2.85)

Multiplying the first equation of (2.85) by s^* or c^* and adding the second equation multiplied by c^* or $-s^*$ gives equations with left-hand sides $(T_2 s^* R - Q_2 c^* R + Vk)^{\cdot}$ and $(T_2 c^* R + Q_2 s^* R)^{\cdot}$, respectively. Integrating, with the additive of integration included in V in the first equation and denoted by T in the second one, and solving the integrals for T_2 and Q_2 gives

$$T_2 R = -Vks^* + T(\xi)c^* + I(\hat{q}_i) ,$$

$$Q_2 R = Vkc^* + T(\xi)s^* + J(\hat{q}_i) .$$

(2.86)

The terms $I(\hat{q}_i)$ and $J(\hat{q}_i)$ designate the operations $I(\ )$ and $J(\ )$ defined in (2.84) over the quantities denoted $\hat{q}_2$ and $\hat{q}_3$ in (2.85).

The next step, necessary to convert the four integrals of the equations to usable formulas, is one of expressing in (2.84) and (2.86) the quantities p_i and $\hat{q}_i$ in terms of ϑ and V. This will be done with the help of the first of the equations (1.36) and (1.56) written in the simple form (to be justified later)

$$(\tau R^2)^{\cdot} = R\vartheta_{,1} , \qquad (SR^2)^{\cdot} = -R\dot{V}_{,1} - q_1 bR^2 . \qquad (2.87)$$

The relations (2.87) together with the estimates of the derivatives from (2.7), the elasticity relations (1.74) and fourth and fifth of equations

(1.36) and (1.56), lead to the following estimates for the "nonaxisymmetrical" stress and strain resultants,

$$\tau \sim \frac{H}{D} \sim j\frac{\vartheta}{R}, \qquad \frac{S}{Gh} \sim \gamma \sim j\frac{\psi h}{R},$$

$$\frac{Q_1}{D} \sim \frac{jn}{bR}\vartheta, \qquad \lambda_2 \sim \frac{jn}{bR}\psi h. \qquad (\star)$$

With ($\star$) and (1.36) and (1.56) it becomes clear that the terms omitted in (2.87) are indeed negligible in accordance with (2.7) if

$$\left|\frac{h}{R_1}\frac{\psi}{\vartheta}\right| \ll 1, \qquad \left|\frac{h}{3R_1}\frac{\vartheta}{\psi}\right| \ll 1. \qquad (2.88)$$

The term $Q_2\tau$ provides an exception: it is small if $|\vartheta|b/(nR) \ll 1$.

The errors introduced by taking $S_1 = S$ and $\tau_2 = \tau$ have the same order of magnitude as the quantities (2.88). The error of assuming $S_2 = S$ and $\tau_1 = \tau$ is of the order of magnitude of $\psi h/(\vartheta R_2)$ and $\vartheta h/(3\psi R_2)$.

Finally, all terms with γ_1, λ_2, Q_1 or $H_{,1}$ may be omitted in the first three equations of compatibility and equilibrium and in their integrals (2.84) and (2.86) with relative errors exceeding neither $(jb/nR)^2$ nor the estimates (2.88). Thus,

$$[p_2 \ p_3] = [\tau_{,1}/R \ \tau^2], \qquad \hat{q}_2 = -q_2 - \frac{1}{R}S_{,1} - \frac{1}{Rb}\dot{V}J(p_i),$$

$$\hat{q}_3 = q - 2S\tau - \frac{1}{Rb}\dot{V}I(p_i). \qquad (2.89)$$

Now, (2.82), (2.84) and (2.86) together with the elasticity equations give expressions in terms of ϑ and V for all stress and strain resultants, except Q_1 and λ_2 which are negligible.

4.2. Reissner-type equations for nonsymmetric problems

Expressing in the fourth of equations of compatibility (1.36) and of equilibrium (1.56) all dependent variables in terms of ϑ and V gives a system of equations for ϑ and V. The system differs from the Reissner equations (2.16) or (2.18) in the *meaning* of the functions ϑ and V (or ψ) and in *additional* terms with p_i, $\gamma_{,1}$, $H_{,1}$ and $\hat{q}_i$. All the simplifications that have led to (2.30) remain applicable in the more general non-axisymmetrical case. Upon omitting also those of the nonaxisymmetrical terms, which are negligible in accordance with the hypothesis (2.7) and conditions (2.88), the two equations may be written in the form

$$\left(\dot{\psi}\frac{r}{t}\right)^{\cdot} + \mu^*c'\vartheta + \mu J\left(\frac{1}{R}\tau_{,1};\tau^2\right) = -ms\ ,$$

$$(\dot{\vartheta}rt^3)^{\cdot} - \mu^*c^*\psi + \mu J\left[\frac{S_{,1}}{KR} + \frac{\dot{\psi}}{bR}J\left(\frac{\tau_{,1}}{R};\tau^2\right); \frac{2S\tau}{K} + \frac{\dot{\psi}}{bR}I\left(\frac{\tau_{,1}}{R};\tau^2\right)\right]$$

$$= \frac{\mu}{K}Ts^* + \frac{\mu}{K}J(q_2;-q)\ ,\quad K = Eh^\circ h_m b\ . \tag{2.90}$$

The twist and the shear-stress resultant (τ, S) are represented in (2.90) in terms of ϑ and ψ (and of additives of integration S_0 and τ_0) as follows from (2.87),

$$\begin{bmatrix} \tau \\ S \end{bmatrix} = \frac{1}{R^2}\int\left[\begin{array}{c} \dot{\vartheta}_{,1} \\ -\dot{\psi}_{,1}K - q_1 bR \end{array}\right]R\ d\eta\ . \tag{2.91}$$

The coinciding notations mean the same as in the symmetrical case (present chapter, §2). But μ^*, c^* and c' depend here also on ξ as $\vartheta(\xi,\eta)$ and $k(\xi)$ do.

Accuracy of (2.84), (2.86), (2.90) and (2.91) describing the deformation of flexible shells can be easily tested once a solution (ϑ, ψ) is found. The estimates (2.7) are simple to apply.

In practice, the conditions (2.88) are fulfilled if that under (2.7) is. When the stress state varies in accordance with the flexible-shell hypothesis (2.7), the functions ϑ and ψ are not too different in their order of magnitude.

On the other hand, an *a priori* assessment of the accuracy of the flexible-shell equations requires an analysis of the specific shell shape and loading (present chapter, §5).

With some additional assumptions, leading in most cases to no significant error, the governing equations (2.90) and (2.91) may be substantially simplified.

The derivatives of the functions ϑ and ψ with respect to ξ are represented in the equations (J terms) by $\vartheta_{,11}$ and $\psi_{,11}$ integrated with respect to η. The differentiation with respect to ξ and integration over η diminish the order of magnitude of functions that describe any deformation corresponding to the conditions (2.5)–(2.7). Thus the $J(\ ;\ )$ terms on the left-hand sides of (2.90), estimated on the basis of the relations (2.6) with $a = R$, have the order of magnitude of $j^2b/(nR)$ or $j^2b/(nR)$, compared to the terms $\mu^*c'\vartheta$ or $\mu^*c^*\psi$ of the respective equation. In the most favourable case the J terms can be omitted altogether. This is possible for the shells of revolution complete in the circumferential direction (Fig. 12, III), i.e. having no edges $\xi = $ const., under the condition

$$j^2b/(nR) \ll 1 . \qquad (2.92)$$

The J terms are indispensable when the shell has edges other than $\eta = $ const. The terms with derivatives with respect to ξ are responsible for any effects of the edges. But a substantial *simplification* of the J terms is often feasible. For, in most cases, the condition

$$|\vartheta| j^2b/(nR) \ll 1 \qquad (2.93)$$

is fulfilled. This allows us to neglect the nonlinear components of the J terms on the left-hand side of (2.90). The equations become

$$\left(\psi\frac{r}{t}\right)^{\cdot} + \mu^*c'\vartheta + \mu J(\tau_{,1}/R; 0) = -ms ,$$
$$(\dot\vartheta rt^3)^{\cdot} - \mu^*c^*\psi + \frac{\mu}{K}J(S_{,1}/R; 0) = \frac{\mu}{K}Ts^* + \frac{\mu}{K}J(q_2; -q) . \qquad (2.94)$$

Equations (2.90) and their simplified version (2.94) present a direct generalization of the Reissner equations (2.11) and of the Schwerin–Chernina equations (2.63).

Inserting expansions of the form

$$[\psi\ \vartheta\ q_i] = \sum_{j=0,1,\dots} [\psi^j(\eta)\ \vartheta^j(\eta)\ q_i^j(\eta)]\cos j\xi \qquad (2.95)$$

into (2.90) or (2.94) yields a system of ordinary differential equations for the amplitudes $\psi^j(\eta)$ and $\vartheta^j(\eta)$. This system falls apart into separate pairs of equations for each of $j = 1, 2, \dots$ only in *linear* approximation. (The axisymmetric deformation $j = 0$ is an exception.)

However, the separation of variables is possible also for one practically interesting *large*-displacement case—that of superposition of small nonsymmetrical deformation $\Sigma \, [\psi^j \, \vartheta^i] \cos j\xi$ upon an unbounded axisymmetrical deformation represented by ψ^0 and ϑ^0. Substituting the expansions (2.95) into equations (2.94) renders for ψ^0 and ϑ^0 two equations identical to (2.30) and a linear system for each harmonic $j = 1, 2, \ldots$, determining the ψ^j and ϑ^j. The coefficients of these linear equations depend on the axisymmetrical stress, deformation and loading (ψ^0, ϑ^0 and q_i^0).

The reader will encounter no serious difficulty in deriving these equations.

4.3. Semi-momentless model of a shell

Equations (2.90) and the deliberations leading to them serve yet another useful purpose. It has, in fact, come out that the hypothesis on the flexible-shell stress state (2.7) implies a *simplified model* of a shell. The equations are mathematically deducible (a restriction is discussed in the present chapter, §5.3) from the following sets of simplified equations of compatibility, of equilibrium and of elasticity,

$$\kappa_{2,1} - (a^2\tau)^{\cdot}/(ab) = 0 \, ,$$

$$\frac{(a\kappa_1)^{\cdot}}{ab} + \frac{\lambda_1}{R_2'} + \frac{\kappa_2}{\rho_1} - \frac{\tau_{,1}}{a} = 0 \, ,$$

$$\frac{\kappa_1}{R_2'} - \frac{(a\lambda_1)^{\cdot}}{ab} + \frac{\kappa_2}{R_1} - \tau^2 = 0 \, , \qquad (2.96)$$

$$\lambda_1 ab = -(a\varepsilon_1)^{\cdot} \, ;$$

$$T_{1,1} + (Sa^2)^{\cdot}/ab = -q_1 a + S\lambda_1 a \, ,$$

$$\frac{(aT_2)^{\cdot}}{ab} + \frac{Q_2}{R_2'} + \frac{T_1}{\rho_1'} + \frac{S_{,1}}{a} = -q_2 \, ,$$

$$\frac{T_2}{R_2'} - \frac{(Q_2 a)^{\cdot}}{ab} + \frac{T_1}{R_1'} + 2S\tau = q \, , \qquad (2.97)$$

$$Q_2 ab = (aM_2)^{\cdot} \, ;$$

$$\varepsilon_1 = T_1/(Eh) \, , \quad S = Gh\gamma \, , \quad M_2 = D\kappa_2 \, , \quad H = (1 - \nu)D\tau \, . \quad (2.98)$$

As in the foregoing, a local geometry identical to that of a shell of revolution is considered: in (2.96) and (2.97) the variables a, R_i and ρ_1 do not depend on the coordinate ξ and b = const. But the coordinate ξ is not necessarily equal to the angle (φ in Fig. 14) between two "meridional" planes. Correspondingly, the parameter a is not necessarily equal to the radius $R(\eta)$ (Fig. 14).

Equations (2.96) and (2.97) can be obtained from the first four of equations (1.36) and (1.56) upon omitting all terms with the resultants

$$\varepsilon_2 , \quad \gamma , \quad \lambda_2(\varepsilon_2, \eta) ; \quad M_1 , \quad H , \quad Q_1(M_1, H) . \qquad (2.99)$$

Actually, only terms with derivatives of these variables are dropped. Equations (2.98) follow from (1.83) and (1.84) of the general theory by dropping the terms with

$$-\nu T_2/(Eh) , \qquad \nu D\kappa_1 . \qquad (2.100)$$

This may appear inconsistent: the parameters T_2 and κ_1 are retained in the equilibrium and compatibility equations and are by no means negligible there. But the inaccuracy introduced into the governing equations is below the contribution of the terms disregarded in (2.98). The derivation of (2.94), (2.63) and (2.30) shows that the contribution of the quantities (2.100) cancels out with terms determined by other parameters (ε_2 and M_1) neglected in this theory. Another justification of (2.98) follows from the present chapter, §4.4.

Thus, the character of deformation, which corresponds to the hypothesis of the flexible-shell theory (2.7), allows us to neglect the terms indicated in (2.99) and (2.100). Specifically, the stress state has a *semi-momentless* character, the deformation is *semi-membrane*: in a section ξ = const. only the *membrane* stress resultants are substantial (Fig. 14). At the same time, in sections η = const. *all* stress resultants have to be accounted for, except H_2 which is set equal to zero together with H_1.

The deformation of flexible shells is dual with respect to the stress state (in the sense of Chapter 1, §8.1). This is demonstrated by the equations (2.96)–(2.98). Specifically, the essential *membrane* stress resultants T_1 and S are dual to those two *bending* strain resultants κ_2 and τ which are substantial in the η-direction. The static-geometric analogy can be observed also between the neglected resultants—the moments M_1 and H and the dual membrane strains ε_2 and γ, respectively.

Among the discarded terms is one with $Q_2\tau$, though both Q_2 and τ are retained in other terms. The quantity $Q_2\tau$ proves to be small as a product of two factors of minor magnitude.

The system of equations (2.96)–(2.98) has eighth order in η. A full complement of eight boundary conditions—four on each edge $\eta =$ const.—can and must be satisfied by the solution. This is true also for the governing system (2.90). Its order in η is reduced to four by integration, while four functions of $\xi\,(k,\ T,\ \tau_0,\ S_0)$ to be determined by conditions on edges $\eta =$ const. are introduced.

The situation on edges $\xi =$ const. is different. The system (2.96)–(2.98), as well as (2.90), is of order four in ξ. This reduction of order results from *omitting* terms. Thus, only two conditions can be satisfied on an edge $\xi =$ const. The semi-momentless or, synonimously, semi-membrane[5] model indicates which two of the boundary conditions can be fulfilled on an edge $\xi =$ const. and which are not eligible.

As the stress state in sections $\xi =$ const. is membrane in character, boundary conditions on an edge $\xi =$ const. may include neither moments $H_1 = H$, M_1 nor the Q_1. Excluded from boundary conditions are also the parameters λ_2, ε_2 and γ, omitted in the equations, as well as displacements directly connected with these parameters, in particular w. Parameters, which may enter the conditions on edges $\xi =$ const., include the membrane stress resultants T_1 and $S_1 = S$, the strain resultants κ_2 and τ and the tangential displacements u and v (cf. §4.4).

Of the stresses caused by forces applied on an edge $\xi =$ const. the semi-momentless theory separates and determines the *main* component—those stresses *slowly fading* with distance from the edge and thus extending over a substantial part of the shell. Omitted is another component of the shell deformation: the one that *rapidly* diminishes away from the edge—the *edge effect*.

The two conditions at an edge $\xi =$ const., excluded from this theory, constitute the two principal (of four) boundary conditions determining the edge effect.

The separation of a stress state into the *main, slowly varying* with ξ,

---

[5] The term *semi-momentless* has proved to be the more unambiguous one and will be used predominantly in the following.

part and the edge effect depends, of course, on the shell geometry and the character of the edge loading. For circular-cylinder shells the conditions leading to the separation are comprehensively discussed in GOL'DENVEIZER's book [48] (cf. [205]). For the variety of flexible shells it can be said that this separation is effective precisely then, when the basic semi-momentless hypothesis holds.

When the "bending type" edge conditions, not accounted for by the semi-momentless theory, are of interest, they can be satisfied by superimposing[6] a boundary-effect deformation. But this correction concerns only the edge zone narrower than $2\sqrt{hR_2}$. In most cases this refinement is of no practical importance as the real edge conditions are not known exactly enough. Particularly uncertain are the "bending" and ε_2 conditions—those decisive for the boundary effect.

The semi-momentless model allows two substantial extensions:

(a) A class of orthotropic and inhomogeneous shells can be encompassed by using instead of (2.98) the elasticity relations based on (1.86) and (1.87),

$$\varepsilon_1 = T_1 B_1' = T_1/(E_1 h) , \quad M_2 = D_2 \kappa_2 ; \quad \gamma = B_G' S , \quad H = D_G 2\tau .$$
$$(2.101)$$

All flexible-shell equations become applicable for these cases if E and $h°$ are respectively replaced by E_1 and $h° = \sqrt{B_1' D_2}/b$ (cf. (2.31)).

(b) The moment H and the shear strain γ can be without substantial complications retained in the theory. They are directly expressed in terms of τ and S with the help of (2.98) or (2.101).

The extended theory is effective for the intermediate class of problems—for shells that are not really flexible, but on the other hand, have a basic stress state that is far from being a membrane one. Consider this extension of the theory using a more general approach to its formulation.

4.4. Semi-momentless theory encompassing shear strain and torsion

The shell shape to be considered is restricted in this section merely by the (rather natural) requirement that it varies less intensively than stress

---

[6] The superposition is possible even for most basically nonlinear problems in which the shell deformation remains small near a stiffened edge. In any case, a small number of iterations must be sufficient.

and strain resultants. That is, for the terms of *secondary* importance it is assumed

$$(r_{,\alpha}F)_{,\beta} \sim r_{,\alpha}F_{,\beta} \quad (r_{,1} = at_1, r_{,2} = bt_2; F = H, M_1; \beta = 1, 2).$$

$$(2.102)$$

The equilibrium equations (1.53) with the Q_α-components written out explicitly and the notation $T_\alpha = N_\alpha + Q_\alpha n^*$ become

$$(bN_1)_{,1} + (aN_2)_{,2} + (bQ_1 n^*)_{,1} + (aQ_2 n^*)_{,2} + qab = 0, \quad (2.103)$$

$$(bM_1 t_2')_{,1} - (bH_1 t_1')_{,1} - (aM_2 t_1')_{,2} + (aH_2 t_2')_{,2}$$

$$+ ab(S_1 n^* - S_2 n^* - Q_1 t_2' + Q_2 t_1') = 0. \quad (2.104)$$

With the expressions of Q_α in terms of M_α, $H_1 = H_2 = H$, following from (2.104) and simplificatons (2.102), the sum of *estimates* of the M_1 and H terms and the *main* M_2 term of (2.103) takes the form

$$\left(\frac{b}{a}M_{1,11} + H_{,12} + H_{,21}\right)n^* + \frac{a}{b}M_{2,22}n^*. \quad (2.105)$$

Hence, the terms with M_1 can be dropped with the accuracy of the relation

$$\left|\frac{b}{a}M_{1,11}\right| \ll \left|\frac{a}{b}M_{2,22}\right|.$$

The static-geometric analogy (Chapter 1, §8) indicates the possibility to neglect under similar conditions all the ε_2 terms in the equations of compatibility (1.33)–(1.36). The conditions allowing us to neglect all the terms with M_1 and ε_2, respectively, are

$$\left|\frac{M_{1,11}}{a^2}\right| \ll \left|\frac{M_{2,22}}{b^2}\right|, \quad \left|\frac{\varepsilon_{2,11}}{a^2}\right| \ll \left|\frac{\varepsilon_{1,22}}{b^2}\right|. \quad (2.106)$$

The indicated simplifications in the description of equilibrium and of deformation, without any relevance to their cause or justification, lead to consequential simplifications in the expressions of strain energy, boundary conditions and elasticity relations. Similar to the derivation in Chapter 1, §6.1 insert into (1.73) the expression of q from (2.103) and (2.104) without the M_1 terms. This gives (cf. (1.81))

$$\int\int\left[\left(T_1-\frac{\partial U}{\partial\varepsilon_1}\right)\delta\varepsilon_1+\left(T_2-\frac{\partial U}{\partial\varepsilon_2}\right)\delta\varepsilon_2+\left(S-\frac{\partial U}{\partial\gamma}\right)\delta\gamma-\frac{\partial U}{\partial\kappa_1}\delta\kappa_1\right.$$
$$\left.+\left(M_2-\frac{\partial U}{\partial\kappa_2}\right)\delta\kappa_2+\left(2H-\frac{\partial U}{\partial\tau}\right)\delta\tau\right]ab\,d\xi\,d\eta+J_s=0,\quad(2.107)$$

$$J_s=\oint[(T_1^B-T_1)\cdot\delta u-(H_1^B-H_1)t_1'\cdot\delta\vartheta]b\,d\eta.\quad(2.108)$$

As all variations $\delta\kappa_1,\dots$ are arbitrary, it must be $\partial U/\partial\kappa_1=0$. Hence U is determined by (1.65) (or (1.68)) *without* the κ_1 *terms*. This expression of the strain-energy density results (instead of (1.83)) in the relation $M_2=\partial U/\partial\kappa_2=D\kappa_2$ identical to the one in (2.98).

In contrast to the general equations (1.84), the relation (2.107) gives no elasticity equation for M_1. The resultant M_1 assumes the role of a *reactive* moment; it follows from $M_2=D\kappa_2$ and (1.84) and (1.83) that $M_1=\nu M_2$.

The static-geometric duality (1.112) and (1.113) (or, for linear approximation, the similar to Chapter 1, §6.1 derivation based on the principle of complementary energy) leads to relations $\varepsilon_1=T_1/Eh$ and $\varepsilon_2=-\nu\varepsilon_1$. Thus, it has been found that absence in the field equations of terms with M_1 and ε_2, respectively, leads to the elasticity relations (2.98) and to

$$M_1=\nu M_2,\qquad\varepsilon_2=-\nu\varepsilon_1.\quad(2.109)$$

Furthermore, the *elastic-energy* expression (1.68) without the κ_1 terms, taking into account the relation $\varepsilon_2=-\nu\varepsilon_1$ and, of course, $U_c=0$ (Chapter 1, §5.2), *becomes*

$$U=\tfrac{1}{2}(Eh\varepsilon_1^2+Gh\gamma^2+D\kappa_2^2+D_G4\tau^2),\quad D_G=Gh^3/12.\quad(2.110)$$

The *contracted*, in consequence of discarding the terms with M_1 and ε_2, *boundary conditions* on an edge $\xi=$ const. follow from (2.107) and (2.108) as equation $J_s=0$. It remains to present $t_1'\cdot\delta\vartheta$ and $\delta u=$

$t_1' \, \delta u + t_2' \, \delta v + n^* \, \delta w$ in terms of mutually independent virtual displacements.

The expression of $t_1' \cdot \delta \boldsymbol{\vartheta} = -\delta \vartheta_2$ in terms of δu, δv and δw is provided by (1.98). Neglecting $\delta \varepsilon_2$ in the equation $\delta w / R_2' + \delta v_{,2}/b + \delta u/\rho_2' = \delta \varepsilon_2$, which follows from (1.76) similarly to (1.40), renders a relation between δu, δv and δw. This results in $J_s = 0$ with (2.108) being transformed to (cf. (1.100))

$$\oint [(T_{(1)}^B - T_{(1)}) \, \delta u + (S_{(1)}^B - S_{(1)}) \, \delta v] b \, \mathrm{d}\eta = 0 , \qquad (2.111)$$

where

$$T_{(1)} = T_1 + \frac{H}{R_{21}'} - \left(Q_1 + \frac{H_{,2}}{b}\right)\frac{R_2'}{\rho_2'} ,$$

$$S_{(1)} = S_1 + \frac{H}{R_2'} + \left(Q_1 \frac{R_2'}{b} + H_{,2}\frac{R_2'}{b^2}\right)_{,2} . \qquad (2.112)$$

Equation (2.111) indicates two conditions on an edge $\xi = \mathrm{const}$. The stress resultants are represented at the edge by the parameters (2.112); the external forces acting on the edge—by the parameters $T_{(1)}^B$ and $S_{(1)}^B$, which are composed of the resultants $T_1^B, \ldots$ in the way $T_{(1)}$ and $S_{(1)}$ are composed of $T_1, \ldots$.

The static-geometric analogy (1.112) indicates for linear approximation boundary conditions dual to (2.111) and (2.112). Namely, there can be stated two *conditions for the strain* at an edge $\xi = \mathrm{const}$., which can be expressed in terms of two parameters (cf. (1.105a)) defined by the formulas

$$\kappa_{(2)} = \kappa_2 + \frac{\gamma}{2R_{21}} + \left(\lambda_2 - \frac{\gamma_{,2}}{2b}\right)\frac{R_2}{\rho_2} ,$$

$$\tau_{(2)} = \tau_2 - \frac{\gamma}{2R_2} + \left(\lambda_2 \frac{R_2}{b} - \gamma_{,2}\frac{R_2}{2b^2}\right)_{,2} . \qquad (2.113)$$

In (2.112) and (2.113) resultants Q_1 and λ_2 are expressed in terms of M_2, H and ε_1, γ through the fourth equation of (1.56) and (1.36), respectively.

The expression (2.105) allows an assessment (however crude) of the gain in accuracy that can be expected by retaining terms with H and γ. With the quantity ε indicating the relative error of the theory based on the relation (2.5), we have $|M_{2,11}/a^2| \sim \varepsilon |M_{2,22}/b^2|$. Hence, the relations (2.109) and (2.105) and a dual to (2.105) expression for strains show the error of dropping the terms with M_1 and ε_1 to be of the order of magnitude of $\nu\varepsilon$.

Dropping the H terms has according to (2.105) the effect of neglecting $2H_{,12}$ compared to $aM_{2,22}/b$. The elasticity relations (2.98) $H = (1 - \nu)D\tau$ and $M_2 = D\kappa_2$, together with the relation $\tau_{,2}/b \sim \kappa_{2,1}/a$ from (1.36), lead to the estimate $H_{,12} = (1 - \nu)D\tau_{,12} \sim (1 - \nu)M_{2,11}b/a$. Recalling (2.105) we find the relative error of neglecting the H terms to be of the order of magnitude of $2(1 - \nu)\varepsilon$. Dual relations for strain lead to the estimate $2(1 + \nu)\varepsilon$ for the error of neglecting the γ terms.

Thus, retaining the terms with H and γ reduces the estimated inaccuracy of the semi-momentless theory from 2.6ε to 0.3ε. This is illustrated by results of calculations in Chapter 3, §7.3.

5. Semi-momentless equations

The equations of the canonical semi-momentless theory, disregarding M_1, ε_2 and H, γ, are reduced to a system of two. Linear approximation is used to look into the applicability of the semi-momentless model. The initial local geometry considered is that of a shell of revolution.

5.1. System of governing equations

We express all the stress and strain resultants in terms of T_1 and κ_2. The first, third and fourth of equations (2.96) and (2.97) give

$$\tau = \frac{b}{a^2} \int \kappa_{2,1} a \, d\eta \,, \qquad \kappa_1 = -\frac{R_2'}{R_1} \kappa_2 - \frac{R_2'(a\varepsilon_1)_{,22}}{a} \frac{}{b^2} + \underline{R_2'\tau^2} \,,$$

$$S = -\frac{b}{a^2} \int (T_{1,1} + q_1 a)a \, d\eta \,, \tag{2.114}$$

$$T_2 = -\frac{R_2'}{R_1'} T_1 + \frac{R_2'(aM_2)_{,22}}{a} \frac{}{b^2} + R_2'q - \underline{R_2'2S\tau} \,.$$

The nonlinear terms underlined in (2.114) are minor, compared to the main terms with κ_2 and T_1, being at most of the order

$$\left| \frac{\partial^2 \kappa_2}{a^2 \partial \xi^2} R_1 \right| \frac{b^2}{n^2} \sim \left| \vartheta \frac{R_1}{a} \right| \frac{j^2 b}{na}. \tag{2.115}$$

These terms will be omitted, as in (2.94).

With (2.114) and expressions for λ_1 and Q_2 from (2.96) and (2.97), the second of these equations of compatibility and of equilibrium contain only two unknowns ε_1, κ_2 and T_1, M_2, respectively. Through appropriate constitutive equations two of these are expressed in terms of the remaining two. Using (2.98) we have for $h = \text{const.}$, the system

$$\left(\frac{\partial^2}{\partial \xi^2} - \frac{\partial}{\partial \eta} r^2 \frac{b}{\rho_1'} + \frac{\partial}{\partial \eta} r^2 \frac{\partial}{\partial \eta} \frac{R_2'}{R_1'} \right) rT_1 - \frac{D}{b} W' r\kappa_2 = q_\Sigma' b \,,$$

$$\left(\frac{\partial^2}{\partial \xi^2} - \frac{\partial}{\partial \eta} r^2 \frac{b}{\rho_1} + \frac{\partial}{\partial \eta} r^2 \frac{\partial}{\partial \eta} \frac{R_2'}{R_1} \right) r\kappa_2 + \frac{1}{Ehb} W' rT_1 = 0 \,. \tag{2.116}$$

Here

$$W' = \frac{\partial}{\partial \eta} r^2 \left(\frac{\partial}{\partial \eta} \frac{R_2'}{b} \frac{\partial^2}{\partial \eta^2} + \frac{b}{R_2'} \frac{\partial}{\partial \eta} \right), \quad r = \frac{a}{b},$$

$$q_\Sigma' = -r^2 q_{1,1} + (r^3 q_2)^{\cdot} + \left[r^2 \left(r \frac{R_2'}{b} q \right)^{\cdot} \right]^{\cdot}, \quad (\,)^{\cdot} = \frac{\partial(\,)}{\partial \eta}. \tag{2.117}$$

The parameters κ_1 and λ_1, implicit in (2.116) through $1/R_1' = 1/R_1 + \kappa_1$ and $1/\rho_1' = 1/\rho_1 + \lambda_1$, are expressed in terms of κ_2 and T_1 using (2.114) and (2.96)–(2.98).

In *linear* approximation, i.e with $R_i' = R_i$ and $\rho_1' = \rho_1$ and thus $W' = W$ equations (2.116) are static-geometric duals of each other (Chapter 1, §8). This allows us to present the system (2.116) as one complex equation for the complex stress resultant $\tilde{T}_1 = T_1 - i(DEh)^{1/2}\kappa_2$ $(i = \sqrt{-1})$. Indeed, the linearized second of equations (2.116) multiplied by $-i\sqrt{DEh}$ and added to the linearized first of equations (2.116) yields (with $\sqrt{D/Eh} = bh°$),

$$\left(\frac{\partial^2}{\partial \xi^2} - \frac{\partial}{\partial \eta} r^2 \frac{b}{\rho_1} + \frac{\partial}{\partial \eta} r^2 \frac{\partial}{\partial \eta} \frac{R_2}{R_1} - ih°W \right) r\tilde{T}_1 = bq_\Sigma \,, \tag{2.118}$$

This equation allows an assessment (however crude) of the *bounds of applicability* of the basic hypothesis (2.5) of the semi-membrane theory. Assume the parameters of shell form to vary no more intensively than the stress state

$$\left|\frac{\partial}{\partial \eta}(A\tilde{T}_1)\right| \sim \left|A\frac{\partial}{\partial \eta}\tilde{T}_1\right| \quad \text{for } A = a, \frac{1}{\rho_1}, \frac{1}{R_i}.$$

With the notation b/n, introduced in (2.6) for the interval of variation of the stress state, equation (2.118) gives

$$\left|\frac{\partial^2 \tilde{T}_1}{a^2 \partial \xi^2}\right| \sim \delta \left|\frac{\partial^2 \tilde{T}_1}{b^2 \partial \eta^2}\right|, \quad \delta = \left|-\frac{b/n}{\rho_1} + \frac{R_2}{R_1} + ih°\left(n^2\frac{R_2}{b} - \frac{b}{R_2}\right)\right|.$$

(2.119)

The hypothesis (2.5) thus means $\delta \ll 1$. This is fulfilled, when

$$\left|\frac{b/n}{\rho_1}\right|, \left|\frac{R_2}{R_1}\right|, \frac{h}{3}\left|\frac{R_2}{(b/n)^2}\right| \ll 1.$$

(2.120)

Obviously, these are only *sufficient conditions*. The actual error of the theory may be *lower* than any single one of the three terms of δ. This is the case for bent circular tubes with thin flanges (Chapter 3, §7).

For circular cylinders the conditions (2.120) reduce to that established by GOL'DENVEIZER [48]: $|hR_2| \ll 3(b/n)^2$ (cf. [205]).

Equations (2.116) or (2.118) are in fact an extension of the linear equations of cylinder shells, due to WLASSOW [36].

5.2. Semi-momentless version of Novozhilov's complex equations

With the estimate (2.119) and $\delta \ll 1$ the semi-momentless model of the shell and the corresponding equations follow directly from the Novozhilov's complex equations. Differences between the new version of the semi-membrane equations and that of (2.118) will indicate a possibility of simplification of (2.116) and (2.118).

In linear approximation the expressions (2.114) for T_2 and κ_1 are static-geometric duals. With the elasticity relations (2.98) the two dual formulas express the complex stress resultant $\tilde{T}_2$,

$$\tilde{T}_2 \equiv T_2 - iEhh'\kappa_1 = -\left(\frac{R_2}{R_1} + i\frac{h^\circ R_2}{ab}\frac{\partial^2}{\partial\eta^2}a\right)\tilde{T}_1 .$$

Comparison with the relation (2.119) gives $|\tilde{T}_2| \sim |\tilde{T}_1|\delta$. Hence, in the *complex* semi-momentless theory, $\tilde{T}_2$ can be neglected relative to $\tilde{T}_1$. This is sufficient to simplify the Novozhilov equations (1.122) to semi-momentless form. It remains to neglect in (1.122) $\tilde{T}_1(ih/3R_1)$ compared to $\tilde{T}_1$ and $\tilde{T}_{1,11}/a^2$ compared to $\tilde{T}_{1,22}/b^2$ (second term in (1.123) for ∇^2).

Equations (1.122) become[7] (for b = const.)

$$b\tilde{T}_{1,1} + (\tilde{S}a^2)_{,2}/a = -abq_1 ,$$

$$(a\tilde{T}_2)_{,2} - \tilde{T}_1 a_{,2} + b\tilde{S}_{,1} + i(ah'/R_2)\tilde{T}_{1,2} = -abq_2 , \qquad (2.121)$$

$$\frac{\tilde{T}_1}{R_1} + \frac{\tilde{T}_2}{R_2} - i\frac{h'}{ab^2}(a\tilde{T}_{1,2})_{,2} = q .$$

The load term of (1.122) with the small factor h' is omitted here. Eliminating the complex stress resultants $\tilde{S}$ and $\tilde{T}_2$ from the second of equations (2.121) with the help of the first and third of equations (2.121) leads to an equation of the semi-momentless theory which coincides with (2.118) when a = const. The difference is as follows: In the equation following from (2.121) there occurs a term $a\tilde{T}_{1,2}$ in a position corresponding to that of $(a\tilde{T}_1)_{,2}$ in the term $ih^\circ Wr\tilde{T}_1$ of (2.118). It is a difference of equations based on nearly equivalent assumptions. Indeed, besides the hypothesis (2.7), equations (2.121) are founded only on assuming $(1 + ih/3R_j)\tilde{T}_k = \tilde{T}_k$, which, as discussed in Chapter 1, §8, only seldom leads beyond the bounds of accuracy of the thin-shell theory.

Hence, the difference between $a\tilde{T}_{1,2}$ and $(a\tilde{T}_1)_{,2}$ should be negligible in the semi-momentless theory. This suggests that without substantial detriment to accuracy of the theory, the function $a(\eta)$ may be treated, with respect to differentiation, as a constant. This is in fact vindicated in what follows by solutions of concrete problems (Chapter 3, §7).

Choosing the coordinate ξ so that $a_m = b$ we have $b\xi$ equal to the

---

[7] Corresponding nonlinear equations are derived in [208].

length of the ξ-line of some appropriate middle-value radius R_m, as shown in Fig. 14.

Equations (2.116) assume with $r = a/b = 1$ the following simplified form

$$\left(\frac{\partial^2}{\partial \xi^2} - \frac{\partial}{\partial \eta} \frac{b}{\rho_1'} + \frac{\partial^2}{\partial \eta^2} \frac{R_2'}{R_1'} \right) T_1 - \frac{D}{b} W' \kappa_2 = bq_\Sigma' ,$$

$$\left(\frac{\partial^2}{\partial \xi^2} - \frac{\partial}{\partial \eta} \frac{b}{\rho_1} + \frac{\partial^2}{\partial \eta^2} \frac{R_2'}{R_1} \right) \kappa_2 + \frac{1}{Ehb} W' T_1 = 0 ; \tag{2.122}$$

$$W' = \frac{\partial^2}{\partial \eta^2} \frac{R_2'}{b} \frac{\partial^2}{\partial \eta^2} + \frac{\partial}{\partial \eta} \frac{b}{R_2'} \frac{\partial}{\partial \eta} , \qquad q_\Sigma' = -\frac{\partial q_1}{\partial \xi} + \frac{\partial q_2}{\partial \eta} + \frac{\partial^2}{b \, \partial \eta^2} (R_2' q) . \tag{2.123}$$

5.3. Displacements

The simplifications introduced in the description of shell deformation by the omission of terms with strain resultants ε_2 and γ are, naturally, reflected in the displacements' pattern. In *linear* approximation this leads to simple relations between components of displacement.

Omitting terms ε_2 and γ in (1.40) gives following relations, representing the semi-membrane features of deformation in terms of displacement components

$$\frac{\dot{v}}{b} + \frac{w}{R_2} = 0 , \qquad \left(\frac{u}{a} \right)^{\cdot} \frac{a}{b} + \frac{v_{,1}}{a} = 0 , \qquad (\ )^{\cdot} = \frac{\partial}{\partial \eta} (\ ) . \tag{2.124}$$

The coordinates ξ and η are those introduced in the present chapter, §5.2 ($b = \text{const.}$, $1/\rho_2 = 0$).

The relations (2.124) together with formulas (1.40) and (1.41) express the tangential components of displacement (u, v) the rotation angle ϑ_2 and the parameter κ_2 in terms of w,

$$v = -\int w \frac{b}{R_2} \, d\eta , \qquad u = a \int \frac{b}{a^2} \int w_{,1} \frac{b}{R_2} \, d\eta^2 ,$$

$$\vartheta_2 = -\frac{w_{,2}}{b} - \frac{1}{R_2} \int w \frac{b}{R_2} \, d\eta , \qquad \kappa_2 = \frac{1}{b} \vartheta_{2,2} . \tag{2.125}$$

With formulas (2.125), elasticity relations (2.98) and expression (1.40) for ε_1, the first equation of (2.116) or of (2.122) can be transformed into a single equation for w. Such an equation of the linearized "second-order theory" (the sort of linearization discussed in the present chapter, §6.3) has provided the first basis for a solution of the bending problem for curved tubes of finite length [95].

6. Linear and nonlinear approximations

A feature of the equations of the semi-momentless theory, which can be useful in fulfilling the conditions of continuity for tubes, and an application of the perturbation method to these equations are discussed briefly.

6.1. Linear equations

Consider a shell with local geometry described by (1.15) and (1.17). The coordinates ξ and η are as shown in Fig. 14: $a = a_m R / R_m$; $a_m = b = \text{const}$. With this, the *linear* approximation of the resolving system (2.116) can be presented in the following nondimensional form

$$W\kappa - LT = -q_\Sigma^\circ, \qquad \kappa \equiv \kappa_2 b,$$

$$L\kappa + WT = 0, \qquad T \equiv T_1/Ehh^\circ = \varepsilon_1/h^\circ; \tag{2.126}$$

$$L = \frac{r}{h^\circ}\frac{\partial^2}{\partial\xi^2} - \mu\frac{\partial}{\partial\eta}r^2 s + \mu\frac{\partial}{\partial\eta}r^2\frac{\partial}{\partial\eta}\frac{R_2}{b}c, \qquad q_\Sigma^\circ \equiv b^3 q_\Sigma/D.$$

The operator W and q_Σ denote respectively the linear approximation of W' and q_Σ' from (2.117): with $R_2' = R_2$.

For the case of *cylinder* shell, i.e. for $1/R = 0$ and $\mu = 0$, $r = 1$, equations (2.126) are substantially simplified. With $L = \partial^2/h^\circ \partial\xi^2$ they become identical to the WLASSOW [36] equations and one of the unknowns is easily eliminated. This gives

$$\left(\frac{1}{h^{\circ 2}}\frac{\partial^4}{\partial\xi^4} + W^2\right)\kappa = -Wq_\Sigma^\circ, \qquad \left(\frac{1}{h^{\circ 2}}\frac{\partial^4}{\partial\xi^4} + W^2\right)T = \frac{\partial^2}{h^\circ \partial\xi^2}q_\Sigma^\circ.$$

$$\tag{2.127}$$

The operator W includes derivatives only in η. For the cases when the W can be used in the form (2.123) (i.e. with $r = 1$), the general integral of the equation $W\kappa_g = 0$ or $WT_g = 0$ is known to have the form

$$[\kappa_g \ T_g] = [C \ C'] + [C_x \ C_x']x + [C_z \ C_z']z + [C \ C']\omega ,$$

$$\omega = \tfrac{1}{2} \int_0^\eta (x\dot{z} - \dot{x}z) \, d\eta , \quad (\ )^{\cdot} = \partial(\ )/\partial\eta , \qquad\qquad (2.128)$$

$$C = C(\xi) , \quad C' = C'(\xi), \dots .$$

It can be verified, by substitution, that $W[x \ z \ \omega] = [0 \ 0 \ 0]$ for $r = 1$ and $x(\eta)$ and $z(\eta)$ being Cartesian coordinates of the η-line in its plane. The function $\omega(\eta)$ is equal to the area crossed by a radius vector from the pole $(x, z) = (0, 0)$, when its end passes from the point $\eta = 0$ along the η-line. Thus, when this coordinate line is closed, the function $\omega(\eta)$ has at the point $\eta = 0$ a discontinuity equal to the value of the area bounded by the η-line. The leap is eliminated if the shear strain (γ) is taken into account (cf. present chapter, §4.4 and [104]).

With the four particular solutions and the arbitrary functions of ξ, introduced by integration, incorporated in T_g and κ_g, equations (2.126) with $r = 1$ give

$$T = T_g - W^{-1}L\kappa , \qquad \kappa = \kappa_g + W^{-1}(LT - q_\Sigma^\circ) . \qquad (2.129)$$

The second terms in these formulas represent arbitrary particular solutions of the corresponding equation (2.126) with κ or T considered as known functions.

Note that the longitudinal stress resultant $T_1 = T_g Ehh^\circ$ and the deformation of the η-line—the curvature change $\kappa_2 = \kappa_g/b$—correspond to two limit cases, namely, the stressed states of thin-walled beams with *undeformable* cross-section $(\kappa_2 = 0)$ and with *freely deformable* cross-section, respectively.

6.2. Continuity conditions for closed shells

In a shell *closed* along coordinate lines ξ or η, stress and strain as well as displacement vary continuously and periodically, without jumps at any value of the coordinates. This demands continuity and periodicity of the eight parameters of stress and strain, listed in (1.106).

Over and above this, a possibility of a relative displacement of congruent edges of a cut through the shell must be eliminated. Such discontinuity is ruled out if, besides the continuity of strain, the displacements are also continuous *at least along one of the closed coordinate lines.*

Continuity of deformation (but not of displacement) along the ξ-lines is assured when the functions T_1, κ_2 or ψ, ϑ are periodic continuous functions of ξ. The stress and strain parameters are expressed in terms of these two dependent variables without integrating over ξ.

The situation is different for the continuity along the η-lines. Indeed, (2.114) expressing S and τ in terms of T_1 and κ_2 include integration. Hence, periodicity of the variables T_1 and κ_2 does not assure the same for S and τ.

We choose in the following the value of b to make $2\pi b$ equal to the perimeter of the η-line. Then the coordinates ξ, η and ξ, $\eta + 2\pi$ denote *the same* point and the periodicity of S and τ means that

$$S(\xi, \eta) = S(\xi, \eta + 2\pi) , \qquad \tau(\xi, \eta) = \tau(\xi, \eta + 2\pi) . \quad (2.130)$$

With (2.114) and $a = R$ these conditions become

$$\frac{d}{d\xi} \int_{\eta}^{\eta+2\pi} T_1 R \, d\eta = - \int_{\eta}^{\eta+2\pi} q_1 R^2 \, d\eta , \quad (2.131)$$

$$\frac{d}{d\xi} \int_{\eta}^{\eta+2\pi} \kappa_2 R \, d\eta = 0 . \quad (2.132)$$

The condition (2.131) has on its left-hand side the derivative of the moment M_z of all internal forces (by $M_1 \approx 0$) acting in the section $\xi = $ const. of the shell. The condition means that the moment, with respect to the axis of z, which is the focus of the centers of curvature of the ξ-lines (Fig. 14), depends only on forces applied on the edge $\xi = $ const. and on q_1. The other components of distributed load (q_2 and q) do not influence the moment in a cross-section $\xi = $ const. as they act within the planes passing through the axis of z.

The condition (2.131) is fulfilled if the internal forces in the section $\xi = $ const. are statically equivalent to the correct values of normal force

N and bending moments M_z and M_x in the plane of the cross-section (Fig. 14). For closed sections, this condition can disregard M_1, H_1:

$$\int_0^{2\pi} [1 \ x \ z] T_1 b \, d\eta = [N \ M_z \ M_x] \quad (x = R - R_m).\tag{2.133}$$

The values of N, M_z and M_x are determined with the help of equations of equilibrium of one of two parts of the shell separated by the section $\xi = $ const. (provided the two are not interconnected).

The conditions of continuity of the shell along a closed η-line demand continuity of the angle of rotation ϑ_2 and of the components of displacement in the directions of axes x and z and along the ξ-line. This requires that the values of ϑ_2 and of the components of displacement at any point, $m(\xi, \eta)$, and at the infinitesimally near point $m(\xi, \eta + 2\pi)$ be equal. These conditions for ϑ_2 and the displacements in the x- and z-directions may be written (neglecting the ε_2 term) in the form

$$\int_\eta^{\eta + 2\pi} [1 \ x \ z] \kappa_2 \, d\eta = [0 \ 0 \ 0].\tag{2.134}$$

The remaining condition of continuity of the displacement along the ξ-line is not given here. It is simpler to write it for each concrete problem.

Together with conditions (2.134) the condition (2.132) is fulfilled automatically.

The conditions of continuity in the form (2.133) and (2.134) can be generalized to include large displacements. In (2.133) this requires replacing x and z with coordinates of *deformed* η-line x^* and z^*. The conditions (2.134) can be considered as a particular case of conditions for a small increment $\Delta\kappa_2$ of shell deformation, when the shell has been previously deformed,

$$\int_\eta^{\eta + 2\pi} [1 \ x^* \ z^*] \Delta\kappa_2 \, d\eta = [0 \ 0 \ 0].\tag{2.135}$$

The continuity of stress and deformation is expressed by conditions (2.131) and (2.132) also for large displacements. As has been mentioned, the dislocation of a shell—discontinuity in displacements—is ruled out if, besides (2.131) and (2.132), displacements are continuous and periodic on at least one η-line.

Usually at least one edge ξ = const. undergoes only small displacements. On such edges the conditions (2.134) remain valid for nonlinear problems.

6.3. Perturbation method and "second-order theory"

An iterative solution of the nonlinear equations (2.96)–(2.98) or (2.116) with appropriate boundary conditions encounters no fundamental difficulties (Chapter 3, §8). But this method has certain shortcomings. It provides results on stress and deformation only for a specific level of loading. Thus, characteristics of a shell—their dependence on geometric and other parameters—can be deduced only through analysis of extensive numerical data.

When nonlinear effects are not too great, namely, do not change the picture of deformation fundamentally, but constitute merely more or less important corrections to it, the solution may be substantially simplified. A way to this simplification is provided by the perturbation method.

The starting point of the solution is constituted by the assumption that the stress and strain resultants (or other dependent variables) may be sought in the form of expansions in powers of a *load parameter* χ,

$$[T_1 \; \kappa_2 \; \cdots] = \sum_{n=1,2,\ldots} \chi^n [T_1^{(n)} \; \kappa_2^{(n)} \; \cdots] . \tag{2.136}$$

Inserting these expansions into the governing equations (2.116) and equating on both sides of each equation sums of terms with factor χ^n for each power $n = 1, 2, \ldots$ results in a linear system of equations for each nth approximation. The same result is obtained with substitution of expansions of ψ and ϑ in powers of χ into the Reissner equations or into their generalization (2.94).

The first-approximation ($n = 1$) equations, following from (2.116) are essentially equivalent to equations (2.126). The difference amounts to the replacement of κ, T by $\kappa^{(1)}$, $T^{(1)}$. For the second and the higher approximations ($n = 2, 3, \ldots$), the equations differ from the first approximation only on the right-hand sides. These depend on the terms of the expansions (2.136) which have been determined in the preceding lower approximations.

The version of the perturbation solution summarized here is the simplest possible. But as mentioned its applicability is restricted to the cases where the nonlinearity does not change the *character* of stress and strain distribution substantially. Such changes cannot be accounted for effectively, since the left-hand sides of the equations obtained do not reflect the changes in geometry and stresses. The convergence of (2.136) can become insufficient.

In Chapter 3, two problems of this sort are discussed.

(a) Bending and axial compression of a cylinder shell or of a tube with small initial curvature.

(b) A cylinder or a curved tube under external or internal pressure, which is not negligible compared to the critical value of external pressure.

The following variant of the small-parameter method allows us to take account of the *nonlinearity* of the relation of a stress state to loading already in the *first-approximation* solution of the two problems. The approach to be used is similar to that applied in stability problems. (An instance is provided by Chapter 1, §9.2.)

The stress state and deformation of the shell is sought in the form of a sum of zeroth approximation and corrections,

$$[T_1 \ \kappa_2 \ T_2 \ \cdots] = \sum_{n=0,1,\ldots} [T_1^{(n)} \ \kappa_2^{(n)} \ T_2^{(n)} \ \cdots]. \qquad (2.137)$$

The zeroth approximation includes values of stress and strain resultants determined in a sufficiently *simple* way. Only those resultants that are of basic importance for the particular problem need be included in the zeroth approximation. Those remaining can be set equal to 0.

The corrections $T_1^{(1)}, \kappa_2^{(1)}, \ldots$ are determined by equations resulting from substitution of the expansion (2.137) into the governing equations of the problem, when only terms *linear* in the first-approximation corrections $T_1^{(1)}, \kappa_2^{(1)}, \ldots$ are retained.

The equations of the second approximation result similarly, when after substitution of expansions (1.137) besides terms of the approximations $n = 0$ and 1 only terms linear in $T_1^{(2)}$ and $\kappa_2^{(2)}$ are retained. In this way equations of any (nth) approximation can be constructed.

Consider the *first-approximation* equations for shells for which the simplified form (2.122) of the resolving equations is applicable. Let the zeroth approximation describe deformation of the beam type determined by axial force N and by the bending moment M_z,

$$T_1^{(0)} = \frac{N}{2\pi b} + \frac{M_z x}{I} h , \qquad \kappa_2^{(0)} = 0 , \qquad I = \int_0^{2\pi} x^2 hb \, d\eta , \quad x = R - R_m .$$

(2.138)

Here and in the following, the coordinate η is defined (by the value of $b = \text{const.}$) to vary from 0 to 2π.

Equations (2.96) and (2.114) with $a = b$ and ε_1, κ_2 set equal to $\varepsilon_1^{(0)} = T_1^{(0)}/Eh, \kappa_2^{(0)} = 0$ give

$$[\lambda_1^{(0)} \ \kappa_1^{(0)}] = [s \ c] M_z / EI .$$

Introduce this into expansions (2.137) retaining only the $n = 0, 1$ terms,

$$T_1 = T_1^{(0)} + T_1^{(1)} , \qquad \frac{b}{R_2'} = \frac{b}{R_2} + \kappa , \qquad \kappa \equiv \kappa_2^{(1)} b ,$$

$$\frac{1}{R_1'} = \frac{c}{R} + \kappa_1^{(0)} + \kappa_1^{(1)} , \qquad \frac{1}{\rho_1'} = \frac{s}{R} + \lambda_1^{(0)} + \lambda_1^{(1)} .$$

(2.139)

We insert these expressions into (2.96)–(2.98) with $a = b$ ($r = 1$) and the identities $\kappa_1/R_2' + \kappa_2/R_1 = \kappa_1/R_2 + \kappa_2/R_1'$ and $\lambda_1/R_2' + \kappa_2/\rho_1 = \lambda_1/R_2 + \kappa_2/\rho_1'$. Retaining only those terms linear in $T_1^{(1)}$ and $\kappa \equiv \kappa_2 b$ we obtain a system similar to (2.126).

$$\left(W + 2\mu' \frac{\partial^2}{\partial\eta^2} \rho^2 c T^{(0)} - \frac{\partial^2}{\partial\eta^2} q^\circ \rho^2 \right) \kappa$$

$$- \left(L' + \frac{\partial}{\partial\eta} T^{(0)} \frac{\partial}{\partial\eta} - \frac{\partial^2}{\partial\eta^2} \rho^2 T^{(0)} \frac{\partial^2}{\partial\eta^2} \right) T^{(1)}$$

$$= L' T^{(0)} - q_\Sigma^\circ , \qquad L'\kappa + W T^{(1)} = 0 ,$$

(2.140)

where

$$L' = \frac{1}{h^\circ}\frac{\partial^2}{\partial \xi^2} - \mu'\frac{\partial}{\partial \eta}s + \mu'\frac{\partial^2}{\partial \eta^2}c\rho \,, \quad \rho = \frac{R_2}{b} \,,$$

$$W = \frac{\partial^2}{\partial \eta^2}\rho\frac{\partial^2}{\partial \eta^2} + \frac{\partial}{\partial \eta}\frac{1}{\rho}\frac{\partial}{\partial \eta} \,, \tag{2.141}$$

$$\mu' = \mu + M^\circ \,, \quad M^\circ = \frac{M_z\, b}{EI\, h^\circ} \,, \quad [T^{(0)} \ T^{(1)}] = \frac{1}{Ehh^\circ}[T_1^{(0)} \ T_1^{(1)}] \,.$$

The coefficients of these linear equations depend on loading (q, N, M_z). The term $L'T^{(0)}$ on the right-hand side of (2.140) depends on load parameter M° nonlinearly.

The linear system (2.140) describes the deformation of shell taking into account its *nonlinear* dependence on loading. Solutions of this sort are known in structural mechanics as "second-order theory". The system has all merits of linear equations. But its accuracy depends in the nonlinear range on the zeroth-approximation stress resultants and curvatures being not too different from the actual values.

Notice that, when the nonlinearity is too pronounced for the "second-order theory", a simple solution cannot be achieved by retaining next $(n = 2, 3, \ldots)$ terms in expansion (2.137). More straightforward is a solution of the basic system by successive approximations (Chapter 3, §8).

7. Matrix form of operations with trigonometric series

Trigonometric series of functions are represented by matrices of the Fourier coefficients. Multiplication, differentiation and integration of functions are effected with matrix operations. Thus, differential equations are transformed to matrix ones.

7.1. Trigonometric-series method

At least since the work of L. Navier and M. Lévy, for more than 100 years, the trigonometric-series method remains the most popular one for analytical solution of plate and shell problems. It is amply illustrated by monographs like those of SEIDE [173] or FLÜGGE [152]. This role of the trigonometric series is not accidental but reflects its remarkable position in mathematics—in the words of LANCZOS [112]: ". . . if we were asked to abandon all mathematical discoveries save one, we would hardly fail to vote for the Fourier series. . . ".

The method is essentially quite straightforward. Trigonometric Fourier series is known to render (under conditions, which are safely fulfilled by the continuous functions to be considered) a true and *unique* representation of a function. Hence, a set of Fourier coefficients of a function provides an adequate image of the function. It is, in fact, a Fourier transform with finite bounds—those of the region of variation of the argument (or two arguments).

Though the number of the coefficients of a function is formally infinite, it remains moderate for functions describing the stressed state and deformation with the accuracy of the thin-shell theory.

The Fourier series render expansion of a function into a spectrum of components in a very *effective basis*. A number of Fourier coefficients is, as a rule, much smaller than the number of values of a function in chosen points, necessary to define the function with the same accuracy. This simplifies the computer solution, particularly the data input and analysis of the results. Moreover, the series solution can serve in many cases as an analytical one, giving the principal relations between parameters of a problem in a simple way.

The trigonometric-series solution amounts to the following:

All known and unknown functions are represented by their trigonometric Fourier expansions. Thereafter, both sides of an equation are reduced to the form of Fourier series. An ordinary differential equation is thus reduced to an algebraic system equating the Fourier coefficients of $\cos n\eta$ and $\sin n\eta$ (or by the same functions of ξ) of both sides of the original equation for *each* of $n = 1, 2, \ldots$. A partial differential equation system determining, for instance, $T_1(\xi, \eta)$ and $\kappa_2(\xi, \eta)$ is in this way reduced to a system of ordinary differential equations determining Fourier coefficients of T_1 and κ_2, which are functions of ξ. The same applies to integral equations or integro-differential equations, for instance, to equations (2.94). An identical procedure, this time with Fourier expansions in the other coordinate, reduces the original two-dimensional equations to a system of *algebraic* equations. (This can, of course, be achieved at a stroke with double Fourier series.)

A simple generalization makes the trigonometric-series solution applicable for *any* boundary conditions on edges $\eta = $ const. or $\xi = $ const.

The procedure of the series solution requires the determination of

Fourier coefficients for all terms in the equations to be solved. Such terms can be products of several functions represented each with its trigonometric series. This involves determination of Fourier coefficients of a product of several series. The technique includes also differentiation and integration of the series. All this can be realized in the most simple, systematic and easily programmable matrix form. The basis is provided by representation of a function with a matrix appropriately composed of the Fourier coefficients. Further multiplication, differentiation and integration of functions are carried out by matrix operations. Finally, a system of algebraic or of ordinary differential equations for the Fourier coefficients of the unknown functions is arrived at. The system is simply written out as a matrix form of the original differential or integro-differential equations.

7.2. Matrix form of algebraic operations with trigonometric series

Consider a function $a(\eta)$ represented by a sum of sufficient number of terms of the trigonometric Fourier series and introduce notation a for a *column matrix* composed of the Fourier coefficients

$$a(\eta) = \tfrac{1}{2}a_0 + \sum_k (a_{ck} \cos k\eta + a_{sk} \sin k\eta) , \qquad a = \begin{bmatrix} a_c \\ a_s \end{bmatrix} = \begin{bmatrix} \tfrac{1}{2}a_0 \\ a_{c1} \\ a_{c2} \\ \dots \\ a_{s1} \\ a_{s2} \\ \dots \end{bmatrix} .$$

$$(2.142)$$

Similar to the matrix form of the vector product of two multidimensional vectors a formula can be composed determining the matrix p of the Fourier coefficients of a product of two functions $p = a(\eta)b(\eta)$ in terms of the coefficients of the cofactors. The formula has two alternative forms

$$p = \breve{a}b = \breve{b}a . \qquad (2.143)$$

Here $\breve{a}$ and $\breve{b}$ denote square matrices composed of the Fourier coefficients of the functions $a(\eta)$ and $b(\eta)$. The starting point for constructing matrices $\breve{a}$ and $\breve{b}$ is provided by the formulas

$$2 \cos m\eta \cos n\eta = \cos(m - n)\eta + \cos(m + n)\eta ,$$

$$2 \cos m\eta \sin n\eta = \sin(m + n)\eta - \sin(m - n)\eta , \tag{2.144}$$

$$2 \sin m\eta \sin n\eta = \cos(m - n)\eta - \cos(m + n)\eta .$$

Multiplication of the series

$$p = ab = \left[\frac{a_0}{2} + \sum_j (a_{cj} \cos j\eta + a_{sj} \sin j\eta) \right]$$

$$\times \left[\frac{b_0}{2} + \sum_k (b_{ck} \cos k\eta + b_{sk} \sin k\eta) \right] ,$$

and use of (2.144) renders, after reshuffling of terms in accordance with the multiplication rule for matrices

$$p = \begin{bmatrix} p_c \\ p_s \end{bmatrix} = \begin{bmatrix} a_c^+ & a_s^- \\ a_s^+ & a_c^- \end{bmatrix} \begin{bmatrix} b_c \\ b_s \end{bmatrix} = \breve{a}b . \tag{2.145}$$

The notations p_c, p_s, b_c and b_s are clear from (2.142) for a_c and a_s: these are *column* matrices composed of the Fourier coefficients of corresponding functions. The matrices denoted a_c^+ and a_c^- are *quadratic*, composed of the coefficients a_{cj}; matrices a_s^+ and a_s^- are *nonquadratic square* ones, composed of coefficients a_{sj}. The four matrices composing $\breve{a}$ are determined by the following formulas for elements of a row number i or m and column number j or k:

$$2a_{cij}^+ = \begin{cases} a_{cj} & (i = 0) , \\ a_{c|i-j|} + a_{c(i+j)} & (i \geqslant 1) , \end{cases}$$

$$2a_{sik}^- = \begin{cases} a_{sk} & (i = 0) \\ a_{s(i+k)} - a_{s(i-k)} & (i \geqslant k) , \\ a_{s(i+k)} + a_{s(k-i)} & (k > i) \end{cases} \tag{2.146}$$

$$a_{s0} = 0 , \qquad j = 0, 1, 2, \ldots ; \quad k = 1, 2, \ldots .$$

$$2a^+_{smj} = \begin{cases} a_{s(m+j)} + a_{s(m-j)} & (m \geqslant j), \\ a_{s(m+j)} - a_{s(j-m)} & (j > m); \end{cases} \quad 2a^-_{cmk} = a_{c|m-k|} - a_{c(m+k)},$$

$$\text{(2.147)}$$

$$a_{s0} = 0, \qquad j = 0, 1, 2, \dots; \quad m, k = 1, 2, \dots.$$

Thus, the four matrix elements of $\check{a}$, written out with rows and columns of numbers under 3 (to save space), are

$$2a^+_c = \begin{bmatrix} a_{c0} & a_{c1} & a_{c2} \\ 2a_{c1} & a_{c0} + a_{c2} & a_{c1} + a_{c3} \\ 2a_{c2} & a_{c1} + a_{c3} & a_{c0} + a_{c4} \end{bmatrix},$$

$$2a^-_s = \begin{bmatrix} a_{s1} & a_{s2} \\ a_{s2} + 0 & a_{s3} + a_{s1} \\ a_{s3} - a_{s1} & a_{s4} + 0 \end{bmatrix},$$

$$\text{(2.148)}$$

$$2a^+_s = \begin{bmatrix} 2a_{s1} & a_{s2} + 0 & a_{s3} - a_{s1} \\ 2a_{s2} & a_{s3} + a_{s1} & a_{s4} + 0 \end{bmatrix},$$

$$2a^-_c = \begin{bmatrix} a_{c0} - a_{c2} & a_{c1} - a_{c3} \\ a_{c1} - a_{c3} & a_{c0} - a_{c4} \end{bmatrix}.$$

Notice that the matrices a^-_s and a^-_c do not include *column* number 0; the matrices a^+_s and a^-_c do not include the *row* number 0. The lack of the column or row number 0 in the matrices a^-_s and a^+_s and of both in a^-_c is a consequence of the lack of a corresponding term $a_{s0} \sin 0\eta$ in the Fourier series and of an element a_{s0} in a column matrix a. (A version including the zeroth columns and rows in all these matrices is presented in [205].)

Formula (2.145) leads directly to a more general matrix formula for the product of more than two trigonometric series. For three factors it is

$$p = \check{a}\check{b}c . \qquad\qquad \text{(2.149)}$$

Here p is a column matrix composed of Fourier coefficients of a function $p(\eta) = a(\eta)b(\eta)c(\eta)$; $\check{a}$ and $\check{b}$—quadratic matrices of the coefficients of the cofactors, composed according to (2.145)–(2.148); c—a column of the coefficients of $c(\eta)$ according to (2.142).

To justify (2.149) it is sufficient to note that according to (2.145) $\check{b}c$ is the column matrix of the coefficients of a function $b(\eta)c(\eta)$ and thus $\check{a}\check{b}c = \check{a}(\check{b}c) = p$ follows from (2.143).

The order of matrices representing functions in (2.143) and (2.149) can be very substantially reduced when the Fourier series of the functions include only terms with $\sin k\eta$ or with $\cos k\eta$. For instance, consider functions

$$a(\eta) = \tfrac{1}{2}a_0 + \sum_k a_{ck} \cos k\eta \,, \qquad b(\eta) = \sum_k b_{sk} \sin k\eta \,,$$

$$c(\eta) = \tfrac{1}{2}c_0 + \sum_k c_{ck} \cos k\eta \,, \qquad d(\eta) = \sum_k d_{sk} \sin k\eta \,. \qquad (2.150)$$

Formulas (2.145) and (2.148) give for the column matrices of Fourier coefficients of functions $a(\eta)b(\eta)$, $b(\eta)c(\eta)$, ... expressions (the symbol $\dot{\rightarrow}$ connects a function and a column matrix of its Fourier coefficients),

$$a(\eta)b(\eta) \dot{\rightarrow} \begin{bmatrix} a_c^+ & 0 \\ 0 & a_c^- \end{bmatrix} \begin{bmatrix} 0 \\ b_s \end{bmatrix} = \begin{bmatrix} 0 \\ a_c^- b_s \end{bmatrix} ,$$

$$b(\eta)a(\eta) \dot{\rightarrow} \begin{bmatrix} 0 & b_s^- \\ b_s^+ & 0 \end{bmatrix} \begin{bmatrix} a_c \\ 0 \end{bmatrix} = \begin{bmatrix} 0 \\ b_s^+ a_c \end{bmatrix} , \qquad \text{etc.}$$

Here 0 denotes matrices, both column and square, composed of only zero-value elements. Omitting the matrices 0, the formulas for matrices of Fourier coefficients of *products* of functions become simplified

$$ab \dot{\rightarrow} a_c^- b_s = b_s^+ a_c \,, \qquad ac \dot{\rightarrow} a_c^+ c_c = c_c^+ a_c \,, \qquad bd \dot{\rightarrow} b_s^- d_s = d_s^- b_s \,.$$

$$(2.151)$$

Notice that the square-matrix cofactors are composed as "minus" matrices when the next cofactor to the right is composed of *sine* coefficients. The "plus" matrices are cofactors by columns of *cosine* coefficients.

In what follows, expansions of the type (2.150) will be used. The indices c and s will be, as a rule, omitted. The cosine or sine character of the corresponding expansion will be indicated at the start of a solution.

The indices $+$ or $-$ will be, when convenient, placed at the foot of the letter denoting a square matrix.

Consider now the determination of the Fourier coefficients of a function $1/b(\eta)$ given the coefficients of $b(\eta)$. For the square matrix $\check{b}^{-1}$ and the column matrix b^{-1} representing the function $1/b(\eta)$ in the formulas

$$\frac{a(\eta)}{b(\eta)} \dot{\rightarrow} \check{b}^{-1} a = \check{a} b^{-1} , \qquad (2.152)$$

one can easily deduce from relations $b^{-1} b a \dot{\rightarrow} a = \check{b}^{-1} \check{b} a$ and $a b b^{-1} \dot{\rightarrow} a = \check{a} \check{b} b^{-1}$ the following:

$$\check{b}^{-1} \check{b} = E , \quad E \equiv \begin{bmatrix} 1 & 0 & 0 \\ 0 & 1 & 0 \\ 0 & 0 & 1 \\ & & & \ddots \end{bmatrix} , \quad \check{b} b^{-1} = \begin{bmatrix} 1 \\ 0 \\ 0 \\ \cdots \end{bmatrix} . \qquad (2.153)$$

A useful formula is obtained by direct integration

$$\frac{1}{\pi} \int_0^{2\pi} a(\eta) b(\eta) \, d\eta = \tfrac{1}{2} a_0 b_0 + a_{c1} b_{c1} + a_{c2} b_{c2} + \cdots + a_{s1} b_{s1}$$
$$+ a_{s2} b_{s2} + \cdots . \qquad (2.154)$$

7.3. Matrix form of derivative and integral

Assume that the Fourier series of all derivatives considered can represent the functions satisfactorily. This justifies representation of a derivative of a function by derivative of its trigonometric series

$$\frac{da}{d\eta} = \sum_k \left(-k a_{ck} \sin k\eta + k a_{sk} \cos k\eta \right) .$$

This means

$$\dot{a}(\eta) \dot{\rightarrow} \dot{a} = \{ 0 \ \ a_{s1} \ \ 2a_{s2} \ \ 3a_{s3} \ \cdots \ -a_{c1} \ -2a_{c2} \ -3a_{c3} \ \cdots \} . \qquad (2.155)$$

The brackets $\{ \ \}$ denote a *column* matrix.

The column matrix $\dot{a}$ can be expressed through the column a and a *differentiation matrix*

$$\dot{a} = \Lambda_d a \equiv \begin{bmatrix} 0 & \Lambda \\ -\Lambda_+ & 0 \end{bmatrix} a \, , \qquad \Lambda = \begin{bmatrix} 0 & 0 & \cdots \\ 1 & 0 & \cdots \\ 0 & 2 & \cdots \\ \cdots & \cdots & \cdots \end{bmatrix} \, ,$$

$$\Lambda_+ = \begin{bmatrix} 0 & 1 & 0 & \cdots \\ 0 & 0 & 2 & \cdots \\ \cdots & \cdots & \cdots & \cdots \end{bmatrix} = \Lambda^{\mathrm{T}} \, , \tag{2.156}$$

$$\ddot{a} = \Lambda_d^2 a \equiv \begin{bmatrix} -\Lambda_+^2 & 0 \\ 0 & -\Lambda^2 \end{bmatrix} a \, , \qquad \Lambda_+^2 = \Lambda\Lambda_+ \, , \qquad \Lambda^2 = \Lambda_+\Lambda \, .$$

The superscript T denotes here and in what follows a *transposed matrix*.

For functions represented by series with only sine terms or only cosine terms, the matrix differentiation formula becomes simpler,

$$\dot{a} = -\Lambda_+ a \quad (\text{for } a = a_c) \, , \qquad \dot{b} = \Lambda b \quad (\text{for } b = b_s) \, . \tag{2.157}$$

(The omission of the indices c and s requires of course knowledge of the art of elements of the matrix in question.)

Any converging Fourier series may be integrated term-by-term. Thus, integration of series (2.142) gives

$$\int_{\eta_0}^{\eta} a(\eta) \, d\eta = \left[\tfrac{1}{2} a_0 \eta + \sum_k \left(\frac{1}{k} a_{ck} \sin k\eta - \frac{1}{k} a_{sk} \cos k\eta \right) \right]_{\eta_0}^{\eta} \, . \tag{2.158}$$

To present the result of integration (2.158) in a matrix form, the nonperiodical term $\tfrac{1}{2} a_0 \eta$ must be substituted with a Fourier series. It is easily done when only the segment $-\pi < \eta < \pi$ is considered, as is in fact the case in the following: Introduce instead of η a *periodic* function η_s equal to η for $-\pi < \eta < \pi$—the "saw-tooth" function,

$$\eta_s = -\sum_k \frac{2}{k} \cos k\pi \sin k\eta \, , \qquad \eta_s \dot{\to} \boldsymbol{\eta}_s = 2\{1 \ -\tfrac{1}{2} \ \tfrac{1}{3} \ -\tfrac{1}{4} \ \cdots\} \, . \tag{2.159}$$

This series reproduces the Gibbs-effect peaks (Fig. 16). Much more

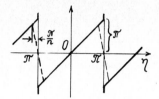

FIG. 16. Saw-tooth functions: η_s—thick line, η_σ—broken line.

rapid convergence is assured for the series modified by the Lanczos [112] σ-factors

$$\eta_\sigma = -\sum_{k=1}^{n-1} \sigma_k \frac{2}{k} \cos k\pi \sin k\eta , \quad \sigma_k = \frac{\sin(k\pi/n)}{k\pi/n} ,$$

$$\boldsymbol{\eta}_\sigma = 2\left\{ \frac{\sigma_1}{1} - \frac{\sigma_2}{2} \frac{\sigma_3}{3} \cdots \right\}. \tag{2.160}$$

Using the relations (2.158) and (2.159) an integral of a function can be written in a matrix form with the help of an integration matrix.

With the lower endpoint η_0 of integration set at 0 or $\pi/2$, the formula of integration becomes

$$\int_0^\eta a(\eta)\,\mathrm{d}\eta = \int_0^\eta \left(\frac{a_0}{2} + \sum_k a_k \cos k\eta \right) \mathrm{d}\eta \dot{\to} \frac{a_0}{2}\boldsymbol{\eta}_s + \Lambda_+^{-1}\boldsymbol{a} ,$$

$$\int_{\pm\pi/2}^\eta b(\eta)\,\mathrm{d}\eta = \int_{\pm\pi/2}^\eta \sum_k b_k \sin k\eta\,\mathrm{d}\eta \dot{\to} -\Lambda^{-1}\boldsymbol{b} . \tag{2.161}$$

The integration matrices are

$$\Lambda_+^{-1} = \begin{bmatrix} 0 & 1 & 0 & 0 & \cdots \\ 0 & 0 & 1/2 & 0 & \cdots \\ 0 & 0 & 0 & 1/3 & \cdots \\ \cdots & \cdots & \cdots & \cdots & \cdots \end{bmatrix} ,$$

$$\Lambda^{-1} = \begin{bmatrix} 0 & 1/2 & 0 & \cdots \\ 1 & 0 & 0 & \cdots \\ 0 & 1/2 & 0 & \cdots \\ 0 & 0 & 1/3 & \cdots \\ \cdots & \cdots & \cdots & \cdots \end{bmatrix} . \tag{2.162}$$

The relation

$$\frac{d}{d\eta} \int a(\eta)\, d\eta = a(\eta)$$

implies defining (contrary to the formal result of multiplication)

$$\Lambda \Lambda_+^{-1} \equiv E, \qquad \Lambda_+ \Lambda^{-1} \equiv E \tag{2.163}$$

The formulas of differentiation and integration (2.156)–(2.161) determine *column* matrices of derivative and integral of a function.

For the *square* matrices corresponding to a column matrix a and its parts a_c and a_s introduce the notation

$$\check{a} = [a]^{\vee}, \qquad a_c^+ = [a]_c^+, \qquad a_s^- = [a]_s^-. \tag{2.164}$$

For functions represented by purely cosine series or purely sine series, as indicated in (2.150), we have according to (2.156) and (2.164),

$$a\dot{b} \dot{\Rightarrow} a_c^+ \Lambda b_s = [\Lambda b_s]_c^+ a_c, \qquad a\dot{c} \dot{\Rightarrow} -a_c^- \Lambda_+ c_c = -[\Lambda_+ c_c]_s^+ a_c,$$

$$b\dot{a} \dot{\Rightarrow} -b_s^- \Lambda_+ a_c = -[\Lambda_+ a_c]_s^- b_s, \qquad b\dot{d} \dot{\Rightarrow} b_s^+ \Lambda d_s = [\Lambda d_s]_c^- b_s. \tag{2.165}$$

8. Trigonometric-series solution

The matrix form of the series solution is presented for boundary and periodic problems of flexible shells.

8.1. Periodic solution of the Reissner equations

Consider a shell of axially symmetric local shape (Fig. 14). With no loss of generality the shell form is assumed *symmetric* with respect to the plane defined by the line $\eta = 0$.

Consider the deformation symmetric to the plane $\eta = 0$. All functions describing the shell form or the stressed state are presented with trigonometric series composed of either cos $j\eta$ terms alone or of sin $j\eta$ terms alone.

The value of the Lamé parameter b will be chosen so that on the boundaries of the shell, $\eta = \pm\pi$ or $0, \pi$. Thus, the form of the shell meridian is described by the expansions

$$\begin{bmatrix} c \\ r \\ t \end{bmatrix} = \frac{1}{2} \begin{bmatrix} c_0 \\ r_0 \\ t_0 \end{bmatrix} + \sum_j \begin{bmatrix} c_j \\ r_j \\ t_j \end{bmatrix} \cos j\eta \,, \qquad s = \sum_j s_j \sin j\eta \,. \tag{2.166}$$

The resolving functions are represented by series of the form

$$\frac{T_1}{Eh^\circ h_m} = \dot\psi = N^\circ + \sum_j g_j \cos j\eta \,,$$

$$\kappa_2 b = \dot\vartheta = \tfrac{1}{2} f_0 + \sum_j f_j \cos j\eta \,, \qquad (\ )^{\boldsymbol\cdot} = \frac{\mathrm{d}(\ )}{\mathrm{d}\eta} \,. \tag{2.167}$$

The term N° is expressed through the resultant force N of the stresses T_1/h in the section $\xi = \mathrm{const.}$ (cf. (2.133)):

$$N^\circ = N/(2\pi bEh^\circ h_m) \,. \tag{2.168}$$

The value of N is statically determinate. Equilibrium of a part of the shell cut off by the plane $\xi = \pm\pi/2$ (Fig. 5) is expressed by

$$2N = \int_{-\pi}^{\pi} b \, \mathrm{d}\eta \int_{-\pi/2}^{\pi/2} (-q_2 s^* + qc^*) \cos \xi R \, \mathrm{d}\xi \,. \tag{2.169}$$

The expansions of the resolving functions ψ and ϑ have the form resulting from integration of the series (2.167). For continuous periodic variation of $\vartheta(\eta)$, i.e. for $f_0 = 0$,

$$\psi = N^\circ\eta + \sum_j b_j \sin j\eta \,, \qquad \vartheta = \sum_j a_j \sin j\eta \quad (j = 1, 2, \ldots) \,. \tag{2.170}$$

The nonperiodic term $N^\circ\eta$ in the function $\psi(\eta)$ leads on the left-hand side of the resolving equations (2.18) to terms exactly equal to the nonperiodic parts of the load terms (with the functions F and F_1) on the right-hand sides of the equations.

Consider the system of algebraic equations determining the Fourier coefficients of the series (2.170)

$$\vartheta = \{a_1 \ a_2 \ \cdots\} \,, \qquad \psi = \{b_1 \ b_2 \ \cdots\} \,.$$

The system can be written directly as a matrix form of the Reissner equations or their simplified form (2.30), while the nonperiodic terms appearing in the loading functions (F_1, F or p) as a result of integration, are accounted for as compensated by the $N°\eta$ term in $\psi(\eta)$. The matrix system, corresponding to (2.30) in accordance with (2.151)–(2.157), is

$$\Lambda_+ r_+ t_+^{-1} \Lambda \psi - \mu^* c'_- \vartheta = ms \, ,$$

$$\Lambda_+ r_+ t_+^3 \Lambda \vartheta + \mu^* c_-^* \psi = -P° s^* - q° p \, ; \tag{2.171}$$

$$p = s_+^* \Lambda^{-1} r_- s^* - c_-^* \Lambda_+^{-1} r_+ c^* \, ,$$

$$s^* = s + c'_- \vartheta \, , \quad c^* = c - s'_- \vartheta \, . \tag{2.172}$$

The formula for p corresponds to the definition of $p(\eta)$ in (2.22) with $\eta_0 = \pi/2$ and $\eta_1 = 0$. (Note that using the integration matrix Λ_+^{-1} from (2.162) amounts to excluding the nonperiodic terms in $p(\eta)$.)

Writing (2.17) in the matrix form and using the well-known expansions

$$\sin \vartheta = \vartheta - \vartheta^3/3! + \vartheta^5/5! - \cdots \, ,$$

$$\cos \vartheta = 1 - \vartheta^2/2! + \vartheta^4/4! - \cdots \tag{2.173}$$

gives for the matrices c' and s' expressions in terms of the unknown ϑ,

$$\begin{bmatrix} c' \\ s' \end{bmatrix} = \begin{bmatrix} c_+ & -s_- \\ s_+ & c_- \end{bmatrix} \begin{bmatrix} e - \vartheta_- \vartheta/3! + \vartheta_+^2 \vartheta_- \vartheta/5! - \cdots \\ \vartheta/2 - \vartheta_+ \vartheta_- \vartheta/4! + \vartheta_+^3 \vartheta_- \vartheta/6! - \cdots \end{bmatrix} , \tag{2.174}$$

$$e \equiv \begin{bmatrix} 1 \\ 0 \\ 0 \\ \cdots \end{bmatrix} .$$

Solutions of the system (2.171) are used for investigation of several problems in the following sections.

8.2. Solution of the Reissner equations for boundary problems

A solution in the form (2.170) gives no way of satisfying concrete boundary conditions on edges $\eta = $ const. Even conditions of continuity and periodicity—for the shell with closed η-lines (tube)—are not necessarily fulfilled by the series (2.170). Only three of these conditions are satisfied. Continuity of the rotation ϑ and of radial displacement u_r is assumed directly by the form of solution (2.170). Displacement u_ξ (along t_1 in Fig. 5) is continuous as a consequence of rotational symmetry of deformation, conserving the η-line plane. The remaining condition—continuity of the displacement u_z—is not automatically satisfied by solution (2.170): the formula (2.25) does *not* render $u_z(-\pi) = u_z(\pi)$ for arbitrary a_j and b_j.

Boundary conditions on the lines $\eta = $ const., as well as the continuity conditions for shells with closed η-lines, can be satisfied by a solution containing *four* arbitrary constants,

$$\psi = \tilde{\psi} + \sum_1^4 C_j \psi_j , \qquad \vartheta = \tilde{\vartheta} + \sum_1^4 C_j \vartheta_j .$$

For a linear problem, $\tilde{\psi}$, $\tilde{\vartheta}$ and ψ_j, ϑ_j are, respectively, particular *solutions* of the governing system and of the same system without the load and distortion terms. The constants $C_1, \ldots, C_4$ are determined to satisfy the boundary conditions.

For *nonlinear* equations a sum of particular solutions does *not* constitute a solution. For any set of boundary conditions, loading and distortion, the solution must be determined individually. Such a solution will be sought in the form

$$\vartheta = \vartheta^0 + \sum_1^4 C_k \vartheta_k^0 , \qquad \psi = \psi^0 + \sum_1^4 C_k \psi_k^0 ,$$

$$\vartheta^0 = \sum_n a_n \sin n\eta , \qquad \psi^0 = \sum_n b_n \sin n\eta + N^0 \eta . \tag{2.175}$$

The functions ϑ^0, ψ^0, ϑ_j^0 and ψ_j^0 introduced here do *not* necessarily satisfy the equations. They constitute a solution only in the combination (2.175) with due values of the constants C_j. The functions ϑ_j^0 and ψ_j^0 are *chosen*. They must be linearly independent and have all those properties that allow us to satisfy both the boundary conditions and equations by duly determining C_j and ϑ^0 and ψ^0 in the expression (2.175). Thus, the procedure of solution requires a *choice* of the functions ϑ_j^0 and ψ_j^0 and *determination* of a particular solution (2.175). As a variant of functions ϑ_i^0 and ψ_j^0, suitable for most cases of flexible-shell deformation—when the shell is predominantly bent in the η-direction, solutions of the following equations (cf. (2.30)) may be used

$$(\dot{\vartheta}_i^0 rt^3)^{\bullet} = 0 , \quad \psi_i^0 = 0; \quad \left(\dot{\psi}_j^0 \frac{r}{t}\right)^{\bullet} = 0 , \quad \vartheta_j^0 = 0 \quad (i = 1, 2; j = 3, 4) . \tag{2.176}$$

For the boundary conditions

$$
\begin{array}{ccccccccc}
& \vartheta_1^0 & \vartheta_2^0 & \psi_3^0 & \psi_4^0 & \vartheta_3^0 & \vartheta_4^0 & \psi_1^0 & \psi_2^0 \\
\eta = \pi & 1 & 0 & 1 & 0 & 0 & 0 & 0 & 0 \\
\eta = 0 & 0 & 1 & 0 & 1 & 0 & 0 & 0 & 0
\end{array} \tag{2.177}
$$

the four solutions of the auxiliary equations (2.176) are

$$[\vartheta_1^0 \quad \vartheta_2^0] = \left[\int_0^\eta \frac{d\eta}{rt^3} \quad \int_\eta^\pi \frac{d\eta}{rt^3}\right]\left(\int_0^\pi \frac{d\eta}{rt^3}\right)^{-1} ,$$

$$[\psi_3^0 \quad \psi_4^0] = \left[\int_0^\eta \frac{t}{r} d\eta \quad \int_\eta^\pi \frac{t}{r} d\eta\right]\left(\int_0^\pi \frac{t}{r} d\eta\right)^{-1} . \tag{2.178}$$

With these functions the constants in the solution (2.175) have a definite mechanical meaning: C_1 and C_2 are the rotation angles ϑ on the lines $\eta = \pi$ and 0, respectively; C_3 and C_4—the radial stress resultant T_r (cf. (2.27)) at $\eta = \pi$ and 0, divided by $Eh^\circ h_m$.

Substituting the expressions (2.175) into (2.30) gives for ψ^0 and ϑ^0 a system of equations, which together with the boundary conditions of the problem fully determines the solution. Thus, the equations for $\vartheta^0 = [a_1 \ a_2 \ \cdots]$, $\psi^0 = [b_1 \ b_2 \ \cdots]$ and $C_1, \ldots, C_4$ following from (2.30) are

$$\Lambda_+ r_+ t_+^{-1} \Lambda \psi^0 - \mu^* c'_- \left(\vartheta^0 + \sum C_k \vartheta_k^0\right) = ms ,$$

$$\Lambda_+ r_+ t_+^3 \Lambda \vartheta^0 + \mu^* c_-^* \left(\psi^0 + \sum C_k \psi_k^0\right) = -P^\circ s^* - q^\circ p . \tag{2.179}$$

Here we take into account (2.176) and that the nonperiodic terms $N°\eta$ in ψ^0, compensate the nonperiodic load terms.

Naturally, the c^*, c', s^* and p, determined by (2.172), (2.171) and (2.174) as functions of ϑ, *depend on the constants* C_k. The system (2.179) can be solved, for any appropriate boundary conditions, in successive approximations.

In *linear* approximation $c^* = c' = c$ and $s^* = s$. A linear solution can be sought in the form (2.175) with the five particular solutions determined by specialized cases of (2.179). For instance,

$$\vartheta_1 = \vartheta_1^0(\eta) + \sum_j a_j \sin j\eta \,, \qquad \psi_1 = \sum_j b_j \sin j\eta \,, \qquad (2.180)$$

with the Fourier coefficients determined by linearized equations (2.179), where the right-hand sides and C_2, C_3 and C_4 are set equal to zero and $C_1 = 1$,

$$\Lambda_+ r_+ t_+^{-1} \Lambda \psi_1 - \mu c_- \vartheta_1 = \mu c_- \vartheta_1^0 \,,$$

$$\Lambda_+ r_+ t_+^3 \Lambda \vartheta_1 + \mu c_- \psi_1 = 0 \,; \qquad \psi_1 = \begin{bmatrix} b_1 \\ b_2 \\ \dots \end{bmatrix}, \quad \vartheta_1 = \begin{bmatrix} a_1 \\ a_2 \\ \dots \end{bmatrix}.$$

$$(2.181)$$

The functions ϑ_i^0 and ψ_j^0 determined by (2.176) are not always optimal as terms of boundary-problem solution (2.175). These functions vary slowly (approximately as η or $1 - \eta$). Therefore in those cases, when the deformation (ϑ, ψ) near an edge has the intensively varying character of boundary effect, the use of these functions ϑ_i^0 and ψ_j^0, leaves in (2.175) the description of the variation to the Fourier series ϑ^0 and ψ^0. It leads to poor convergence of these series. This is the case, when near the edges ($\eta = \pm\pi$ or $\eta = 0$) there is in (2.30),

$$\mu\left(\frac{|c|}{r}\right)_{\eta=n} \equiv \mu_{(n)} \gg 1 \,, \quad n = \pi, 0 \,. \qquad (2.182)$$

On the other hand, for $\mu_{(n)} \gg 1$ deformation caused by edge loading decreases very rapidly with the distance from the edge. Often the zone of this boundary effect is so small that (2.30) can be approximately represented by equations with *constant* coefficients (the Geckeler–

Staerman method [205]):

$$\ddot{\psi} + \mu_{(n)} t(n) \vartheta = 0 , \qquad \ddot{\vartheta} - \mu_{(n)} t^{-3}(n) \psi = 0 , \quad n = \pi, 0 . \qquad (2.183)$$

The required solutions of the system (2.183) are easily written out in both explicit form and as trigonometric series. Indeed the system is readily reduced to the well-known form

$$\psi^{IV} + 4\beta^4 \psi = 0 , \quad \vartheta^{IV} + 4\beta^4 \vartheta = 0 ,$$

$$2\beta^2 = \mu_{(n)}/t(n) , \quad n = \pi, 0 . \qquad (2.184)$$

Hence, one obtains with the functions K_k defined in Table 4 (Chapter 3, §6.5),

$$\begin{bmatrix} \vartheta_1^0 \\ \psi_1^0 \end{bmatrix} = \begin{bmatrix} \psi_3^0 \\ -\vartheta_3^0 \end{bmatrix} = \frac{1}{A_2^2 + 4A_4^2} \begin{bmatrix} A_2 & 4A_1 \\ 2A_4 & -2A_2 \end{bmatrix} \begin{bmatrix} K_2(\beta\eta) \\ K_4(\beta\eta) \end{bmatrix} , \qquad (2.185)$$

$$A_k = K_k(\beta\pi) .$$

The series technique described is also directly applicable to the Reissner–Meissner equations in their full form (2.18) as well as to the Schwerin–Chernina equations (2.59) or (2.63).

8.3. Solution of the semi-momentless equations

Consider boundary problems for the equations stated in the present chapter, §5. It is a system of partial differential equations of eighth order in η. The solution must satisfy eight conditions on two boundaries $\eta = \text{const}$. Similar to the expressions (2.175), a series solution will be sought in the form

$$\kappa = \kappa_c + f , \quad f = \frac{f_0}{2} + \sum_n [f_{cn}(\xi) \cos n\eta + f_{sn}(\xi) \sin n\eta] ,$$

$$T = T_c + g , \quad g = \frac{g_0}{2} + \sum_n [g_{cn}(\xi) \cos n\eta + g_{sn}(\xi) \sin n\eta] . \qquad (2.186)$$

The terms κ_c and T_c must each be a combination of four linearly independent chosen functions F_k and G_k

$$\kappa_c = \sum_1^4 C_k(\xi)F_k(\eta), \qquad T_c = \sum_1^4 C_k'(\xi)G_k(\eta). \qquad (2.187)$$

The functions $F_k(\eta)$ and $G_k(\eta)$ must be such as to make possible satisfying all of the boundary conditions by duly determining the eight variables $C_k(\xi)$ and $C_k'(\xi)$. In many cases, κ_c and T_c may be taken identical in form to κ_g and T_g defined in (2.128).

Substitution of expressions (2.186) into the equations and edge conditions (with $F_k(\eta)$ and $G_k(\eta)$ also represented by trigonometric series) gives a system of equations for the $f_{cn}(\xi), \ldots, g_{sn}(\xi), C_k(\xi), C_k'(\xi)$. This system can be written out directly in the matrix form. Taking, for instance, the simplest linear form of the governing equations (see (2.126)) leads to the matrix equations

$$\begin{bmatrix} W & -L \\ L & W \end{bmatrix}\begin{bmatrix} f + \kappa_c \\ g + T_c \end{bmatrix} = \begin{bmatrix} -q_\Sigma^\circ \\ 0 \end{bmatrix}, \qquad \begin{aligned} \kappa_c &= \sum_1^4 C_k F_k, \\ T_c &= \sum_1^4 C_k' G_k, \end{aligned} \qquad (2.188)$$

which, together with the boundary conditions, constitute a closed system for the functions $f_{cn}(\xi), \ldots, C_k'(\xi)$.

In (2.188), f, g, $F_{k'}$, G_k and q denote column matrices of the Fourier coefficients of the corresponding functions.

The square operator matrices W and L are composed according to the rules described in the present chapter, §7.

As discussed in the present chapter, §8.1, a solution can be sought with the help of trigonometric series containing either only sine terms or only the cosine terms. When the plane $\eta = 0$ is a plane of symmetry of deformation, the functions f, g, F_k, G_k, r, R_2, c and q_Σ expand in cosine series; s and q_2 are represented by sine series.

The matrix formulas of the present chapter, §7 applied to the linearized equations (2.116) with $a/b = r$ yield

$$W = \Lambda r_-^2(\Lambda_+ \rho_+ \Lambda - \rho_-^{-1})\Lambda_+ r_+ \quad (\rho = R_2/b),$$

$$L = r_+ \frac{d^2}{h^\circ \, d\xi^2} - \mu\Lambda r_-^2(s_+ + \Lambda_+ \rho_+ c_+), \qquad (2.189)$$

$$q_\Sigma^\circ = \frac{b^3}{D}\left(-\Lambda r_-^2 \Lambda_+ r_+ \rho_+ q + \Lambda r_-^3 q_2 - r_+^2 \frac{d}{d\xi} q_1\right).$$

Note an important circumstance concerning the differentiation of functions κ_c and T_c in η. To satisfy the boundary conditions the functions κ_c and T_c and their derivatives may have to attain *boundary* values, which *cannot* be represented exactly by the kind of series chosen. In such cases the series of the functions κ_c and T_c cannot be differentiated with respect to η without obtaining divergent series. Consequently, the terms $W\kappa_c$, WT_c, $L\kappa_c$ and LT_c cannot be computed through multiplication of matrices W and L with κ_c and T_c. The Fourier coefficients of the derivatives of κ_c and T_c must be computed on the basis of differentiated *functions* $F_k(\eta)$ and $G_k(\eta)$, which constitute κ_c and T_c.

Consider now the case of closed η-lines, i.e. tubes.

8.4. *Semi-momentless solution for tubes*

For tubes the role of boundary conditions at $\eta = $ const. goes over to conditions of continuity and periodicity. These conditions are presented in (2.131)–(2.135).

For the deformation *symmetric* to the plane of $\eta = 0$, the nonsymmetric terms of (2.186), disappear. Taking $\kappa_c = \kappa_g$ and $T_c = T_g$ we have to retain only the constant and the x terms in (2.128). The expressions (2.186) become

$$T = N^\circ + C'_x(\xi)x + \sum_n g_n(\xi)\cos n\eta,$$

$$\kappa = C_x(\xi)x + \sum_n f_n(\xi)\cos n\eta.$$

(2.190)

The constant term in T is equal to the parameter N° of the resultant of normal forces acting on the plane section $\xi = $ const. For a tube the constant term of κ is equal to zero according to (2.132). (This term reflects a jump in the angle of rotation at some value of η.)

We fix the value of R_m, giving the expansion of $x(\eta)$ a form

$$x(\eta) = R - R_m = b \sum_n x_n \cos n\eta.$$

(2.191)

Thus, expansions (2.190) and (2.191) satisfy all the conditions of continuity (2.133) and (2.134) except for two. The remaining conditions define C_x and C'_x in terms of $g_n(\xi)$, $f_n(\xi)$ and x_n:

$$C'_x b = M^\circ - \left(\sum_n g_n x_n\right)\Big/\sum_n x_n^2 \,,$$

$$C_x b = -\left(\sum_x f_n x_n\right)\Big/\sum_n x_n^2 \quad (n=1,2,\ldots)\,.$$

Now, the governing functions (2.190) can be presented (in the equations determining f_n and g_n) by matrices of Fourier coefficients κ, T, where C_x and C'_x are defined through f_n and g_n

$$\kappa = E_x f\,,$$
$$T = E_x g + T_M\,, \qquad E_x = E - \frac{xx^{\mathrm{T}}}{x^{\mathrm{T}} x}\,,$$

$$[f \; g \; T_M] = \begin{bmatrix} 0 & 0 & N^\circ \\ f_1 & g_1 & M^\circ x_1 \\ f_2 & g_2 & M^\circ x_2 \\ \cdots & \cdots & \cdots \end{bmatrix}. \tag{2.192}$$

Recall that κ, T and x denote column matrices of Fourier coefficients of the corresponding functions κ, T and x; x^{T} denotes the transposed matrix x, i.e. $x^{\mathrm{T}} = [0 \; x_1 \; x_2 \; \cdots]$; E is the unit matrix (2.153). The values of the normal force and bending moment in the section $\xi = $ const. (represented by N° and M°) are determined by conditions of equilibrium of a part of the shell.

With formulas (2.192), the resolving matrix system (2.188) for a shell with closed η-lines becomes

$$WE_x f - LE_x g = -q_\Sigma^\circ + LT_M\,, \qquad LE_x f + WE_x g = -WT_M \equiv 0\,. \tag{2.193}$$

It is necessary to retain in the series representation of $x(\eta)$ *more* terms than in the representations of the unknowns f and g. Correspondingly, the number of rows in the matrix E_x and the number of columns in the matrices L and W must be substantially greater than the number of elements in the unknowns f and g.

The case of *circular tubes* is the most interesting one. In this case, when $\alpha = \eta$, $\rho = 1$ and $x = b\cos\eta$, equations (2.192) and (2.193) are much simpler and determine the series of κ and T directly,

$$W\kappa - LT = -q_\Sigma^\circ,$$

$$L\kappa + WT = 0;$$

$$[\kappa \ \ T] = \begin{bmatrix} 0 & N^\circ \\ 0 & M^\circ \\ f_2 & g_2 \\ f_3 & g_3 \\ \cdots & \cdots \end{bmatrix},$$

$$T = N^\circ + M^\circ \cos\eta + \sum_{2,3,\ldots} g_n(\xi)\cos n\eta,$$

$$\kappa = \sum_{2,3,\ldots} f_n(\xi)\cos n\eta.$$

(2.194)

Notable further simplification becomes possible when $r = 1$ can be assumed in the equations. (This possibility, as discussed in the present chapter, §5.3 and Chapter 3, §7.3, is available in most cases.) With $r = 1$ the operator W has the form (2.122) and $Wx = 0$ (for large displacements $W'x^* = 0$). Hence, $WT_c = 0$ and $W\kappa_c = 0$.

For a circular tube the simplification through the assumption $r = 1$ is particularly substantial. Namely, for $\rho = 1$ and $r = 1$ the matrix W is diagonal with elements $W_{ii} = i^4 - i^2$.

In this case half of the unknowns are readily eliminated from (2.194) and the equations become

$$T = \begin{bmatrix} N^\circ \\ M^\circ \\ 0 \\ 0 \\ \cdots \end{bmatrix} - W^{-1}L\kappa, \quad W^{-1} = \begin{bmatrix} 0 & 0 & 0 & 0 & \cdots \\ 0 & 0 & 0 & 0 & \cdots \\ 0 & 0 & 12^{-1} & 0 & \cdots \\ 0 & 0 & 0 & 72^{-1} & \cdots \\ \cdots & \cdots & \cdots & \cdots & \cdots \end{bmatrix};$$

(2.195)

$$[W + LW^{-1}L]\kappa = -q_\Sigma^\circ + L \begin{bmatrix} N^\circ \\ M^\circ \\ 0 \\ \cdots \end{bmatrix}.$$

One feature of the equations (2.193)–(2.195) must be mentioned. Two of the equations—those corresponding to the zeroth rows of the matrices W and L become identities if the resultant N is known (determined by the statics). The corresponding zeroth terms of the Fourier expansions of κ and T are equal to 0 and N°, respectively. For the *linear* approximation the first rows and columns of the matrices W and L may also become unnecessary—the respective terms of the Fourier series of κ and T are equal to $0 \cos \eta$ and $M° \cos \eta$.

The series solution is directly applicable to the "*second-order-theory*" equations (2.140). We define the zeroth approximation $T^{(0)}$, $\kappa^{(0)}$ in accordance with (2.138) and (2.140) and seek the solution in the series form

$$T = T^{(0)} + T^{(1)}, \quad T^{(0)} = N° + M° \sum x_n \cos n\eta,$$

$$T^{(1)} = \sum g_n \cos n\eta, \quad \kappa^{(0)} = 0, \quad\quad (2.196)$$

$$\kappa = \sum f_n \cos n\eta \quad (n = 0, 1, \ldots; f_0 = 0).$$

Substitution of these expansions into (2.140) gives for the $g_n(\xi)$ and $f_n(\xi)$ equations, which can be written out in the matrix form

$$(W - 2\mu' \Lambda_+^2 \boldsymbol{\rho}_+^2 c_+ T_+^{(0)} + \Lambda_+^2 q_+° \boldsymbol{\rho}_+^2) f$$

$$- (L' - \Lambda T_-^{(0)} \Lambda_+ - \Lambda_+^2 \boldsymbol{\rho}_+^2 T_+^{(0)} \Lambda_+^2) g = L' T^{(0)}, \quad\quad (2.197)$$

$$L' f + W g = 0; \quad f = \{0 \ f_1 \ f_2 \ \cdots\}, \quad g = \{g_0 \ g_1 \ \cdots\}.$$

The deformation determined with the aid of (2.196) and (2.197) satisfies the continuity conditions of the closed cross-sections identically, provided these conditions are satisfied for one of the edges (as in most problems of practical interest). As the second-order-theory solution with $T^{(0)}$ determined by the "beam" stresses (2.196) is usable only for small curvature tubes, equations (2.197) are intended for $r = 1$ (cf. (2.189) for W and L).

For the important case of *circular* tubes the equations are simplified by $\rho = 1$. With $r = 1$ and $\rho = 1$ the matrix W becomes diagonal.

For linear problems (without the $T^{(0)}$ and $q°$ terms), (2.196) and (2.197) apply for tubes with any μ-values. These equations are a (simpler) case of those in (2.193).

Double Fourier-series solutions are discussed, for concrete problems, in Chapter 3, §§7, 9.

Commentary

To §1: Recent rigorous investigations of specific problems of large deformations of shells by small strain (e.g. [206, 221, 222]) provide confirmation of the basic hypothesis (2.5) as representing the immanent feature of such stress states (those of flexible shells).

To §2.1, 2: A governing system of two second-order differential equations of the axisymmetric problem was first proposed by H. REISSNER [5] for the case of spherical shell. These equations were extended by MEISSNER [6] to other shapes of meridian.

The next stage of development of the Reissner–Meissner equations (and of their applications) was introduced by E. REISSNER [35, 38]. The 1949 work generalized the equations of shells of revolution to include the Saint-Venant *linear* problem of pure *flexure* of curved *tubes* (problem M_z in Fig. 13). In 1950 REISSNER [38] extended the equations in another respect—to include unboundedly *large axisymmetric*, displacements (problem P_z in Fig. 13).

AXELRAD [73] used the REISSNER [35] idea to generalize the nonlinear equations of shells of revolution to include unboundedly *large flexure* of tubes or any other thin-walled sections. This has resulted in (2.16) and (2.18) (where the stress function V, or ψ, is the one introduced by LUR'E [31]).

This has been made possible by the use of purely *intrinsic* equations. The necessary step of deriving the *nonlinear* integral (2.11) of the compatibility equations has been previously held back by the description of the axisymmetric flexure in terms of displacements. It has been achieved only recently by REISSNER [199]. LIBAI and SIMMONDS [206] returned to the intrinsic derivation of of [73].

KOITER [195] established an important link between the nonlinear equations of axisymmetric deformations, which he called Reissner–Meissner–Reissner (R–M–R) equations, and the general tensor-form thin-shell equations. The R–M–R equations extended [73] to include *nonlinear flexure* are referred to in the present book as "Reissner equations".

To §2.3: The problem of getting rid of the lower-order terms of the Reissner–Meissner equations has been discussed in many investigations of the axisymmetric deformation (cf. reviews [193, 216]).

Recently, SIMMONDS [174] achieved a rigorous proof of the insignificance of the ν terms in these equations. KOITER [195] has shown that the ν terms need not appear, when the equations are derived from the general tensor-form simplified equations of [155]. The relevant eqs. (5.6) of [195] are similar to (2.30) with $t = k = 1$, $q = 0$.

To §2.3 and §4.3: As mentioned in §4.3, the semi-momentless model disregards a set of terms (2.99) and (2.100). The simplifications leading to the Reissner equations (2.30) are actually equivalent to dropping just these quantities. However, if retained, these terms would, for the most part, cancel each other in (2.30), (2.63) or (2.116). The recurrent in the recent work on Reissner equations and on tube analysis assumption, setting $\varepsilon_2 = 0$ and $M_1 = 0$ also in the *elasticity* relations, upsets the balance. It leads to relations $\varepsilon_1 = T_1(1 - \nu^2)/(Eh)$ and $M_2 = \kappa_2(1 - \nu^2)D$, corrected already in [25, 29] (cf. footnotes on pp. 162, 239).

To §2.4: The relatively simple analysis of the middle-thickness-range shells is, of course, due to the boundary conditions being left out.

To §3: For the nonsymmetric Saint-Venant problem the equations of the Reissner–Meissner type were first obtained by SCHWERIN [7], who considered spherical shells. A nearly equivalent equation of nonspherical shells was derived by NOVOZHILOV [30] on the basis of the complex version of the shell equations (Chapter 1, §8). CHERNYKH [67] extended this result to shells of variable thickness. CHERNINA [66] derived the equations without resort to the somewhat approximate complex version. The reader will also recognize in §3 the influence of the work of REISSNER and WAN [125], LARDNER and SIMMONDS [97].

To §3.5: The discussion leans on the work of CLARK and REISSNER [39]. Further asymptotic analysis and references are given by CHERNINA [114] and SEIDE [173].

To §§4, 5 and §1.3: The Reissner–Meissner and Schwerin–Chernina equations provide a standard of accuracy and simplicity as well as a motivation to extend the nonlinear Reissner equations to nonsymmetrical stress. The possibility of such an extension suggests itself on inspection of the equations of §§2, 3, the equations of the two types being strikingly similar. It leaves an impression of relative insignificance of the variation of stress around the shell of revolution.

Retrospectively, it is now clear that the basic hypothesis, as expressed in (2.99) and (2.100) or in the semi-momentless model of §4.3, was indicated by the work of KÁRMÁN [4], KARL [25] and BESKIN [29] on tubes (discussed in Chapter 3, §1.1)). However, the hypothesis has been enunciated for two-dimensional problems [95] on a different basis—as an extension of the *linear* semi-momentless theory of the *cylinder* shells. This theory was originated by G. Ehlers, P.L. Pasternak and W.S. Wlassow in 1930–1932 as a set of simplifications for evaluation of shed-type roofs. WLASSOW [36] developed the theory in the period 1932–1936 on the basis of neglecting the moments M_1 and H_1 and the strains ε_2 and γ in the analysis of equilibrium and deformation, respectively. Besides, the elasticity equations were used in the form (2.98)—without the T_2 and κ_1 terms. Wlassow's theory acquired a clearer formulation, using *dual* equations, after the invention (GOL'DENVEIZER [20]) of the compatibility equations.

NOVOZHILOV [40] noted that this theory follows from the general linear theory of the cylinder shell as a consequence of one single assumption: "variation of the displacements and stresses in the direction along the cylinder shell is much smoother than variation in the direction of the arc of the cylinder's cross-section."

GOL'DENVEIZER [48] showed by way of the asymptotic analysis that the semi-momentless theory describes the main, slowly varying in the longitudinal direction, stress state of the cylinder shell.

SCHNELL [53] developed a linear theory of cylinder shells, which comes in fact near to the semi-momentless theory but takes into account additionally the shear deformation. This theory is useful for reinforced shells that are less rigid with respect to shear.

It may not be out of place to remark that in its further development the linear semi-momentless theory of cylinder shells has in fact become the main tool of applied analysis of these shells (GOL'DENVEIZER [178]).

It was GOL'DENVEIZER's [48] version of the semi-momentless theory that was extended in [95] to noncylinder shells and to large displacements. This theory found considerable application (cf. [117, 118, 143, 153]) before it has been [105, 166, 176] supplemented by the system of intrinsic equations (2.96)–(2.98) and (2.114)–(2.116). Avoiding the use of displacements, the intrinsic equations are particularly effective for nonlinear problems.

The internal consistency has been assured for the semi-momentless theory of *cylinder* shells by both its derivation from a single principle (NOVOZHILOV [40]) and by analysis of the equations (GOL'DENVEIZER [48]). But these ways cannot be extended to noncylinder shells. The derivation presented in §4.1 becomes possible through the use of the integrals of the compatibility and equilibrium equations and the resulting Reissner-type equations (2.90) obtained in [166, 193].

To §4.4: The discussion extends that in [208, 216].

To §6.1: The operator W for $r = 1$ was investigated by WLASSOW [36] in the framework of the theory of beams with undeformable cross-section.

CHAPTER 3

TUBES

CHAPTER 3

TUBES

1. Basic problems of tube analysis

The basic problems are discussed in the form of a review of the development of the theory,[1] in particular of that part of it presented below in §§2–10.

1.1. Linear bending of long tubes

The work of KÁRMÁN [4] (1911) laid the basis for at least 40 subsequent years of tube analysis. KÁRMÁN has set forth five basic assumptions:

(i) The stresses were assumed not to vary along the tube.

(ii) The extension ε_2 of the cross-section profile (of the η-line in Fig. 17) was neglected compared to the derivative of displacement $\partial v/b\,\partial\eta$ in the expression (1.40) for ε_2. This gave a relation between the components of displacements for a circular tube ($R_2 = b$) and $u = 0$,

$$w = -\frac{\partial v}{\partial \eta} + \varepsilon_2 b \approx -\frac{\partial v}{\partial \eta}. \tag{3.1}$$

(iii) The tubes were assumed to be "slender". That is, the curvature $1/R$ of any of the ξ-lines was set equal to the middle value $1/R_m$ (Fig. 17):

$$R = R_m + b \cos \eta \approx R_m. \tag{3.2}$$

---

[1] Further references and discussion can be found in [216] and in *Appl. Mech. Rev.* **37** (1984) 891–897.

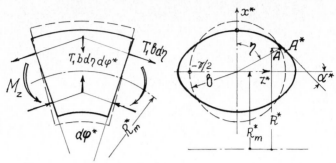

FIG. 17. Kármán effect.

(iv) The elasticity relations were used (in the indirect way—through the elastic energy expression) in the form

$$\varepsilon_1 = T_1/Eh , \qquad M_2 = \kappa_2 Eh^3/12 . \tag{3.3}$$

These relations mean neglecting the terms with T_2 and M_1, respectively in the relations (1.84), or assuming

$$\varepsilon_2 = -\nu\varepsilon_1 , \qquad M_1 = 0 . \tag{3.4}$$

(v) Finally, the bending moment in a cross-section of a tube was determined without taking into account the wall-bending moment M_1 (Figs. 14 and 17),

$$M_z = \int_0^{2\pi} T_1(b\cos\eta)b\,\mathrm{d}\eta . \tag{3.5}$$

Kármán solved the problem with the aid of the Ritz method setting $v = \Sigma\, C_n \sin 2n\eta$ and explained the interdependence of the longitudinal stress resultants T_1 with the deformation of the cross-section: The radial components $(T_1 b\,\mathrm{d}\eta)\,\mathrm{d}\varphi^*$ flatten the cross-sections (Fig. 17) causing substantial wall-bending stresses $\sigma_{M2} = M_2/(h^2/6)$ and decrease of the flexural stiffness from EI to KEI. Thus,

$$\frac{1}{R_m^*} - \frac{1}{R_m} \equiv \frac{\mathrm{d}\varphi^* - \mathrm{d}\varphi}{R_m\,\mathrm{d}\varphi} = \frac{M_z}{KEI}, \quad I = \pi b^3 h . \tag{3.6}$$

In the course of flexure *decreasing* the tube curvature $(\mathrm{d}\varphi^* < \mathrm{d}\varphi)$ the cross-section is not flattened but extended in the direction of the curvature plane (the stiffness-reduction factor K is equal to that for $d\varphi^* > d\varphi$).

KÁRMÁN [4] determined the K and the stresses. The tube geometry was thereby represented by a single parameter $\lambda_k = Rh/b^2$. The first approximation (one term in the series solution) gave

$$K = \frac{1 + 12\lambda_k^2}{10 + 12\lambda_k^2}. \tag{3.7}$$

THULOUP [10] solved the Kármán problem taking into account the normal *pressure q*. Formula (3.7) was extended to

$$K = \frac{1 + 12\lambda_k^2 + 48\lambda_k^2 qb^3/Eh^3}{10 + 12\lambda_k^2 + 48\lambda_k^2 qb^3/Eh^3}. \tag{3.8}$$

Later, THULOUP [10] investigated the bending of *oval* tubes.

KARL [25] considered, besides the plane flexure, bending *out of the plane* of curvature. Using the principle of minimum potential energy and the principle of least work, he obtained both the upper and lower bounds for the rigidity of the tube. Karl also made more realistic the assumptions (3.3) and (3.4). He took into account the reactive moment $M_1 \approx \nu M_2$ occurring in the cross-section of tube as a consequence of $|\kappa_1| \ll |\kappa_2|$ and used (in an energy form) the relations

$$\varepsilon_1 = \frac{T_1}{Eh}, \qquad M_2 = \kappa_2 \frac{Eh^3}{12(1 - \nu^2)}. \tag{3.9}$$

As a consequence, the Kármán formulas were improved by replacing λ_k by

$$\frac{1}{\sqrt{1 - \nu^2}} \frac{R_m h}{b^2}.$$

Further, Karl showed assumption (3.2) to be acceptable when

$$1 \pm (b/R_m)^2 \approx 1.$$

Almost simultaneously the space bending of tubes was considered by VIGNESS [26].

For the space-bending analysis BESKIN [29] formulated and used, in fact, the full set of the semi-membrane assumptions (Chapter 2, §4). For both plane and space bending he evaluated the Kármán coefficient K and the stress factors for a broad range of tubes.

The investigations of KARL [25] and BESKIN [29] have substantially solved the Saint-Venant linear problems of long circular tubes. But also the shortcomings of each of the two works have had a lasting influence. For a tube bent out of plane, Karl stated: "... The cross-section is deformed into an oval, as ... for the plane bending, only the axes of symmetry are turned through 45°...". The correct picture is discussed in present chapter, §4.2, and Fig. 29.

For his part, BESKIN [29] argued that the accuracy of the assumption (3.2) is the same as in the theory of curved beams and thus "not satisfactory for determination of stress distribution, when R_m/b is less than 10". However, this reasoning is not applicable for *deformable* cross-sections. The error is much smaller. This question is discussed in present chapter, §2.1 and §3.1.

The results referred to so far were obtained by the use of minimum principles, and approximate expressions for the displacements or stresses. The next stage of tube analysis was introduced by REISSNER [34, 35]: the problem of bending of long tubes was formulated in terms of the differential equations of the thin-shell theory.[2] CLARK and REISSNER [39] exhaustively investigated the linear pure bending of long circular tubes. KÁRMÁN's [4] and BESKIN's [29] results were confirmed (in the face of repeated attempts to refine them).

CLARK, GILROY and REISSNER [45] extended the 1951 [39] analysis to elliptic tubes. The works [39, 45] provided also the effective *asymptotic solutions* of the tube-bending problems (presented in a somewhat more general form in Chapter 2, §3.5 and present chapter, §3.3).

The *influence of* inner or outer normal *pressure* on the stresses and stiffness of tubes under bending was evaluated theoretically and experimentally by KAFKA and DUNN [56] who, independently of THULOUP [10], obtained an energy solution. CRANDALL and DAHL [54] investigated this problem through generalizing the equations of REISSNER [34] and the series solution of CLARK and REISSNER [39]. Starting from the

---

[2] This was also achieved in the work of TUEDA [16].

nonlinear equilibrium and large displacements relations this work simplifies them by dropping terms of the order of magnitude of ϑ. Further, linearization is achieved in [54] by the use of a "second-order theory" (discussed in Chapter 2, §6.3). The basic equations were $(\ddot{\psi} = d^2\psi/d\eta^2)$,

$$\ddot{\psi} + \mu\vartheta \cos \eta = -m \sin \eta \,,$$

$$\ddot{\vartheta} - q^\circ \vartheta - \mu\psi \cos \eta = 0 \,. \tag{3.10}$$

The pressure term is here very approximate. Equation (3.10) gives for a cylinder ($\mu = 0$) $q_{cr} = -4D/b^3$, instead of the well-known correct result $q_{cr} = -3D/b^3$ (given also by (3.20)).

For *space* bending the influence of normal pressure was investigated first by RODABAUGH and GEORGE [60] who obtained an energy solution of this problem.

Advantages of the shell-theory methods were made available for the lateral and space tube bending problems (Fig. 13(e) and (f); Figs. 27 and 28) by the work of CHERNYKH [74]. A comprehensive investigation of these problems is due to SEAMAN and WAN [164].

In so far as it concerns the *linear* theory, §§2, 3 of the present chapter are based on the two last named works and those of CLARK and REISSNER [39] and CLARK, GILROY and REISSNER [45].

The analysis of the normal-pressure influence on the space flexure, set forth in present chapter, §4, follows [176]. The asymptotic formulas (3.79) have not been published previously.

A very considerable effort was spent since 1911 on attempts to improve on the last four of the Kármán hypotheses (3.1)–(3.5). This effort has, however, conclusively demonstrated that these assumptions (with the modification replacing (3.3) and (3.4) by (3.9)) are as exact[3] as the first one—the Saint-Venant setting of the boundary conditions (cf. present chapter, §2.1). Most valuable data in this respect are supplied by the work of JONES [109], DODGE and MOORE [145] and SEAMAN and WAN [164].

---

[3] This invariably follows from the data obtained by the investigations. However, some of the respective authors consider their results as a refinement of the accuracy.

1.2. Nonlinear flexure of long tubes

The problem of nonlinear bending of long tubes was set by BRAZIER (1927) [9] who considered pure bending of a cylinder retaining the Kármán assumptions (including the most important one—of the invariable along the tube stress state). With the one-term expression $v = C \sin 2\eta$, Brazier obtained a simple energy solution. This gave the nonlinear relation between the bending moment and elastic curvature of the tube. With some value of the curvature the moment reaches a maximum. Obviously, the allowable loading cannot exceed this maximum, which was therefore assumed to be the *collapse load*.

There were numerous attempts at improving on the BRAZIER [9] result by means of more complicated energy solutions. All of the energy-method solutions were unsuccessful, producing higher critical moments than Brazier's. REISSNER [70] proposed for the nonlinear tube bending an integro-differential equation and obtained (by perturbation solution) for a cylinder $M_{max}^{\circ} = 1.002$. REISSNER [70] succeeded also in solving for the first time the *nonlinear* flexure problem for tubes with slight *initial curvature*.

But these equations apparently proved complicated and REISSNER [82] in 1961 derived another system especially for cylinders ($\ddot{\psi} = d^2\psi/d\eta^2$),

$$\ddot{\psi} + m \sin(\alpha + \vartheta) = 0 \,,$$

$$\ddot{\vartheta} - m\psi \cos(\alpha + \vartheta) = 0 \,. \tag{3.11}$$

A year earlier AXELRAD [73] developed REISSNER's [34, 38] ideas to obtain the system[4] (3.17), which, written without the load term, is

$$\ddot{\psi} + (\mu + m) \sin(\alpha + \vartheta) - \mu \sin \alpha = 0 \,,$$

$$\ddot{\vartheta} - (\mu + m)\psi \cos(\alpha + \vartheta) = 0 \,. \tag{3.12}$$

For a cylinder ($\mu = 0$) this gives (3.11).

Expanding $\sin(\alpha + \vartheta)$ and $\cos(\alpha + \vartheta)$, with accuracy $1 \pm \vartheta^2/2 \approx 1$

---

[4] The background of these equations was discussed in the Commentary to Chapter 2, §2.1.

and using the perturbation method AXELRAD [79, 84] solved the Brazier problem both for cylinders and for initially curved long tubes, including the effect of internal or external pressure and noncircular profile. REISSNER and WEINITSCHKE [90] obtained an exact numerical solution of (3.11). By a different method the nonlinear bending of cylinder was exactly investigated in the work of THURSTON [184] (where four other numerical treatments of the problem are referred to). A numerical solution of equations (3.11) was obtained by WEINITSCHKE [137] for nonlinear pure flexure of *elliptic* cylinders. The elliptic orthotropic tubes under bending and normal pressure were studied numerically and experimentally by SPENCE and TOH [190].

For initially *curved*, circular and elliptic, tubes the nonlinear bending was investigated by EMMERLING [198, 201], with the aid of the Fourier-series solution of equations (3.12), the results are presented in Figs. 21, 25 and 26. For circular curved tubes these results confirm the *perturbation* solution [79, 84] for the precritical range (cf. the review [193]).

There is a case—flexure of circular cylinder—for which results of different investigations (more than a dozen in all) are available for comparison thus throwing some light on the methods used.

After the classic result obtained by BRAZIER [9], namely $M^\circ_{max} =$ 1.089, the calculated snap-through moment varied upwards to $M^\circ_{max} =$ 1.25 (CHWALLA, *ZAMM* **13** (1933) 48) and even 1.33 [19]. Subsequent results fell to $M^\circ_{max} = 1.002$ (REISSNER [70]) and 1.025 (AXELRAD [78, 79]) and rose to 1.063 (REISSNER and WEINITSCHKE [90]). Following the value $M^\circ_{max} = 1.057$ (of THURSTON [184]) the energy-method solution of SPENCE and TOH [190] gave $M^\circ_{max} = 1.10 \ldots$.

The early overestimates $M^\circ_{max} = 1.25, 1.33$ can be definitely traced to the geometrical assumptions made. Dropping some and retaining others among the Brazier simplifications (which had caused mutually compensating errors) increased the inaccuracy. The conservative values 1.002 and 1.025 resulted mainly from the simplification $1 \pm \vartheta^2/2 \approx 1$. The causes of the aberrations in the other results—gained by computer solutions—are not as easy to ascertain. The Fourier-series solution of equations (3.12), obtained by EMMERLING [198] as described in present chapter, §2.4, corroborates the result $M^\circ_{max} = 1.057$. The series solution allows a check with a pocket calculator (cf. Table 2).

Evidently, the large effort invested in the nonlinear flexure problem of tubes was hampered by duplication and adopting unclear or already discredited hypotheses.[5]

1.3. Tubes with end constraints

The work discussed so far is concerned with no given conditions on tube ends. The stress state was assumed either constant along the tube or varying as $\cos \xi$ and $\sin \xi$. (In the second case, only linear solutions were possible.) There was in 1911–1965 no attempt to assess the range of applicability of this Saint-Venant type theory. On the other hand, experiments displayed substantial influence of conditions on tube ends. As early as 1951 PARDUE and VIGNESS [41] came to the conclusion that ". . . the flexibility factors should be taken from the experimental data".

The first analysis of the flexure of tubes of *finite* length was carried out with the aid of the semi-momentless theory in 1965 [95]. The nonlinear part of this analysis was restricted to the tubes of small initial curvature. But within the bounds of the linear approximation, the theory describes flexure of tubes of *any* curvature compatible with the assumption of slenderness (3.2).

The semi-momentless (semi-membrane) theory was extensively used in publications, summarized in the book [143], where the tube elbows with adjoining straight tubes and the space flexure of tubes with end flanges were discussed. IL'IN [153] extended some of these results to bimetallic tubes.

CHENG and THAILER [131] considered a tube with end flanges, assuming the cross-sections to stay plane and the shear stiffness of the tube wall to be zero. These, neglected, items are of first-order importance in the deformation of tubes (cf. present chapter, §7).

Nonslender tubes—those with $b \sim R$—were discussed in several publications indicated in [143]. An estimate of the error of the semi-momentless theory, relevant to large values of curvature b/R_m, was discussed in [166] (cf. Chapter 2, §5.1).

A solution of the flexure problem of a tube with flanges, calculated

---

[5] For instance, there reappears the uncertainty whether the assumption (3.3), that of (3.9) or yet another version should be used. In a work of 1980 it is stated (with resignation) about a version of formula (3.8), that it is "the same except for the factor $(1 - \nu^2)$, which haunts pipe-bend analysis".

by N.N. Ometova by means of the full equations of the shell theory in 1976 [176], allowed a check on the accuracy of the corresponding semi-momentless solution (cf. present chapter, §7).

The progress in computer technology made it possible to solve the tube problems numerically. The first computer solution of the boundary problem of a curved tube is due to KALNINS [123]. Later, the numerical solutions of these problems were obtained by the method of finite elements. A comprehensive investigation of the bending of curved tube with straight extensions was made using this method by NATARAJAN and BLOMFIELD [170].

A broad computer analysis of a series of tube bends with straight-tube extensions was carried out by RODABAUGH, ISKANDER and MOORE [187]. Properties of flanged tubes are described by W. HÜBNER [229].

WHATHAM and THOMPSON [192] investigated tube bends with flanged straight tangents. Far-reaching results were achieved in this work and its development [214] by way of solution of full thin-shell equations.[6]

The mentioned work together with results presented (mostly in internal reports) by L.H. Sobel, L.M. Habip, H.D. Hibitt, A. Hoffmann, G.E. Findlay and some other investigators make it possible to compute the stresses and displacements of any particular tube bend. A review of the tube analysis may be found in [216] and in *Appl. Mech. Rev.* **37** (1984) 891–897.

From the standpoint of practical applications it is no less important to have an overall view of the dependence of design stress and stiffness of a tube bend on its geometry parameters. This remains the aim of the theory.

1.4. *Collapse of tubes under bending*

The collapse of bent tubes invariably starts with local buckling in the compressed zone. This was indicated already by BRAZIER [9]. Later, the *limit-point* concept of instability was intensively developed (as reviewed in the foregoing) for infinitely long tubes, *without* any reference to *buckling*. The collapse load was in this theory identical with *maximum bending moment,*

---

[6] Further treatment of the problem is given by WHATHAM in *J. Nucl. Eng. Des.* **65** (1981) 77, **72** (1982) 175; *Trans. Inst. Engrs. Austr.* **CE25**(1) (1983).

$$\frac{M_{max}}{W} \equiv M^\circ_{max} Eh^\circ = 0.5285\,\sigma_{cl}\,,$$

$$\sigma_{cl} \equiv 2Eh^\circ = \frac{Eh}{b\sqrt{3(1-\nu^2)}} \quad (W = \pi b^2 h)\,. \tag{3.13}$$

The bifurcation instability of a cylinder tube in bending was first investigated by FLÜGGE [12]. Part of the relevant later work is referred to in present chapter, §9.

However, until 1965 this analysis took little or no notice of the precritical deformation of the tube. The critical bending moment M_{cr} was thus determined (SEIDE and WEINGARTEN [83]) independently of the length of a cylinder tube to be

$$M_{cr} = \sigma_{cl} W\,. \tag{3.14}$$

The value of M_{max} in (3.13) is lower and thus agrees better with experiments than does (3.14). But it could not be accepted as a collapse load, experiments having shown that the tubes collapse in *local buckling*. Besides, the maximum moment (3.13) was determined only for *unlimitedly long* cylinders.

The two approaches developed independently. It was this, apparently, that caused the presumption—widely propagated in the literature despite the experimental evidence—of the existence of the two different possibilities of collapse.

The two concepts are unified, as proposed in 1965 [95]—by investigating the buckling for the *actual* shell *shape* and *stress state*, which result from the previous nonlinear deformation.

(From this point of view, the Brazier approach determines only the *prebuckling* deformation; it lacks a check of stability with respect to bifurcation buckling. The other approach is a bifurcation buckling analysis without taking into account the real shape and stresses, produced by the preceding elastic deformation.)

The curvatures and stresses of the precritical state vary in both surface coordinates. Investigation of the bifurcation stability of such a state of (nonlinear) deformation is a difficult task. It is drastically simplified and solved in present chapter, §9 for bending of cylinders and initially curved tubes with the help of the *local buckling approach* [95, 193, 218].

The buckling modes of cylinders and curved tubes under *external pressure* are basically different from the buckling modes by flexure. Therefore, the influence of the normal pressure on the collapse by flexure is, in the main, indirect—through influencing the precritical deformation (present chapter, §9).

The situation is different when the external pressure acts alone, without the bending load, or is the predominant factor.

For a *long* cylinder, when the influence of the bottoms is negligible, the critical external pressure q is determined by the well-known formula[7]

$$q_{cr} = -3\frac{D}{b^3}.$$ (3.15)

Circular ribs or flanges at the ends of a cylinder increase the critical pressure. The corresponding critical pressure was found by MISES [11]. The buckling problem of *cylinder* shells is now comprehensively investigated. The references can be found for instance in the monograph of BRUSH and ALMROTH [167], and in the review of GRIGOLJUK and KABANOV [179].

The stability of *curved* tubes and closed *toroidal* shells is a more general problem and has attracted attention comparatively recently.

The stability of curved tubes under external pressure was first investigated under the assumption that the buckling-mode deformation is *constant* along the tube (KOSTOVETSKY [81] and AXELRAD [84]). It was clarified later, by way of semi-momentless analysis [105], that the first of the two solutions gives the lower bound for the critical pressure of long tubes, whereas the second solution provides the upper bound.

For any lengths and longitudinal curvatures of tubes, the buckling deformation varies much less intensively in the longitudinal direction than in the cross-section, thus satisfying the condition (2.5). The semi-momentless theory is fully appropriate to describe the buckling mode of cylinders and torus shells under external pressure.

The semi-momentless analysis of [105] gives an explicit determinantal equation for the critical pressure of tubes and toroidal shells. Almost simultaneously SOBEL and FLÜGGE [113] obtained an effective *numerical*

---

[7] BRESSE, M., *Course de mecanique appliqué*, I, Paris, 1859.

solution of the buckling problem of torus shells. The results (the q_{cr} values) nearly coincide with those of [105]. But the starting points of the two were different. Namely, the semi-momentless equations were used with the simplification $R = R_m$ (Figs. 14 and 17), which eliminates the precritical deformation of a circular meridian by the normal pressure (Chapter 5, §5). The numeric solution disregarded both the transverse stress resultants Q_1 and Q_2 in the first two of the equilibrium equations (similar to the Donnell equations discussed in Chapter 1, §9). This excludes the analysis of long tubes and torus shells with $R \gg b$.

JORDAN [154] derived for the buckling pressure of toroidal shells an effective asymptotic formula (presented in what follows in a different form (3.226)). Results of series of experiments with toroidal shells have been reported by SOBEL and FLÜGGE [113], ALMROTH, SOBEL and HUNTER [121] and NORDELL and CRAWFORD [156]. The tests confirm the accuracy of the semi-momentless solution, which for $\mu > 1$ gives q_{cr} converging to the Jordan asymptotic value. Later, the semi-momentless solution was developed [176, 185] for tubes and torus shells with different kinds of ribs or end flanges. This theory, somewhat extended by new results, is discussed in present chapter, §10.

2. Bending of long curved tubes

Flexure by forces applied on the tube ends is considered. The forces on each end are statically equivalent to a moment acting in the plane of curvature of the tube (Fig. 17). It is assumed that the end forces are arrayed in a way assuring stress distribution uniform along the tube. In other words, the variation of stress and deformation along the tube is disregarded. This gives a solution that must be correct for sufficiently long tubes. But its main objective is to provide a simple reference solution of the limit case. (A description of the tube deformation for specific tube length and end conditions follows in present chapter, §§6–8.)

The stress state and displacements are first determined in linear approximation—the Kármán problem. Next, large displacements and the maximum bending moment are investigated—the Brazier problem.

The bending is considered taking account of the influence of the uniform normal pressure.

2.1. Governing equations

Consider the Saint-Venant's problem of pure bending (illustrated in Fig. 13(b)). This problem is described by the equations (2.16) or (2.30). In solving it, the equations will be simplified by the restriction that the tubes be slender (Fig. 17),

$$r = R(\eta)/R_m \approx 1 , \qquad (3.16)$$

which for circular cross-section, i.e. when $R = R_m + b \cos \eta$, means

$$1 \pm b/R_m \approx 1 .$$

Now (2.30) for homogeneous tubes of *uniform* wall-thickness[8] become

$$\ddot{\psi} + \mu^* c' \vartheta = -ms ,$$

$$\ddot{\vartheta} - \mu^* c^* \psi = P^\circ s^* + q^\circ p , \qquad (3.17)$$

where c', c^* and s^* are functions of η and $\vartheta(\eta)$ defined in (2.17) and

$$p = -c^* \int_0^\eta c^* \, \mathrm{d}\eta - s^* \int_{\pi/2}^\eta s^* \, \mathrm{d}\eta , \qquad P^\circ = 0 . \qquad (3.18)$$

According to these equations the functions ψ and ϑ and consequently the stresses and displacements depend on only one geometric parameter $\mu^* = \mu + m$ defined by (2.20).

The other geometric parameter of the tube, b/R_m, is of no independent influence. This makes the description of the stress and strain much simpler, *without detriment* to the accuracy. Moreover, the Saint-Venant "long-tube" solution *can* be consistent *only* for tubes satisfying the condition (3.16). Indeed, the stress may be constant along the tube

---

[8] The wall-thickness of real curved tubes is variable. The analysis of present chapter, §§2–8 can be extended to take this into account. In many cases the variation of the wall-thickness can be expected to follow the relation similar to (5.1).

only in its part which is distant enough from the tube ends. The distance and therefore the tube length φR_m must be large compared to the tube diameter $2b$. (Recall the principle of Saint-Venant.) But, as a rule $\varphi < \pi$ (Fig. 17). Thus $\pi R_m \gg 2b$, just what is required by the condition (3.16) for circular tubes.

Moreover, calculations taking full account of the η-dependence of $R(\eta)$ invariably yield for a given value of μ practically the same design stress and strain for any values of b/R_m. The cause of this is a localization of stress and strain in the parts of the tube where $\mu |\cos \alpha| \ll 1$ (cf. present chapter, §2.3). For large values of μ these zones around points $\alpha = \pm \pi/2$ (Fig. 17) are narrow and inside them the condition (3.16) is actually satisfied. It is, in particular, the case when b is *not* small compared with R_m. Indeed, for $b \sim R_m$ and a thin shell ($h \ll b$) the μ *must* be large,

$$\mu = b/(R_m h^\circ) \approx 3.3(b/R_m)(b/h) \gg 1 .$$

2.2. Linear Kármán problem

With the representation of c^* and s^* in terms of ϑ according to (2.17) we have from (3.18) expressions for the linear in ϑ part of the load function

$$p = p^0 + p^1(\vartheta), \quad p^0 = -c \int_0^\eta c \, d\eta - s \int_{\pi/2}^\eta s \, d\eta ,$$

$$p^1(\vartheta) = \vartheta s \int_0^\eta c \, d\eta + c \int_0^\eta \vartheta s \, d\eta - \vartheta c \int_{\pi/2}^\eta s \, d\eta - s \int_{\pi/2}^\eta \vartheta c \, d\eta .$$

$$(3.19)$$

For the circular tube this gives $p^0 \equiv 0$.

The linear approximation of (3.17) becomes, using (3.19)

$$\ddot{\psi} + \mu c \vartheta = -ms ,$$

$$\ddot{\vartheta} - q^\circ p^1(\vartheta) - \mu c \psi = q^\circ p^0 .$$

$$(3.20)$$

The solution of simplified equations (3.20) possesses properties of symmetry with respect to the lines $\eta = 0$, π and $\pm\pi/2$ when the cross-section of the tube possesses such properties (Fig. 17).

Suitable expansions (2.170) are (for $p^0 = 0$) of the form

$$\psi = N^\circ\eta + \sum_{j=1,3,\ldots} b_j \sin j\eta \,, \qquad \vartheta = \sum_{n=2,4,\ldots} a_n \sin n\eta \,, \qquad (3.21)$$

where

$$N^\circ = qb/(2Ehh^\circ) \,.$$

The cosine terms are omitted taking into account the symmetry of stress and strain $(\dot\psi, \dot\vartheta)$ with respect to the plane $\eta = 0, \pi$. The lack of even-integer terms in the ψ-series and of odd-integer terms in the ϑ-series is a consequence of the line passing through the points $\eta = \pm\pi/2$ (Fig. 17) being the neutral axis of bending of the tube.

The series (3.21) satisfy the conditions of continuity of the closed double-symmetric section of the tube (Chapter 2, §6.2).

The dimensionless parameter N° presents the axial force produced in the cross-section of the tube by the normal pressure q, according to (2.133) and (2.168). The term $N^\circ\eta$ compensates the aperiodic part of the load term $q^\circ p$ of (3.17). These aperiodic terms will not be discussed further.

To determine the Fourier coefficients of the functions ψ and ϑ, introduce the expansions (3.21) (or more generally (2.170)) into equations (3.20). This renders a system of linear algebraic equations for a_n and b_j. The system may be written as a matrix form of (3.20) or as a particular case of (2.171):

$$\begin{bmatrix} \Lambda^2 & -\mu c_- \\ \mu c_- & \Lambda^2 + q^\circ F \end{bmatrix} \begin{bmatrix} \psi \\ \vartheta \end{bmatrix} = \begin{bmatrix} ms \\ -q^\circ p^0 \end{bmatrix}, \qquad p^0 = s_+ \Lambda^{-1}s - c_-\Lambda_+^{-1}c \,,$$

$$F = [\Lambda_+^{-1}c]_+ s_- + [\Lambda^{-1}s]_- c_- + c_-\Lambda_+^{-1}s_- + s_+\Lambda^{-1}c_- \,. \tag{3.22}$$

Turning now to the tubes of *circular* cross-section ($\alpha = \eta$) we have according to (2.148) and (2.165):

$$c = \begin{bmatrix} 0 \\ 1 \\ 0 \\ 0 \\ 0 \end{bmatrix}, \quad s = \begin{bmatrix} 1 \\ 0 \\ 0 \\ 0 \end{bmatrix}, \quad c_- = \tfrac{1}{2}\begin{bmatrix} 0 & 1 & 0 & 0 \\ 1 & 0 & 1 & 0 \\ 0 & 1 & 0 & 1 \\ 0 & 0 & 1 & 0 \end{bmatrix},$$

$$s_+ = \tfrac{1}{2}\begin{bmatrix} 2 & 0 & -1 & 0 & 0 \\ 0 & 1 & 0 & -1 & 0 \\ 0 & 0 & 1 & 0 & -1 \\ 0 & 0 & 0 & 1 & 0 \end{bmatrix}, \quad s_- = \tfrac{1}{2}\begin{bmatrix} 1 & 0 & 0 & 0 \\ 0 & 1 & 0 & 0 \\ -1 & 0 & 1 & 0 \\ 0 & -1 & 0 & 1 \\ 0 & 0 & -1 & 0 \end{bmatrix},$$

$$F = \begin{bmatrix} 0 & 0 & 0 & 0 \\ 0 & F_{22} & 0 & 0 \\ 0 & 0 & F_{33} & 0 \\ 0 & 0 & 0 & F_{44} \end{bmatrix}, \quad F_{nn} = \frac{n^2}{n^2-1}; \quad p^0 = 0. \qquad (3.23)$$

To save space, only those rows and columns number 0–4 or 1–4 are written.

Using (3.23) the system (3.22) falls apart in two separate systems. One of these gives zero values for all those Fourier coefficients that are set equal to zero in (3.21). The other system determines the solution (3.21). For the comparatively simple case of circular tube it can be deduced directly, in the form

$$j^2 b_j - \tfrac{1}{2}\mu(a_{j-1} + a_{j+1}) = \begin{cases} m, & j=1; \\ 0, & j=3,5,\ldots; \end{cases}$$

$$\left(n^2 + q^0 \frac{n^2}{n^2-1}\right)a_n + \tfrac{1}{2}\mu(b_{n-1} + b_{n+1}) = 0 \quad (n=2,4,\ldots). \qquad (3.24)$$

Obviously half of the unknowns are easily eliminated from the system. Elimination of b_j leads to

$$A(\mu)\begin{bmatrix} a_2 \\ a_4 \\ a_6 \\ \cdots \end{bmatrix} = \begin{bmatrix} -\tfrac{1}{2}m\mu \\ 0 \\ 0 \\ \cdots \end{bmatrix}, \quad A(\mu) = \begin{bmatrix} A_{22} & \mu^2/36 & 0 & \cdots \\ \mu^2/36 & A_{44} & \mu^2/100 & \cdots \\ 0 & \mu^2/100 & A_{66} & \cdots \\ \cdots & \cdots & \cdots & \cdots \end{bmatrix}$$

$$(3.25)$$

with the nonzero elements of A

$$A_{nn} = n^2 + \frac{q^{\circ}n^2}{n^2 - 1} + \tfrac{1}{2}\mu^2 \frac{n^2 + 1}{(n^2 - 1)^2}, \qquad A_{n,n \pm 1} = \frac{\mu^2}{4(n \pm 1)^2}.$$

After eliminating a_n, the system (3.24) becomes

$$B(\mu) \begin{bmatrix} b_1 \\ b_3 \\ b_5 \\ \cdots \end{bmatrix} \equiv \begin{bmatrix} 1 + \mu_2^2 & \mu_2^2 & 0 & \cdots \\ \mu_2^2 & 9 + \mu_2^2 + \mu_4^2 & \mu_4^2 & \cdots \\ 0 & \mu_4^2 & 25 + \mu_4^2 + \mu_6^2 & \cdots \\ \cdots & \cdots & \cdots & \cdots \end{bmatrix} \begin{bmatrix} b_1 \\ b_3 \\ b_5 \\ \cdots \end{bmatrix} = \begin{bmatrix} m \\ 0 \\ 0 \\ \cdots \end{bmatrix},$$

$$\mu_n^2 = \frac{\mu^2}{4n^2} \bigg/ \left(1 + \frac{q^{\circ}}{n^2 - 1}\right). \tag{3.26}$$

Having solved (3.25) or (3.26) and calculated the remaining Fourier coefficients with (3.24), one has defined the solution (3.21). The full picture of the stresses and deformation of the tube can be calculated with the functions ψ and ϑ and formulas of Chapter 2, §2.

The stresses may be determined through (1.89) and (1.90). The maximum through the wall-thickness stress in the longitudinal and circumferential direction respectively, is given by the formulas

$$\sigma_1 = Ek' \frac{b}{R_m} \left(\frac{\dot{\psi}}{m} \pm \nu \sqrt{\frac{3}{1 - \nu^2}} \frac{\dot{\vartheta}}{m}\right) + \frac{qb}{h2} \quad \left(k' = \frac{\varphi^* - \varphi}{\varphi} = k - 1\right),$$

$$\sigma_2 = Ek' \frac{b}{R_m} \left(\pm \sqrt{\frac{3}{1 - \nu^2}} \frac{\dot{\vartheta}}{m} - \frac{b}{R_m} \frac{s}{r} \frac{\psi}{m}\right) + \frac{qb}{h}. \tag{3.27}$$

Formulas for maximum stress and the relation between the angle of flexure $\varphi^* - \varphi$ and the applied bending moment M_z may be presented in a form similar to those well known in the Strength of Materials for beams with *undeformable* cross-sections. For a circular section of radius b and the length of the beam φR_m the beam formulas are

$$\frac{\varphi^* - \varphi}{\varphi R_m} \equiv \frac{k'}{R_m} = \frac{M_z}{EI}, \qquad \sigma_B \equiv \frac{M_z b}{I} = E \frac{k'}{R_m} b \quad (I = \pi b^3 h). \tag{3.28}$$

The relation between the applied moment and the elastic change of curvature of the tube k'/R_m is provided by (2.32). For a tube, the M_1 term in (2.32) is of the order of magnitude of h/b compared to the T_1 term. It must be neglected. With expression (2.24) for T_1, equation (2.32) becomes

$$M_z = Ehh°b \int_{-\pi}^{\pi} (R^* - R_m^*)\dot{\psi}\, d\eta\,, \quad R_m^* = \frac{1}{2\pi} \int_{-\pi}^{\pi} R^*\, d\eta\,.$$

Integrating by parts and introducing $\dot{R}^* = -(1 + \varepsilon_2)bs^* \approx -bs^*$ gives the formula

$$M_z = Ehh°b^2 \int_{-\pi}^{\pi} \psi_s s^*\, d\eta\,, \quad \psi_s = \psi - N°\eta = \sum b_j \sin j\eta\,, \quad (3.29)$$

(where we take account of the continuity of $R^*(\eta)$ and $\psi_s(\eta)$ for a tube).

Upon substitution of the expansion (3.21) for ψ_s and $s^* = s = \sin \eta$ we obtain for linear bending of circular tubes,

$$M_z = KEI\frac{\varphi^* - \varphi}{\varphi R_m}\,, \quad K = \frac{b_1}{m}\,, \quad (3.30)$$

where, as in (3.28) I denotes the moment of inertia of the cross-section of the tube (with respect to the axis $R = R_m$); KEI is the *effective stiffness*.

For a straight tube ($\mu = 0$) we have from (3.26) and (3.30) $K = 1$, and formula (3.30) coincides with (3.28). For curved tubes the *stiffness factor* K is less than 1. The larger the μ, the smaller is the effective stiffness KEI. The values of $K = K(\mu)$ may be taken from the graphs of Figs. 18 and 30. For moderate curvatures, when $\mu < 30$, it is sufficient to retain three terms in the expansion (3.21) of ψ. The solution of the system of first three equations of (3.26) yields a simple formula

$$K = \frac{1 + \mu^2/144 + q°/3 - \Delta}{1 + 10\mu^2/144 + q°/3 - \Delta}\,, \quad \Delta = \frac{(\mu/12)^4}{4 + 4q°/15 + 17\mu^2/1800}\,. \quad (3.31)$$

Taking into account the Kármán effect by means of stress factors we

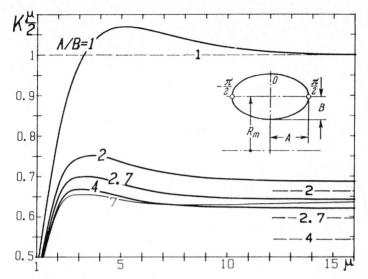

FIG. 18. Stiffness factor for circular and elliptic tubes. Broken lines: (3.34) or (3.63).

write instead of the beam formulas (3.28),

$$\sigma_{1\,max} = \sigma_1^\circ \frac{M_z b}{KI}, \qquad \sigma_{2\,max} = \sigma_2^\circ \frac{M_z b}{KI} \quad \left(\frac{M_z b}{KI} = E \frac{k'}{R_m} b \right). \qquad (3.32)$$

As mentioned, the functions ψ and ϑ and therefore the factors K, σ_1° and σ_2° depend on only one parameter of tube geometry—on μ. The graphs of $K(\mu)$ and σ_1°, σ_2° are plotted from the series solution in Figs. 18 and 19, and in Fig. 30.

The larger the value of the curvature parameter μ, the more terms must be retained in the expansions (3.21). The calculations show sufficient accuracy of stress and stiffness factors to be obtainable, when the series of ψ and ϑ retain terms with numbers

$$j, n \leq \sqrt{\mu} + 2. \qquad (3.33)$$

Thus, for the range of tube curvatures of practical interest ($\mu < 100$), the systems (3.25) or (3.26) contain, at most, five equations.

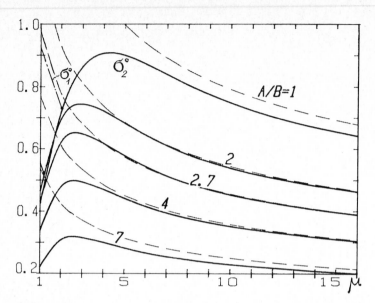

FIG. 19. Stress factors, circular and elliptic tubes. Broken lines: (3.35), (3.61).

2.3. Asymptotic analysis

The cause of the decrease of the stiffness factor and of the less favourable convergence of the series for larger tube-curvatures is the localization of stresses in the zone adjacent to the neutral line. This is illustrated in Fig. 20. In that part of the cross-section, remote from the neutral layer, the longitudinal stress T_1/h acts even in the direction opposite to the bending moment (zone cc in Fig. 20). But for the localized deformation the asymptotic solution of Chapter 2, §3.5 becomes applicable. Substitution of the asymptotic expression (2.79) of ψ into the relation (3.29) with $s^* = s$ gives after integration,

$$K = \frac{2}{\mu}. \tag{3.34}$$

(This elegant formula was discovered by BESKIN [29] as an analytical form for the results of the calculations with Fourier series. It has been derived by way of asymptotic solution by CLARK and REISSNER [39].)

Formula (3.34) is illustrated by a straight-line graph $K\mu/2 = 1$ in Fig.

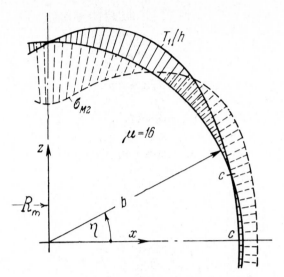

FIG. 20. Stresses in a curved tube. $\sigma_{M2} = -M_2/(h^2/6)$.

18. The accuracy of the asymptotic formula appears to be quite good even for moderately curved tubes $\mu \geqslant 12$.

Asymptotic formulas for maximum circumferential stress follow from the formulas (3.27) and the definition (3.32) of σ_2^o. After inserting the expression (2.79) for $\vartheta(\pi/2)$ (cf. present chapter, §3.3), we obtain

$$\sigma_2^o = \sqrt{\frac{3}{1-\nu^2}}\, \frac{0.939}{\mu^{1/3}}. \qquad (3.35)$$

2.4. Nonlinear flexure

The form (3.21) of the series solution remains valid also for large displacements. Indeed, the symmetry with respect to the cross-section axes z and x (Fig. 17) is fully retained. Therefore, the conditions of continuity of the cross-section of the tube,

$$R^*(\pi) - R^*(-\pi) = -b \int_{-\pi}^{\pi} \sin(\alpha + \vartheta)\, d\eta = 0\,,$$

$$z^*(\pi) - z^*(-\pi) = b \int_{-\pi}^{\pi} \cos(\alpha + \vartheta)\, d\eta = 0\,,$$

are satisfied identically with $\vartheta(\eta)$ in the form of Fourier series (3.21). This is easily checked upon substitution of the expansions (2.17) for $\sin(\alpha + \vartheta)$ and $\cos(\alpha + \vartheta)$, $\alpha = \eta$.

The Fourier coefficients of the solution (3.21) are determined by a system of (nonalgebraic) equations, which follows from substitution of the series (3.21) into the equations (3.17). This system may be written as a matrix form of equations (3.17) or as a particular case of equations (2.171) for $r = t = 1$.

$$\begin{bmatrix} \Lambda^2 & -\mu^*c'_- \\ \mu^*c_-^* & \Lambda^2 \end{bmatrix} \begin{bmatrix} \psi \\ \vartheta \end{bmatrix} = \begin{bmatrix} ms \\ -q^\circ p \end{bmatrix}. \tag{3.36}$$

After elimination of ψ we have

$$(\Lambda^2 + \mu^{*2}c_-^*\Lambda^{-2}c'_-)\vartheta = -q^\circ p - \mu^* mc_-^*\Lambda^{-2}s,$$

$$\psi = \Lambda^{-2}(\mu^*c'_-\vartheta + ms). \tag{3.37}$$

The system is easily solved in successive approximations. Starting with $c' = c^* = c$ and $s^* = s$, one calculates ϑ solving the linear equation (3.37).

The next approximation begins with calculations of c', c^* and s^* by substituting the ϑ of the preceding approximation into (2.172) and (2.174). With the new values of c', s^* and c^*, equations (3.37) yield the current approximation of ϑ. The process is continued until the difference between the two successive approximations of ϑ becomes small enough. The coordinates x^* and z^* of the deformed profile of the tube (Fig. 7) can be determined in terms of $c^* = \cos \alpha^*$ and $s^* = \sin \alpha^*$ according to (2.25) with $1 + \varepsilon_2 \cong 1$, $R_m^* = R^*(\pi/2)$,

$$x^* = R^* - R_m^* = -\int_{-\pi/2}^{\eta} s^*b \, d\eta = \sum_j \frac{b}{j}s_j^* \cos j\eta ;$$

$$z^* = \int_0^{\eta} c^*b \, d\eta = \sum_j \frac{1}{j}bc_j^* \sin j\eta . \tag{3.38}$$

With the Fourier coefficients of ϑ and of ψ evaluated, the stress and strain corresponding to the given change of curvature of tube are easily calculated according to (3.21) and (3.27).

The applied bending moment is determined by (3.29), which can be written in a matrix form with the aid of (2.154)

$$M_z = E\pi b^2 hh° \sum b_j s_j^* = E\pi b^2 hh° \psi^\mathrm{T} s^* . \tag{3.39}$$

The results of the solution just described will be presented both for the circular and noncircular tubes later in this section.

For (a) tubes of *small* initial curvature, which are the most markedly nonlinear in their bending, and (b) the tubes of *large* initial curvature and their not-too-prominent nonlinear effects, respective simpler solutions are available.

(a) In more detail, (3.36) rewritten with $\vartheta c' = s^* - s$ (cf. (2.30)) yield

$$\psi = \Lambda^{-2}(\mu^* s^* - \mu s) ,$$
$$\vartheta = -\Lambda^{-2}(\mu^* c_-^* \psi + q° p) , \tag{3.40}$$

$$p = s_+^* \Lambda^{-1} s^* - c_-^* \Lambda_+^{-1} c^* . \tag{3.41}$$

Equations (3.40) can be solved by direct iteration which converges in a few steps, provided the initial curvature of the tube is not large.

As starting values in the equations $s^* = s$ and $c^* = c$ are used. (This approximation sets the longitudinal stress (ψ) equal to that of a straight beam, which worsens convergence when $\mu > 2$. Thus, for the larger initial curvature values ($\mu > 2$), equations (3.36) should be preferred to those in (3.40).) Introducing the first approximation ψ into the second of matrix equations (3.40) we calculate the first approximation of ϑ. With ϑ, the c^* and s^* describing the deformed profile of the tube are calculated according to (2.172) and (2.174). Using the c^* and s^*, the next approximation can be started. The measure of accuracy of an approximation (ϑ, ψ) is provided by the difference compared to the preceding approximation.

The calculations are made still simpler by the form (3.21) of the series of ψ and ϑ. With $\psi = [b_1 \ b_3 \ \cdots]$ all even-numbered columns and odd-numbered rows of the matrix c_-^* may be omitted. The ϑ, s^*, s_+^* and c_+^* are also abridged.

The following data (Table 2) characterize the solution just outlined for the case of pure bending of a *cylinder* ($\mu = 0$) to the curvature of $\mu^* = 1.6$, which very nearly corresponds to the maximum value of the bending moment—the Brazier collapse load (cf. Fig. 21).

Sufficient accuracy is achieved in the *third* iteration. Only three, at most four, terms in each of the series (3.21) of ψ and ϑ have to be retained. The calculations are easily made with a pocket calculator.

We now turn to

(b) A simple and sufficiently exact (for not too large nonlinearities) correction to the linear solution (of present chapter, §2.3) is provided by the *perturbation method*.

For the circular tubes, the normal-pressure term independent of ϑ (p^0 in (3.19)) vanishes. The expansion of the solution in powers of the bending parameter m may be written in the form not displaying the pressure q explicitly

$$\psi - N^\circ \eta = \sum_{j=1,3,\dots} (mb_j^{(1)} + m^2 b_j^{(2)} + \cdots) \sin j\eta ,$$

$$\psi = m\psi_1 + m^2\psi_2 + \cdots ; \tag{3.42}$$

$$\vartheta = \sum_{n=2,4,\dots} (ma_n^{(1)} + m^2 a_n^{(2)} + \cdots) \sin n\eta , \quad \vartheta = m\vartheta_1 + m^2\vartheta_2 + \cdots .$$

TABLE 2.

	Approximation	1	2	3
	$b_1 10^5$	160000	129797	129712
ψ	$b_3 10^5$	0	−3007	−2993
	$b_5 10^5$	0	122	124
	$b_7 10^5$	0	−4	−4
	$a_2 10^5$	−32000	−32687	−32759
ϑ	$a_4 10^5$	0	1191	1262
	$a_6 10^5$	0	−47	−65
	$a_8 10^5$	0	1	2
$M^\circ = \dfrac{M_z}{E\pi b^2 hh^\circ}$		1.6	1.0581	1.0574

Substituting (3.42), (2.172) and (2.174) into equations (3.36) and equating the sums of terms with factor m on both sides of each equation gives a system of linear algebraic equations for ψ_1 and ϑ_1. A similar procedure gives equations for ψ_2 and ϑ_2 and for other terms of expansions (3.42). (Of course, these equations represent corresponding differential ones that follow in a similar way from (3.17) and (3.42).)

The equations can be presented in the form

$$\begin{bmatrix} \Lambda^2 & -\mu^*c_- \\ \mu^*c_- & \Lambda^2 + q^\circ F \end{bmatrix} \begin{bmatrix} \psi_N \\ \vartheta_N \end{bmatrix} = \begin{bmatrix} u_N \\ v_N \end{bmatrix} \quad (N = 1, 2, \ldots). \quad (3.43)$$

The left-hand sides of these equations differ from the linear equations (3.22) only in the replacement of the quantity μ with $\mu^* = \mu + m$.

The right-hand sides of (3.43) for the first two terms of the expansions (3.42) are

$$u_1 = \begin{bmatrix} 1 \\ 0 \\ 0 \\ \cdots \end{bmatrix}, \qquad v_1 = 0 ; \qquad u_2 = -\frac{\mu^*}{2} s_+ \vartheta_1^- \vartheta_1 ,$$

$$v_2 = \mu^* s_+ \psi_1^- \vartheta_1 + q^\circ F_2 \vartheta_1^- \vartheta_1 . \tag{3.44}$$

The formula for F_2 can be deduced[9] by expanding the expression (3.18) for p in powers of ϑ

$$p = p^0 + F\vartheta + F_2 \vartheta_- \vartheta + \cdots .$$

As indicated, there are cases when the nonlinearity is comparatively weak and then the first two terms of the expansions (3.42) give a good approximation. In the case of pure bending this is assured for $m \leq 1$ (cf. [84], where the coefficients of ϑ_N and ψ_N are presented for $N \leq 3$). In those situations, where the m and m^2 terms of expansions (3.42) are insufficient, it is advisable to turn to the iteration solution of equations (3.36) or (3.40).

---

[9] The term with F_2 may be dropped in most cases. Exceptions occur in the cases of large external pressure and small m and μ^*.

The perturbation solution does not require the determination of s^*. Therefore, it may be convenient to have for the bending moment a formula containing only ψ and s. Such is obtained by substituting into (3.39) the expression of s^* from the first of equations (3.40). For the circular cross-section, when $\alpha = \eta$, $s_1 = 1$ and $s_2 = s_3 = \cdots = 0$, the formula becomes

$$\frac{M_z b}{EIh^\circ} \equiv M^\circ = \frac{1}{\mu^*} \psi^{\mathrm{T}}(\mu s + \Lambda^2 \psi) = \frac{\mu}{\mu^*} b_1 + \frac{1}{\mu^*} \sum_j j^2 b_j^2$$

$$(I = \pi b^3 h, \; b_j = m b_j^{(1)} + m^2 b_j^{(2)} + \cdots) .$$

(3.45)

For a *cylinder* ($\mu = 0$) there follow from (3.45) and (3.40) two versions of a formula for the dimensionless bending moment:

$$M^\circ = \frac{1}{\mu^*} \psi^{\mathrm{T}} \Lambda^2 \psi = \mu^*(s^*)^{\mathrm{T}} \Lambda^{-2} s^* \quad (\mu^* = m) .$$

(3.46)

The nonlinear relationships between the bending moment and the change of curvature of a tube, resulting from the solution described, are presented in Fig. 21 [201, 210]. There are two different cases. The bending *increasing* the curvature of the tube ($m > 0$, $\mu^* > \mu$) leads to the flattening of the cross-sections as shown in Fig. 17. The slope of the curve $M(m)$ declines with m (Fig. 21(a)). The maximum bending moment decreases with the initial curvature parameter μ. When the initial curvature is substantial, the curve $M^\circ(m)$ is monotonously rising with m, there is no maximum bending moment.

The situation is entirely different for the bending which *diminishes* the curvature of the tube—for $m < 0$ and $\mu^* < \mu$ (Fig. 21(b)). The maximum moment exists in this case for *any* initial curvature. It is reached *after* the tube is bent to become straight—after $\mu^* = \mu + m = 0$. The tube becomes straight under the bending moment (corresponding to $M^\circ = m = -\mu$), which would completely straighten a tube with undeformable cross-sections, without the Kármán effect.

The curves of Fig. 21 indicate the nonlinearity to be substantial for $m > 0.5$ and also a comparatively satisfactory accuracy of the perturbation solution [78, 84]—the dotted curves—up to the $M = M_{\max}$. The good convergence of the expansions (3.42) in powers of m is due to the

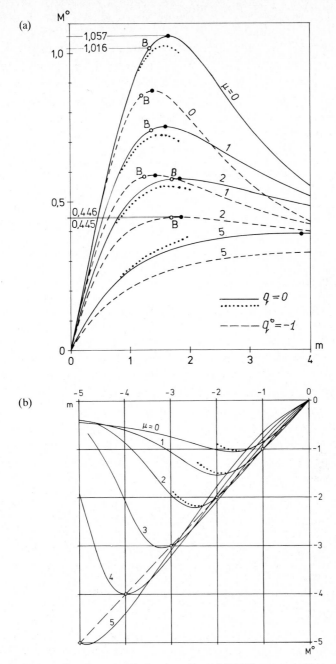

FIG. 21. Nonlinear bending of cylinder ($\mu = 0$) and initially curved tubes. Buckling points denoted by B.

use in (3.43)–(3.46) of the curvature parameter μ^* corresponding to the *deformed* state of the tube (cf. [90]). Owing to this, the first terms of the solution (3.42) (with $\psi = m\psi_1$ and $\vartheta = m\vartheta_1$) take into account no less than 70% of the deviation of the bending moment $M°(m)$ from its linear value even for m up to 1.5. This allows a simple assessment of the nonlinearity.

The characteristic curves $M°(m)$ are shown in Fig. 21 not far beyond $M° = M°_{max}$. The bending of thin tubes is cut short by a *local buckling* by $M° < M°_{max}$. This phenomenon is discussed in present chapter, §9.

The influence of the normal pressure on the nonlinear bending of tubes is also illustrated in the Fig. 21 by the results of the solutions just described.

3. Noncircular long tubes

Retaining the setting of the bending problem of the preceding section, we consider a more general form of the cross-section.

Solution of the Reissner equations in the trigonometric-series form is used to describe linear bending of tubes under normal pressure (the Kármán problem) and for the analysis of large displacements and limit-point instability (the Brazier problem). For linear bending, asymptotic formulas of maximum stress and bending stiffness are derived.

3.1. Basic formulas and equations

The influence of the noncircularity of a tube manifests itself mainly in the effect of normal pressure. In the noncircular tube the pressure causes considerable bending of tube wall, whereas in a circular tube, even though its curvature be high, the pressure can lead to no significant bending of the wall (cf. Chapter 5, §5).

The linear deformation of a tube may be considered as a sum of two parts, which are proportional to the parameters of tube bending (m) and normal pressure ($q°$);

$$\psi = m\psi_m + q°\psi_q , \qquad \vartheta = m\vartheta_m + q°\vartheta_q . \qquad (3.47)$$

The stresses can be determined with the help of formulas (1.90) and (2.24). The part of stresses, connected with the curvature-change

k'/R_m, is thus defined as (where b is the cross-section perimeter divided by 2π)

$$\sigma_1 = Ek' \frac{b}{R_m} \left(\dot{\psi}_m \pm \nu \sqrt{\frac{3}{1-\nu^2}} \dot{\vartheta}_m \right),$$

$$\sigma_2 = Ek' \frac{b}{R_m} \left(\pm \sqrt{\frac{3}{1-\nu^2}} \dot{\vartheta}_m - \frac{b}{R_m} s\psi_m \right).$$

(3.48)

These are the maximum through the wall-thickness stresses in the longitudinal (σ_1) and in the circumferential (σ_2) directions.

The remaining part of the stresses, directly connected with the normal pressure q, is determined by the functions ψ_q and ϑ_q and $p_1(\eta)$ from (2.22),

$$\sigma_1 = q\frac{b}{h} \left[\frac{6b}{h} \left(\dot{\psi}_q \sqrt{\frac{1-\nu^2}{3}} \pm \nu\dot{\vartheta}_q \right) + \frac{1}{2\pi b^2} F_0 \right],$$

$$\sigma_2 = q\frac{b}{h} \left[\pm 6\frac{b}{h} \dot{\vartheta}_q - \mu\frac{s}{r}\psi_q + \frac{p_1}{r} \right], \quad p_1 = s\frac{z}{b} + c\frac{x}{b}.$$

(3.49)

Here F_0 is the area of the tube opening. The small terms of the order of magnitude of $\nu\vartheta/\dot{\vartheta}$ are in (3.48) and (3.49) omitted.

The factors $Ek'b/R_m$ and qb/h are the stresses caused in a circular tube with undeformable cross-section by a change of curvature $k'/R_m = 1/R_m^* - 1/R_m$ and by the normal pressure q, respectively.

A noncircular cross-section, for instance an elliptical one, may have parts where the radius R_2 of the middle-surface curvature is small—of the same order of magnitude as the wall-thickness h. In such zones of elliptical and similar-form tubes the maximum stresses often occur. In this situation the stresses cannot be accurately defined by the thin-shell formulas. Here, formulas (2.39) of *middle-range-thickness* shells may be used. This leads to formulas for the σ_1 and σ_2, which follow from (3.48) and (3.49), when $\dot{\vartheta}_m$, $\dot{\vartheta}_q$, ψ_m and $-\mu s\psi_q + p_1$ are replaced, respectively, by

$$\dot{\vartheta}_m \frac{1 \pm h/6R_2}{1 \pm h/2R_2}, \quad \dot{\vartheta}_q \frac{1 \pm h/6R_2}{1 \pm h/2R_2},$$

$$\psi_m \left(1 + \frac{1+1.2\nu}{2} \frac{h}{R_2} \right), \quad (-\mu s\psi_q + p_1)\left(1 + \frac{1+1.2\nu}{2} \frac{h}{R_2} \right).$$

(3.50)

In the case of tubes of moderate curvature (cf. §2.1), the governing equations may be used in the simplified form (3.20). For noncircular tubes the part $q°p^0$ of the pressure-load function constitutes the main term in the function p. The other pressure term of (3.20) $(q°p^1)$ is less important for noncircular tubes and, in linear approximation, can mostly be omitted.

Introducing the expressions (3.47) into equations (3.20) gives

$$\ddot{\psi}_m + \mu c\vartheta_m = -s\,, \qquad \ddot{\psi}_q + \mu c\vartheta_q = 0\,,$$
$$\ddot{\vartheta}_m - \mu c\psi_m = 0\,, \qquad \ddot{\vartheta}_q - \mu c\psi_q = p^0\,. \tag{3.51}$$

The solutions to (3.51), apart from their being affected by the form of the cross-section represented by $s = \sin\alpha(\eta)$ and $c = \cos\alpha(\eta)$, depend on one parameter of curvature and wall-thickness $-\mu$.

But the equations are applicable not only for tubes of curvature so small that $b/R_m \ll 1$. As was already shown for circular tubes, the maximum values of stress and the bending stiffness depend quite insignificantly on the second parameter of the geometry b/R_m. For the so-called *flat-oval section* this aspect is illustrated by examples in Fig. 22.

Here the stress states of two tubes with the widest difference in values of curvature parameter $\lambda = b/R_m$ are compared. The graphs of Fig. 22 present solutions (produced as described in Chapter 2, §7) to the linearized equations (2.18). For each tube two cases are described: (a) bending, without the normal pressure, $q = 0$; (b) deformation by internal pressure q without any change of curvature of the tube axis, $k' = 0$.

In the case of *flexure* (a) the highest stress is σ_2. It is determined by $\max|\dot{\vartheta}|$ and occurs near the points $\eta = \pm\pi/2$ (Fig. 22(a)). This maximum stress value varies with the variation of b/R_m (from 0 to 0.975) insignificantly—less than 4%. The influence of the "slenderness" parameter b/R_m on the bending stiffness—on the factor K—is even weaker. The longitudinal stress T_1/h, determined by $\dot{\psi}$, is, in the case of tube flexure, much lower than σ_2 and has no substantial influence on the strength of the tube.

Consider now the case of tube *deformation* caused by normal *pressure q*, as presented in Fig. 22(b). The graphs of Fig. 22(b) and formulas (3.49) indicate for this case momentous longitudinal

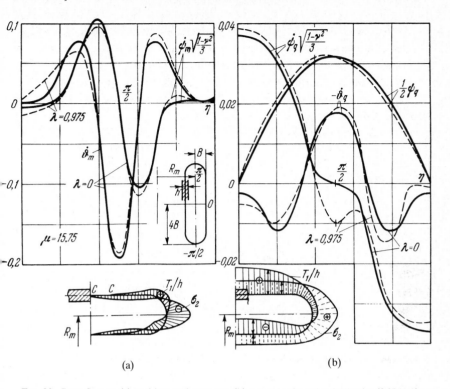

(a)   (b)

FIG. 22. Pure flexure (a) and internal pressure (b) cause in the "slender" tube ($b/R \ll 1$) and in the tube with $b/R = 0.975$ (the broken line graphs) nearly equal maximum stresses by equal values of μ ($\mu = 15.75$).

stress σ_1 with maximum values determined by the values of $|\dot{\psi}|$ at the points $\eta = 0, \pi$. These values vary insignificantly, as b/R_m changes from 0 to 0.975.

Even the utmost variation of b/R_m influences $|\sigma_1|_{max}$ less than 4%. The circumferential stress σ_2 varies at the points $\eta = 0$ relatively more but a summary influence of the parameter b/R_m remains also for the q-loading insignificant for any tube with $h/b < 0.1$.

The examples show the potential for the use of the simplified equations (3.51) and for disregarding the influence of the parameter b/R_m.

The change of curvature (k'/R_m) of the noncircular tube depends directly not only on the bending moment M_z, but also on the normal pressure q. This relation follows from formulas (3.29) after substitution of the expression (3.47) for ψ. Introduce the notation M_q for the moment acting in a cross-section of the tube under internal pressure q, when the change of curvature of the tube is made impossible by fixing its ends. Now the relation discussed may be written in the form

$$M_z = KEI\frac{k'}{R_m} + M_q \quad \left(\frac{k'}{R_m} \equiv \frac{\varphi^* - \varphi}{\varphi R_m} = \frac{1}{R_m^*} - \frac{1}{R_m} \right). \tag{3.52}$$

3.2. Linear bending

The solution of the problem will be presented in the trigonometric-series form

$$\psi = m\psi_m + q^\circ\psi_q = \sum b_j \sin j\eta , \quad b_j = mb_j^m + q^\circ b_j^q ,$$
$$\vartheta = m\vartheta_m + q^\circ\vartheta_q = \sum a_n \sin n\eta , \quad a_n = ma_n^m + q^\circ a_n^q . \tag{3.53}$$

The linear approximation of (3.39),

$$M_z = E\pi b^2 hh^\circ \sum_j b_j s_j , \tag{3.54}$$

yields, after introduction of the series (3.53) and (3.38), the parameters of (3.52)

$$I = \int\limits_{-\pi}^{\pi} x^2 bh \, d\eta = \pi b^3 h \sum_j \frac{s_j^2}{j^2}, \qquad K = \sum_j b_j^m s_j \left(\sum_j \frac{s_j^2}{j^2} \right)^{-1} ,$$
$$M_q = q\frac{b^3}{h^\circ}\pi \sum_j b_j^q s_j . \tag{3.55}$$

The equations for the Fourier coefficients of the solution (3.53), arranged in column matrices $\psi_m, \ldots, \vartheta_q$, follow from (3.51). Using the matrix operations described in Chapter 2, §7, we obtain (cf. (3.22)):

$$\begin{bmatrix} \Lambda^2 & -\mu c_- \\ \mu c_- & \Lambda^2 \end{bmatrix} \begin{bmatrix} \psi_m & \psi_q \\ \vartheta_m & \vartheta_q \end{bmatrix} = \begin{bmatrix} s & 0 \\ 0 & p^0 \end{bmatrix}. \tag{3.56}$$

Half of the unknowns—either those comprising the matrices ψ_m and ψ_q

or those comprising ϑ_m and ϑ_q—can be easily eliminated from (3.56). This leads to equations similar to (3.25) and (3.26).

Consider in more detail the case of the cross-section symmetrical with respect to axes x and z, shown in Fig. 17.

The profile of the tube is determined by the series

$$s \equiv \sin \alpha = s_1 \sin \eta + s_3 \sin 3\eta + \cdots ,$$
$$c \equiv \cos \alpha = c_1 \cos \eta + c_3 \cos 3\eta + \cdots . \tag{3.57}$$

The Fourier series of the functions ψ_m and ψ_q contain only $\sin j\eta$ terms with $j = 1, 3, \ldots$; the series of ϑ_m and ϑ_q consist of only the $\sin n\eta$ terms with $n = 2, 4, \ldots$. The solution has the form (3.21). When those matrix elements which vanish and the corresponding equations are dropped, equations (3.56) may be reduced to the form

$$\left\{ \begin{bmatrix} 1^2 & 0 & \cdots \\ 0 & 3^2 & \cdots \\ \cdots & \cdots & \cdots \end{bmatrix} + \mu^2 \check{c} \begin{bmatrix} 2^{-2} & 0 & \cdots \\ 0 & 4^{-2} & \cdots \\ \cdots & \cdots & \cdots \end{bmatrix} \check{c}^T \right\} \begin{bmatrix} b_1^m & b_1^q \\ b_3^m & b_3^q \\ \cdots & \cdots \end{bmatrix}$$

$$= \begin{bmatrix} s_1 & u_1 \\ s_3 & u_3 \\ \cdots & \cdots \end{bmatrix}, \qquad \begin{bmatrix} u_1 \\ u_3 \\ \cdots \end{bmatrix} = -\check{c} \begin{bmatrix} \frac{1}{4}\mu p_2 \\ \frac{1}{16}\mu p_4 \\ \cdots \end{bmatrix};$$

$$\check{c} = \tfrac{1}{2} \begin{bmatrix} c_{1,2} & c_{1,4} & \cdots \\ c_{3,2} & c_{3,4} & \cdots \\ \cdots & \cdots & \cdots \end{bmatrix}, \qquad c_{i,j} = c_{|i-j|} - c_{i+j}; \tag{3.58}$$

$$\begin{bmatrix} p_2 \\ p_4 \\ \cdots \end{bmatrix} = \tfrac{1}{2} \begin{bmatrix} s_3 + s_1 & s_5 - s_1 & \cdots \\ s_5 + s_3 & s_7 + s_1 & \cdots \\ \cdots & \cdots & \cdots \end{bmatrix} \begin{bmatrix} s_1 \\ \frac{1}{3}s_3 \\ \cdots \end{bmatrix} - \check{c}^T \begin{bmatrix} c_1 \\ \frac{1}{3}c_3 \\ \cdots \end{bmatrix};$$

$$\begin{bmatrix} 4a_2^m \\ 16a_4^m \\ \cdots \end{bmatrix} = -\mu \check{c}^T \begin{bmatrix} b_1^m \\ b_3^m \\ \cdots \end{bmatrix}, \qquad \begin{bmatrix} 4a_2^q \\ 16a_4^q \\ \cdots \end{bmatrix} = -\mu \check{c}^T \begin{bmatrix} b_1^q \\ b_3^q \\ \cdots \end{bmatrix} - \begin{bmatrix} p_2 \\ p_4 \\ \cdots \end{bmatrix},$$

where $p_2, p_4, \ldots$ are elements of p^0 from (3.22).

Obviously, the solution just discussed can be effective only when the

Fourier series of $\cos \alpha(\eta)$ and $\sin \alpha(\eta)$ converge rapidly enough. This is in fact assured for the elliptic and the flat-oval (Fig. 22) profiles.

The Fourier coefficients are determined by the classic formulas

$$\begin{bmatrix} c_j \\ s_j \end{bmatrix} = \frac{2}{\pi} \int_0^{\pi} \begin{bmatrix} \cos \alpha(\eta) \cdot \cos j\eta \\ \sin \alpha(\eta) \cdot \sin j\eta \end{bmatrix} d\eta . \tag{3.59}$$

Elliptic profiles are described for any axis ratio A/B (Fig. 23) with the aid of elliptic functions (e.g. [100]).

The characteristics of elliptic profiles, obtained by the numerical integration in (3.59), are presented in Fig. 23.

For the flat-oval profiles, formulas (3.59) are integrated in the closed form

$$s_j = (-1)^{(j+1)/2} \frac{4}{\pi} \frac{n}{j^2 - n^2} \cos \frac{\pi j}{2n} ,$$

$$c_j = \frac{4}{\pi j} \frac{n^2}{n^2 - j^2} \sin\left(\pi j \frac{n-1}{2n}\right) ,$$

$$\frac{b}{A} = \frac{2}{\pi} + \frac{B}{A} \frac{\pi - 2}{\pi} ,$$

$$n = 1 + \frac{2}{\pi} \frac{A - B}{B} \quad (j = 1, 3, \ldots) .$$

The nearer a profile shape is to circular, the better is the convergence of the series for $c(\eta)$ and $s(\eta)$. But a small number of terms is usually sufficient for a satisfactory description of a profile by equations (Fig. 17),

$$\frac{x}{b} = - \int_{-\pi/2}^{\eta} s \, d\eta = \sum_j \frac{1}{j} s_j \cos j\eta , \qquad \frac{z}{b} = \int_0^{\eta} c \, d\eta = \sum_j \frac{1}{j} c_j \sin j\eta .$$

For instance, an ellipse with the axis ratio as large as $A/B = 2.7$ can be represented by only three terms ($j \leqslant 5$) of the c- and s-series with accuracy of the order of the usual values of $h/b \sim 1/20$.

There are some observations to be made on the systems (3.56) and (3.58).

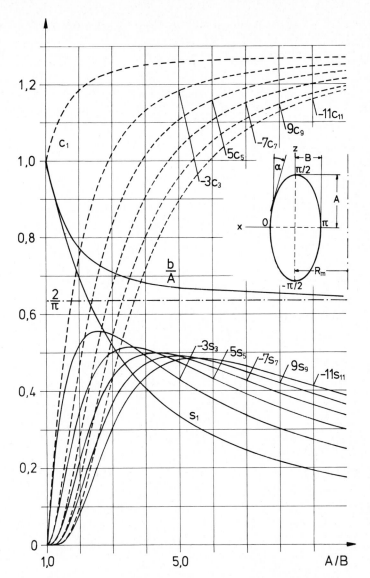

Fig. 23. Fourier coefficients and b for elliptic profiles [201].

The systems are favourably conditioned: for any value of μ the diagonal elements of the matrix of coefficients of the system (3.58), beginning with some row, exceed not only the absolute value of any other element of the row but also the sum of all these values. For

$\mu = 0$ the coefficient matrix is a diagonal one and the series (3.53) converge as s_j/j^2 or c_j/j^2. With larger values of μ and a profile increasingly different from the circular form, the convergence of the series becomes less rapid and more equations have to be retained. But a small number of terms in the series of ψ and ϑ remains sufficient for most practically interesting cases. The calculation is possible even with a pocket calculator.

Numerical examples allow a rough estimate

$$ j \leqslant \sqrt{\mu}(c_1 - c_3) + 2 , \qquad (3.60) $$

for j, the order of the last harmonic, $\sin j\eta$, to be retained in the series of ψ and ϑ.

For a circular tube, i.e. $c_1 - c_3 = 1$, the condition (3.60) is identical with (3.33).

The stress state caused by *internal pressure* is substantially different from that of pure bending. A glance at the graphs of Fig. 22 shows that the internal-pressure case is characterized by a predominance in the stress function ψ_q of the *first* term of the Fourier series ($b_1 \sin \eta$). This is observed not only for circular tubes, but also for a wide range (in μ and A/B) of elliptic and flat-oval tubes. For instance, in an elliptic tube with $A/B = 4.05$ and $\mu = 32$, the term $b_1^q \sin \eta$ accounts for about 80% of the function ψ_q (Table 3).

According to their stress state and flexural stiffness, as illustrated in Figs. 18, 19 and 24, the tubes fall in three ranges.

Small curvature tubes have a stress state, which similar to the straight beams, is dominated by the longitudinal stress T_1/h. These are tubes with $\mu < 1.5$.

In the *middle curvature* tubes, the flexure causes substantial lateral stress σ_2 determined by the moment $M_2 = D\dot{\vartheta}/b$. Transition to the kind of stress that is characteristic for *large curvature* tubes is signalled by localization of the stresses in the zone where $\cos \alpha \ll 1$ and by the

TABLE 3.

j	1	3	5	7	9	11	13
$10^4 b_j^m$	114	-121	90	-59.6	23	-9.4	3
$10^4 b_j^q$	382	-48	20.8	-41	35	-19.6	5

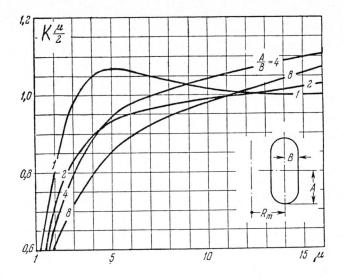

FIG. 24. Stiffness factor for flat-oval tubes.

appearance of the zone where the stress state is nearly membrane. The stress state of the "transitional" type is shown in Fig. 20.

The more localized the stress state is, the more terms must be retained in the trigonometric series solution. (This is reflected in formulas (3.33) and (3.60).) But the localized stressed state can be described by the simple asymptotic solution of Chapter 2, §3.5. For the noncircular tubes the asymptotic formulas are presented in the next section.

3.3. Asymptotic analysis

The asymptotic solution of Chapter 2, §3.5 applied to the *flexure problem* of tubes with the double-symmetric profiles yields simple formulas for the maximum stress and the flexural stiffness.

At points on the symmetry axis z of the cross-section (Fig. 20), where $\alpha = \pm \pi/2$, both the angular displacement ϑ and the resultant T_1 vanish. With the accuracy accepted for (3.51), this means at these points $\psi = 0$. Therefore, on the quarter of the cross-section from $\eta = 0$ to $\pi/2$ (Fig. 20), the functions ψ and ϑ may be identified with the asymptotic solution (2.79). In the remaining three-quarters of the cross-section the functions ψ and ϑ are symmetric or antisymmetric with respect to the first sector.

For the tubes with large values of μ the maximum stress occurs near points $\alpha = \pm\pi/2$. With $g_{kr} = 0$ and $g_{ki} = -ms$ according to (2.73) and (3.20) and μ_a from (2.75) the maximum stress is defined by (2.79) and (2.80) as

$$\sigma_{2\max} = \sigma_2^\circ Ek' \frac{b}{R_m}, \quad \sigma_2^\circ = 1.705\left(\frac{R_{20}}{\mu b}\right)^{1/3}, \quad R_{20} \cong (R_2)_{\eta = \pm\pi/2}. \quad (3.61)$$

Consider now the *bending stiffness* of the tube. Formula (3.29) with $\psi(\eta)$ of (2.79) and $x(\eta)$ of (2.77) gives

$$\frac{M_z}{Ehh^\circ b^2} = 4 \int_{-\pi/2}^{0} \psi s \, d\eta = \frac{m}{\mu_a^2} \int_{0}^{x_{\max}} e_r(x) \sin^2 \alpha \, d\eta, \quad x = x_{\max} c(\eta).$$

Assume the value of

$$x_{\max} = \frac{R_{20}}{b}\mu_a = \left(\mu \frac{R_{20}^2}{b^2}\right)^{1/3}$$

to be large enough to make $e_r(x) \cong 0$ outside the zone around the point $\alpha = \pi/2$, where $\sin\alpha = 1$. Thus, the integral is reduced to $\int_0^\infty e_r(x)\,dx$. This integral is equal to $\pi/2$ and the bending moment is given by the formula

$$M_z = E\pi b^3 h \frac{k'}{R_m \mu} \frac{2 R_{20}}{b}. \quad (3.62)$$

With the bending stiffness KEI, defined in (3.52) and I from (3.55) we have

$$M_z = KEI \frac{k'}{R_m}, \quad K = \frac{2 R_{20}}{\mu b}\left(\sum_j \frac{s_j^2}{j}\right)^{-1}. \quad (3.63)$$

For the circular tube this gives formula (3.34).

The asymptotic formulas are very simple and they are formally applicable to a variety of cross-section forms. But the simplifications made do limit, of course, the applicability range.

The limitations are particularly dependent on the curvature of the profile around the points $\alpha = \pm\pi/2$, i.e. on $1/R_{20}$.

The necessary control is achieved by a comparison with the results of the series solution. For the stress formula (3.61), the comparison (Fig. 19) shows satisfactory accuracy when $A/B < 4$ and $\mu > 10$. The deviations grow for elliptic tubes with the axis ratio. But even for A/B as large as 4, formula (3.61) is applicable for $\mu \geqslant 16$. This might have been expected. The model equation (2.75) offers the best approximation of the Reissner equations in the very zone (around the points $\alpha = \pm\pi/2$) where the maximum stress, determined by formula (3.61), occurs.

The bending-moment formula (3.63) is more restricted in its application. Figure 18 shows sufficient accuracy of the K-value from (3.63) only for circular tubes. This has a simple explanation. The assumption $e_r(x) \ll 1$ for the values[10] of x near to x_{max} requires at least $x_{max} > 2.15$ (Fig. 15). For circular tube this means $\mu > 10$. But for elliptic profile with axis ratio $A/B = 2$, formula (3.63) is applicable only for $\mu > 60$. For $A/B \geqslant 4$ this threshold rises to $\mu \geqslant 1200$.

The flat-oval profile allows a somewhat wider application range than the elliptic one. For instance, when $A/B = 4$, formula (3.62) is valid for $\mu > 80$.

3.4. Nonlinear bending. Nearly circular tubes

For the analysis of nonlinear bending there is no substantial difference between the circular and noncircular profiles. Indeed, the analysis takes into account the *deformed* shape of the tube which is noncircular anyway.

The conditions of continuity of a closed cross-section are identically satisfied by the series solution of the form (3.21) for any double-symmetric profile that can be represented by the series (3.57) for c and s.

The solutions of equations (3.36), obtained for elliptic tubes in the way described in present chapter, §2.4, are presented in Figs. 25 and 26. (These results are due to EMMERLING [198, 201, 210].)

The nondimensional bending moment M° is defined by an expression, which is somewhat more general than in (3.45). Taking account of the moment of inertia I of noncircular cross-section according to (3.55) we

---

[10] The argument x has nothing in common with the coordinate x.

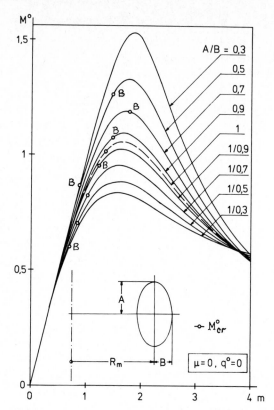

FIG. 25. Flexure of elliptic tubes. Effects of profile shape. Circles denote buckling points.

use (3.39) in the form:

$$M° \equiv \frac{M_z b}{EIh°} = \frac{\sum b_j s_j^*}{\sum (s_j/j)^2} = \frac{\psi^T(\Lambda^2\psi + \mu s)}{\mu^* s^T \Lambda^{-2} s}. \tag{3.64}$$

The curves $M°(m, q°)$ are presented in Fig. 26 only for $M, m \geq 0$—for the flexure *increasing* the initial curvature. The *unbending* curves would be similar to those in Fig. 21(b). The graphs for $M°(m, q°)$ are drawn not far beyond the point of maximum bending moment as, even *before* M_{max} is reached, a *bifurcation* local *buckling* is expected (present chapter, §9) to appear. The corresponding points of the characteristics $M°(m, q°)$ are denoted on the graphs by circles and letters B.

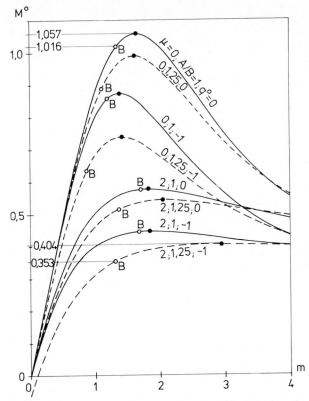

FIG. 26. Flexure of cylindrical and initially curved tubes. Effect of external pressure and ellipticity.

Small deviations from circularity of the cross-sections substantially influence the flexure of tubes[11] under normal pressure. Therefore, tubes with *nearly circular* profile constitute a class of particular interest. Consider briefly a simplified analysis of these tubes, i.e. of tubes satisfying the noncircularity limitations

$$c_1 \pm c_j \approx 1 , \qquad s_1 \pm s_j \approx 1 \quad (j \neq 1) . \tag{3.65}$$

Under these conditions it is possible to take all quadratic matrices on the

[11] Discussion of this problem and additional references are given by SPENCE and BOYLE [191] and KOSTOVETSKI [69].

left-hand side of (3.22) as they are for the circular tube. The bending of
nearly circular tubes is described by equations differing from (3.43) only
in the right-hand sides of linear approximation. This leads to equations, where the matrix of the right-hand sides is the linear combination of the right-hand sides of (3.58) (B is defined in (3.26))

$$B(\mu^*)\begin{bmatrix} b_1^{(1)} \\ b_3^{(1)} \\ \cdots \end{bmatrix} = \begin{bmatrix} s_1 \\ 0 \\ \cdots \end{bmatrix} - \frac{q^\circ}{m2}\begin{bmatrix} 1 & 0 & 0 & \cdots \\ 1 & 1 & 0 & \cdots \\ 0 & 1 & 1 & \cdots \\ \cdots & \cdots & \cdots & \cdots \end{bmatrix}\begin{bmatrix} \alpha_2^{-2}p_2 \\ \alpha_4^{-2}p_4 \\ \cdots \end{bmatrix},$$
$$(3.66)$$
$$\alpha_n^2 = n^2 + \frac{n^2 q^\circ}{n^2 - 1}.$$

After calculating $b_j^{(1)}$ the further evaluation of ψ through equations
(3.43) does not differ from the case of circular cross-section. (Some
examples on slightly noncircular tubes are presented in [84].)

4. Lateral and out-of-plane bending

The linear Saint-Venant problems of tube bending and torsion with
stress varying along the tube as cos ξ or sin ξ (Fig. 13(e) and (f)) are
discussed on the basis of the Schwerin–Chernina equations of Chapter
2, §3. The equations of small displacements are supplemented by terms
that take into account the influence of the normal pressure q.

As in present chapter, §§1, 2, the analysis is restricted to constant
wall-thickness and to cross-sections with double symmetry (with respect
to the axes x and z of Fig. 17).

4.1. Transverse bending in the plane of symmetry

The linear problem of lateral plane bending of tubes (Fig. 27) will be
solved with the help of (2.63). For constant wall-thickness the equations
have the form

$$(r\dot\psi)^\cdot + \mu\vartheta c = c\Omega^\circ - \frac{s}{r}U^\circ,$$
$$(r\dot\vartheta)^\cdot - \mu\psi c = cP_x^\circ - \frac{s}{r}L^\circ.$$
$$(3.67)$$

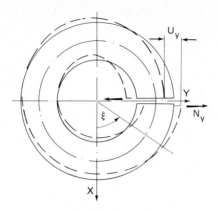

FIG. 27. Plane cross-bending.

As was discussed in present chapter, §§1, 2, the equations can be simplified for the Saint-Venant problems by assuming $r = R/R_m \approx 1$. These considerations apply also for the transverse bending of tubes. In all those cases when the length of the tube is large enough to allow the end conditions to be fulfilled merely in the sense of the Saint-Venant principle, equations (3.67) may be used in the form

$$\ddot{\psi} + \mu c\vartheta = c\Omega° - sU° ,$$
$$\ddot{\vartheta} - \mu c\psi = cP_x° - sL° .$$

(3.68)

In the plane bending being considered (Fig. 27) the angle Ω_x of spatial distortion (shown in Figs. 13 and 28) is equal to zero. It is also not difficult to deduce from the scheme of the plane bending that the component P_x of the stress resultant acting in any section $\eta = $ const. is equal to zero. The same applies for the moment L_y of stresses acting over the section $\eta = $ const. (Fig. 13): $L_y = 0$.

With $\Omega°$, $P_x°$ and $L° = 0$, equations (3.68) are identical in form to the equations of pure flexure (3.51). Hence, the functions ψ and ϑ determined for the cross-bending by equations (3.68) with $U° = m$ and $\Omega°$, $P_x°$, $L° = 0$ are identical to the functions ψ and ϑ describing the *pure* flexure (cf. Chapter 2, §3.4). In the two cases the stresses σ_1 and σ_2 are distributed in the same way. For equal values of the change of curvature of a tube the stresses σ_1 and σ_2 are equal. Indeed, according to formula

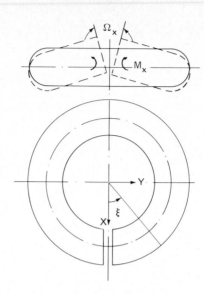

Fig. 28. Spatial bending.

(2.48), U° can be expressed in terms of the change of curvature $\kappa_1 = t_z \kappa_1^1$ of the line $\eta = 0$, lying in the plane of symmetry of the tube (the plane xy in Fig. 27). Neglecting in (2.48) the term with ε_1, representing the direct influence of the extension of the ξ-line, we obtain

$$U^\circ = -\frac{\mu}{\pi R} U_y = -\frac{\mu}{\pi R} \int_0^{2\pi} \kappa_1^1(0) \cos^2\xi \, R^2 \, \mathrm{d}\xi = \mu R \kappa_1^1(0) \, . \qquad (3.69)$$

Comparison of (3.69) with the definition (2.20) of m shows that U° and m represent the change of curvature of the tube in a similar way. This leads to a conclusion synthetizing the solutions of the problems of the pure and of the cross-flexure. The stresses and displacements can be calculated for *any* case of Saint-Venant's plane linear bending with the help of the formulas and graphs of the *pure* bending (presented in present chapter, §§1, 2).

This is done by replacing the value of k'/R_m or m in the formulas of pure bending (present chapter, §1) by the value $\kappa_1^1(0)$ of the change of curvature κ_1 for $\xi = 0$, $\eta = 0$ or respectively by U°. The resulting formulas are

$$\kappa_1^1(0) = \frac{1}{\pi R_m^2} U_y \quad \text{and} \quad \kappa_1^1(0) = \frac{M_z}{KEI}, \tag{3.70}$$

where M_z is a bending moment in the cross-section $\xi = 0$ (Fig. 27) about the z-axis passing through the centroid of the section.

4.2. Spatial bending

Consider a tube bent *out of* the *plane* of its circular axis. In this case, as seen in the schemes of Fig. 28 and Fig. 13(f), both the bending and the twisting resultant moments in the sections $\xi = $ const. are nonzero. The deformation is determined by a solution of equations (3.67) or (3.68) and is proportional to the distortion angle Ω_x (fig. 28) represented by Ω°. In this case $U^\circ = 0$, while the right-hand side of the second equation of (3.67) or (3.68) $(cP_x^\circ - sL^\circ)$ is to be determined together with the functions ψ and ϑ.

The solution of the governing equations will be sought in the form of trigonometric series. For symmetric cross-sections, the series of $\psi(\eta)$ and $\vartheta(\eta)$ satisfying (3.67) or (3.68) must contain for the problem $\Omega^\circ \neq 0$ and $U^\circ = 0$, only cos $j\eta$ terms. The system of algebraic equations determining the Fourier coefficients (arranged in the column matrices $\boldsymbol{\psi}$ and $\boldsymbol{\vartheta}$) may be written as a matrix form (Chapter 3, §7) of (3.67)

$$\boldsymbol{\Lambda}_+ \boldsymbol{r}_- \boldsymbol{\Lambda}\boldsymbol{\psi} - \mu \boldsymbol{c}_+ \boldsymbol{\vartheta} = -c\Omega^\circ ,$$

$$\boldsymbol{\Lambda}\boldsymbol{r}_- \boldsymbol{\Lambda}\boldsymbol{\vartheta} + \mu \boldsymbol{c}_+ \boldsymbol{\psi} = -\boldsymbol{c}_+ \boldsymbol{P}_x^\circ + \boldsymbol{r}_+^{-1} \boldsymbol{s}_- \boldsymbol{L}^\circ . \tag{3.71}$$

For profiles symmetric with respect to both axes x and z (Fig. 17), the Fourier series of the functions $c(\eta)$ and $s(\eta)$ contain only odd-number terms $c_j \cos j\eta$, $s_j \sin j\eta$, $j = 1, 3, \ldots$, as in (3.57).

When the governing equations are used in the form (3.68), the series solution for the double-symmetric profiles may be shown to have the form

$$\psi = \tfrac{1}{2} b_0 + b_1 \cos \eta + b_3 \cos 3\eta + \cdots ,$$

$$\vartheta = \tfrac{1}{2} a_0 + a_2 \cos 2\eta + a_4 \cos 4\eta + \cdots . \tag{3.72}$$

These series can satisfy (3.71) with $r = 1$, provided

$$-cP_x^\circ + sL^\circ = \tfrac{1}{2}\mu(b_1 c_1 + b_3 c_3 + \cdots) + \tfrac{1}{2}\mu b_0 c(\eta). \tag{3.73}$$

The system (3.71) is easily transformed by eliminating the unknown ϑ or ψ and thus becomes (for $r = 1$)

$$\left\{ \begin{bmatrix} 1^2 & 0 & \cdots \\ 0 & 3^2 & \cdots \\ \cdots & \cdots & \cdots \end{bmatrix} + \mu^2 \check{c}_+ \begin{bmatrix} 2^{-2} & 0 & \cdots \\ 0 & 4^{-2} & \cdots \\ \cdots & \cdots & \cdots \end{bmatrix} \check{c}_+^{\mathrm{T}} \right\} \begin{bmatrix} b_1 \\ b_3 \\ \cdots \end{bmatrix}$$

$$= \left(-\Omega^\circ + \mu\frac{a_0}{2} \right) \begin{bmatrix} c_1 \\ c_3 \\ \cdots \end{bmatrix},$$

$$\begin{bmatrix} 4a_2 \\ 16a_2 \\ \cdots \end{bmatrix} = -\mu \check{c}_+^{\mathrm{T}} \begin{bmatrix} b_1 \\ b_3 \\ \cdots \end{bmatrix}, \tag{3.74}$$

$$\check{c}_+ = \tfrac{1}{2} \begin{bmatrix} c_1 + c_3 & c_3 + c_5 & c_5 + c_7 & \cdots \\ c_1 + c_5 & c_1 + c_7 & c_3 + c_9 & \cdots \\ c_3 + c_7 & c_1 + c_9 & c_1 + c_{11} & \cdots \\ \cdots & \cdots & \cdots & \cdots \end{bmatrix}.$$

For the circular profile, this system differs from (3.26) and (3.24) only in the replacement of m by $-\Omega^\circ + \mu a_0/2$ and q° by 0. Hence, the *coefficients* of the series solution (3.72) related to $-\Omega^\circ + \mu a_0/2$ coincide with a_n/m, b_n/m of the *pure* plane bending discussed in present chapter, §2.

The bending moment M_R applied in a cross-section about an axis lying in the plane xy (Fig. 28) can be determined by a formula similar to (3.30). Evaluating the resultant moment M_R of the normal stresses T_1/h in the cross-section and expressing T_1 in terms of ψ with formulas (2.19) and (2.57) for $r = 1$, we obtain

$$M_R = \int_0^{2\pi} T_1 zb \, d\eta, \quad T_1 = Ehh^\circ \psi \cos \xi.$$

Integrating by parts gives for a continuous function $z\psi$, i.e. for closed cross-sections (tubes),

$$M_R = M_x \cos \xi , \quad M_x = Ehh°b^2 \int_0^{2\pi} \psi \cos \alpha \, d\eta . \qquad (3.75)$$

With $\psi(\eta)$ obtained from (3.72) and using (3.74), formula (3.75) yields the expression of the applied moment M_x in terms of the angle of the tube deformation Ω_x and of the quantity a_0 to be determined,

$$\frac{\Omega_x}{\pi R_m} = \frac{M_x}{KEI} + \frac{a_0}{2R_m} . \qquad (a)$$

For a circular tube the stiffness factor has the value

$$K(\mu) = -b_1/(\Omega° - \mu a_0/2) , \qquad (3.76)$$

which coincides with the value of K in (3.31).

It remains to determine the quantity a_0 which represents according to (3.72) the rigid-body rotation of the cross-sections of the tube. In the relation (a) the term $a_0/2R_m$ indicates the contribution to the angle Ω_x of the *membrane torsional* deformation of the tube, one corresponding to the *constant* terms in $\psi(\eta)$ and $\vartheta(\eta)$. This contribution may be evaluated most simply as the elastic angular displacement Ω_x of a *beam* caused by a torsional moment $M_t = M_x \sin \xi$ (Fig. 28). The Maxwell–Mohr formula gives

$$\pi \frac{a_0}{2} = \int_0^{2\pi} \frac{1}{GI_t} M_t \frac{M_t}{M_x} R \, d\xi = \frac{M_x}{GI_t} \pi R_m .$$

Formula (a) now becomes (for a circular tube, when $I_t = 2I = 2\pi b^3 h$)

$$\frac{\Omega_x}{\pi R} = \frac{M_x}{KEI} + \frac{M_x}{GI_t} = \frac{M_x}{EI}\left(\frac{1}{K} + 1 + \nu\right) . \qquad (3.77)$$

This formula is derived for the tube and the loading of Fig. 28. But the foregoing leads to a conclusion valid for any Saint-Venant's problem. The tube deforms as a beam with the *reduced flexural* stiffness[12] KEI

---

[12] In the work of SEAMAN and WAN [164] the stiffness-reduction factor is related to the entire value of Ω_x including the torsional contribution. This makes $K(\mu)$ different from K of the in-plane bending and dependent on the value of the tube angle φ (Fig. 17).

and the *full torsional* stiffness GI_t, which is determined by Bredt's formula.

The stresses are expressed in terms of ψ and ϑ with the aid of relations of Chapter 1, §6 and Chapter 2, §2. The dominant stresses correspond to the longitudinal resultant $T_1 = Ehh°\psi$ and the bending moment M_2 in the circumferential direction. The respective formulas can be written similarly to those under (3.27)

$$\sigma_1 = Eh°\left(\dot{\psi} \pm \nu\sqrt{\frac{3}{1-\nu^2}}\dot{\vartheta}\right), \qquad \sigma_2 = Eh°\left(\pm\sqrt{\frac{3}{1-\nu^2}}\dot{\vartheta} - \frac{b}{R}s\psi\right).$$

$$(3.78)$$

The torsional moment M_t (in the problem of Fig. 28, $M_t = M_x \sin \xi$) causes the shear stresses $\sigma_{12} \approx S/h$ in a cross-section. Here S is determined in terms of ψ by formulas (2.57), (2.19), (3.72) or, more directly, by the elementary beam-torsion formulas. However, these shear stresses do not influence, as a rule, the strength of the tube. The level of the σ_{12} is much lower than that of σ_2. Besides, the peak values of σ_{12} do not occur in the same parts of the tube as those of σ_2 and σ_1.

The distribution of the stresses over the cross-section is presented for the example $\mu = 56$ in Fig. 29. It is quite different from the case of plane

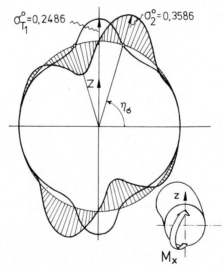

FIG. 29. Stresses by out-of-plane bending. $\mu = 56$. $[\sigma_{T_1 \max} \quad \sigma_{2 \max}] \equiv [\sigma_{T_1}° \quad \sigma_2°]M_x b/KI$.

4.3. Influence of the normal pressure

The normal pressure q, uniform along the tube surface, causes the stress resultants T_1 and T_2, which according to (3.27) have for $b \ll R_m$ nearly constant values (cf. (5.70))

$$T_1 = \tfrac{1}{2} qb , \qquad T_2 = qb .$$

Influence of these stress resultants on the fraction of the tube deformation, which varies as $\cos \xi$ and $\sin \xi$, is determined by the nonlinear terms of the equilibrium equations (1.56). Substituting into the nonlinear terms the constant values $T_1 = qb/2$ and $T_2 = qb$ and the expressions $[\kappa_1 \ \kappa_2 \ \lambda_1] = [\kappa_1^1 \ \kappa_2^1 \ \lambda_1^1] \cos \xi$ and $\lambda_2 = \lambda_2^1 \cdot \sin \xi$ results in additional terms, varying as $\cos \xi$ and $\sin \xi$. These terms account for the influence of the normal pressure. Thus, in the equations describing the deformation varying as $\cos \xi$ and $\sin \xi$ (cf. Chapter 2, §1.2), there appear, instead of the components of loading $q_1^1 \sin \xi$, $q_2^1 \cos \xi$ and $q^1 \cos \xi$, the following "reduced" load terms:

$$(q_1^1 - qb\lambda_2^1) \sin \xi , \quad (q_2^1 - \tfrac{1}{2} qb\lambda_1^1) \cos \xi ,$$

$$(q^1 - qb\kappa_2^1 - \tfrac{1}{2} qb\kappa_1^1) \cos \xi . \tag{3.80}$$

These quantities must be taken into account in the stress resultants P_x and L_y. The values of the resultants follow from the conditions of equilibrium of a part of the shell enclosed between a section $\eta = \eta'$ or $\eta = \eta''$ and another section $\eta = \text{const.}$:

$$P_x = P_x(\eta') - \int_{\eta'}^{\eta} b \, d\eta \int_0^{2\pi} q_x R \, d\xi ,$$

$$L_y = L_y(\eta'') - \int_{\eta''}^{\eta} b \, d\eta \int_0^{2\pi} (zq_x - xq_z) R \, d\xi . \tag{3.81}$$

Determining the projections of the distributed load on the axes x and z (with the help of (2.42) for the tangent vectors of the coordinate lines) we find

$$q_x = -(q_1^1 - qb\lambda_2^1)\sin^2 \xi - (q_2^1 - \tfrac{1}{2}qb\lambda_1^1)s \cdot \cos^2 \xi$$

$$+ (q^1 - qb\kappa_2^1 - \tfrac{1}{2}qb\kappa_1^1)c \cdot \cos^2 \xi , \tag{3.82}$$

$$q_z = (q_2^1 - \tfrac{1}{2}qb\lambda_1^1)c \cdot \cos \xi + (q^1 - qb\kappa_2^1 - \tfrac{1}{2}qb\kappa_1^1)s \cdot \cos \xi .$$

With (3.80)–(3.82) and the expressions of Chapter 2, §3 determining κ_i^1 and λ_i^1 in terms of ϑ and ψ, we obtain the part of the right-hand side of (3.68), which represents the influence of the normal pressure q,

$$cP_x^\circ - sL^\circ = q^\circ \left(c \int_{\eta'}^{\eta} \dot\vartheta c \, d\eta + s \int_{\eta''}^{\eta} \dot\vartheta s \, d\eta \right) . \tag{3.83}$$

Terms of the order of magnitude of b/R, ϑ and ψ compared to unity are omitted in (3.83). The values η' and η'' can be freely chosen.

Substituting into (3.83) a Fourier series

$$\vartheta = \sum_n (a_{cn} \cos n\eta + a_{sn} \sin n\eta) \tag{3.84}$$

leads for circular tubes to the simple expression

$$cP_x^\circ - sL^\circ = q^\circ \sum_n \frac{n^2}{n^2 - 1}(a_{cn} \cos n\eta + a_{sn} \sin n\eta) . \tag{3.85}$$

The normal-pressure terms are identical to those of (3.22) and (3.23) completing the similarity of (3.68) and (3.20). This allows a general conclusion on the influence of pressure.

For both the plane bending and the out-of-plane bending (Figs. 27 and 28) the Fourier-series solution of equations (3.68) retains its form (3.21) or (3.72), respectively. Also in the case of substantial normal pressure (taken into account by the terms (3.85) in equations (3.68)) the coefficients of the series are determined by equations that differ from those in (3.25) and (3.26) only in the replacement of m by U° or P_x°, respectively.

Thus, for the plane cross-bending as well as for the spatial bending the stiffness factor $K(\mu, q^\circ)$ is equal to that of pure bending (present chapter, §2). The influence of the normal pressure is identical.

To determine with the help of superposition the deformation of a curved tube caused by *arbitrary* forces and moments (distributed on the tube's ends according to the principle of Saint-Venant), it remains for us to consider the axisymmetric *torsion* problem of Fig. 13(c) and (d).

5. Torsion

Consider the two cases of torsion constituting the second Saint-Venant problem of axisymmetric deformation, stated in Chapter 2, §1: (i) Torsion of a closed shell of revolution about the symmetry axis z (Fig. 13(c)) by moments applied on the edges $\eta = $ const. (ii) Torsion of a tube, or of an open-section beam, with the axis of circular form, loaded on each end by forces statically equivalent to an axial force N_z, as indicated in Fig. 13(d). According to the Saint-Venant's semi-inverse method, the distribution of the load along the edges is determined together with the stress state of the shell.

5.1. Solution of the Saint-Venant problem

Following the pattern of the semi-inverse method let us assume the values of some of the stress and strain resultants. We set those of the resultants not necessarily connected with torsion equal to zero. The remaining resultants—describing the torsion and shear—are assumed to vary only with the coordinate η. Thus, we prescribe as values of the stress resultants

$$T_1 = T_2 = M_1 = M_2 = 0 , \qquad S = S(\eta) , \qquad H = H(\eta) \qquad (3.86)$$

and in static-geometric analogy the strain resultants

$$\kappa_2 = \kappa_1 = \varepsilon_2 = \varepsilon_1 = 0 , \qquad \tau = \tau(\eta) , \qquad \gamma = \gamma(\eta) . \qquad (3.87)$$

Substitution of these expressions satisfies identically all but the first of the linearised equilibrium equations (1.59) and of the analogous compatibility equations (1.36). The four resultants (3.86) and (3.87) must satisfy the remaining first of equations (1.59) and, respectively, of (1.36). For the present problem these equations have integrals, which follow from (2.41) and (2.48).

The unknowns S, H, τ and γ will be determined with the help of the integrals of the equilibrium and compatibility equations (2.41) and (2.48) and of the elasticity relations (1.86).

The second equations of (2.41) and (2.48) projected on the z-axis, after consideration of (3.86), (3.87) and expressions (3.42) for $\boldsymbol{t}_1$, $\boldsymbol{t}_2$, $\boldsymbol{n}$ and $\boldsymbol{R} \times \boldsymbol{t}_1 = R\boldsymbol{t}_z$, yield equations[13]

$$
\begin{aligned}
&H_2 c + RS_2 = \frac{1}{2\pi R} L_z \quad (c \equiv \cos\alpha)\,, \\
&-\frac{\gamma}{2} c + R\tau_1 = \frac{1}{2\pi R}(u_z + R\vartheta_2) = \frac{1}{2\pi R} U_z\,.
\end{aligned} \tag{3.88}
$$

Similarly, projecting on the z-axis the integrals of $\boldsymbol{T}_1$ and $\boldsymbol{\kappa}_2$ over the section $\xi = \text{const.}$ gives

$$
\begin{aligned}
&\int_{\eta_1}^{\eta_2} (-S_1 c - Q_1 s) b \, \mathrm{d}\eta = Q_z \quad (s \equiv \sin\alpha)\,, \\
&\int_{\eta_1}^{\eta_2} (\tau_2 c - \lambda_2 s) b \, \mathrm{d}\eta = [\vartheta_z]_{\eta_1}^{\eta_2}\,.
\end{aligned} \tag{3.89}
$$

Expressing all resultants in the last four equations in terms of S, H, τ and γ, and using (1.58) and (1.35) provides the basic equations

$$
2cH + RS = \frac{1}{2\pi R} L_z\,, \qquad -c\gamma + R\tau = \frac{1}{2\pi R} U_z\,; \tag{3.90}
$$

$$
\begin{aligned}
&\int_{\eta_1}^{\eta_2} \left[-Sc + 2\frac{(HRs)^{\bullet}}{Rb}\right] b \, \mathrm{d}\eta = N_z + [cH_2]_{\eta_1}^{\eta_2}\,, \\
&\int_{\eta_1}^{\eta_2} \left[\tau c - \frac{(\gamma Rs)^{\bullet}}{Rb}\right] b \, \mathrm{d}\eta = \left[\vartheta_z - \frac{s\gamma}{2}\right]_{\eta_1}^{\eta_2}.
\end{aligned} \tag{3.91}
$$

[13] The first of (3.88) can be easily recognized as a condition of static equivalence on inspection of Figs. 13, 14 and 15.

Equations (3.90) were derived by REISSNER and WAN [125]. Together with the elasticity relations (1.86),

$$H = 2D_G \tau , \qquad S = B_G \gamma , \tag{3.92}$$

(3.90) and (3.91) solve both torsion problems. The resolving system is, moreover, a nondifferential one. Consider some specific cases.

5.2. Torsion of shells of revolution

In this case (Fig. 13(c)) the continuity of the closed ξ-line demands

$$U_z = 0 . \tag{3.93}$$

Solution of the four equations (3.90) and (3.92) gives formulas for the twist and the shear-stress resultant

$$\tau = \frac{c}{B_G 2\pi R^3} L_z , \qquad S = \frac{1}{2\pi R^2} L_z , \tag{3.94}$$

where small terms of the order of magnitude of $4D_G/(B_G R^2) \sim h^2/(3R^2)$ are omitted.

We may further determine the angle θ of twist of the shell—the angle of rotation of the parallel $\eta = \eta_1$ with respect to the parallel line $\eta = \eta_2$. The rotation of a parallel will be measured as rotation of its tangent vector t_1.

The angle of twist is expressed by formulas of Chapter 1, §3.3 in terms of the rotation ϑ of the tangent plane and the shear angle γ,

$$\theta = [\vartheta_z + \tfrac{1}{2}\gamma s]_{\eta_1}^{\eta_2} .$$

Using equations (3.91) with the term $(\gamma s R)^{\cdot}/R$ integrated by parts, and substituting τ and γ according to formulas (3.92) and (3.94) gives finally

$$\theta = \int_{\eta_1}^{\eta_2} \left[\tau c + \gamma \frac{s^2}{R} \right] b \, d\eta = \frac{1}{2\pi} C_B L_z , \qquad C_B = \int_{\eta_1}^{\eta_2} \frac{b \, d\eta}{B_G R^3} . \tag{3.95}$$

5.3. Twisting of curved beams

Consider the problem represented by the scheme of Fig. 13(d). In this case the lines $\eta = $ const. are not closed. Therefore, the condition (3.93) is inapplicable: the distortion displacement $U_z(\eta)$ is an unknown. Using the semi-inverse method, set $U_z = $ const., pending a subsequent verification. The constant U_z characterizes, in accordance with (3.88), a relative displacement of the beam edges in the direction of axis z. It has to be expressed in terms of the forces N_z applied to the edges (Fig. 13(d)). This is made with the help of the first of equations (3.91). The second of the equations renders a condition of continuity of the section $\xi = $ const. When the section is closed, the points η_1 and η_2 coincide yielding

$$\int_{\eta_1}^{\eta_2} \left[\tau c - \frac{(\gamma R c)^{\displaystyle\cdot}}{Rb} \right] b \, d\eta = 0 , \qquad [cH_2]_{\eta_1}^{\eta_2} = 0 . \tag{3.96}$$

Solution of the system (3.90)–(3.92) and the omission of terms of the order of magnitude of $4D_G/(B_G R^2) \sim h^2/3R^2$ leads to the formulas

$$\tau = \frac{U_z}{R^2 \varphi} \left(1 - \frac{c}{R} \frac{C}{C_B B_G} \right) ,$$

$$S = -\frac{U_z}{R_2 \varphi} \left(\frac{C}{C_B} + \frac{c}{R} 4D_G \right) , \qquad C = \int_{\eta_1}^{\eta_2} \frac{c}{R^2} b \, d\eta , \tag{3.97}$$

$$N_z = \frac{U_z}{\varphi} \left(\int_{\eta_1}^{\eta_2} 4D_G \frac{b}{R^3} \, d\eta + \frac{C^2}{C_B} \right) ,$$

where φ is the angle between the planes of edges $\xi = $ const.

Formulas (3.97) with (3.92) define stress and strain resultants, as well as the displacement U_z of the edges in relation to each other, in terms of the forces N_z applied to those edges.

Formulas (3.97) can be substantially simplified for each of two main types of beams.

For the case of *closed* cross-section (tube), the terms with D_G in the formulas (3.97) are to be neglected. Their relative order of magnitude is h^2/R^2.

For beams of *open* cross-section the terms with $1/C_B$ disappear. Indeed, an open section may be considered as a limiting case of a closed one, having on some segment a zero wall-thickness.

On such a segment $B_G = 0$. (Recall the definition of B_G in (1.69).) According to (3.95), this means $1/C_B = 0$.

As discussed in present chapter, §§2–4, the Saint-Venant's problems make practical sense only for "slender" beams—when it may be assumed $R/R_m \approx 1$. Consider curved beams satisfying the less restrictive condition

$$\left(\frac{x}{R_m}\right)^2 \ll 1 , \quad x = R - R_m , \quad \frac{1}{R^2} \cong \frac{1}{R_m^2}\left(1 - \frac{2x}{R_m}\right). \qquad (3.98)$$

For these cases, formulas (3.97) may be written in the simplest form coinciding with the well-known formulas of the Strength of Materials for the torsion of thin-walled beams. With this as goal, we express the load resultant N_z and the distortion displacement U_z in terms of new parameters—of the twisting moment M_t in the cross-section and of the twist measure τ_m,

$$M_t = N_z R_m , \qquad \tau_m = U_z/(\varphi R_m^2) . \qquad (3.99)$$

Let B_G and D_G be constants. (Recall that for homogeneous isotropic material $B_G = Gh$ and $D_G = Gh^3/12$.) Consider again beams with *closed cross-sections*—tubes. In this case we set the coordinate η to make $\eta_1 = 0$, $\eta_2 = 2\pi$ and $2\pi b$—the perimeter of the cross-section. Formulas (3.95), (3.97) and (3.98) give, within a relative error of the order of magnitude of $(x/R_m)^2$,

$$C_B = \frac{2\pi b}{B_G R_m^3} , \qquad C = -\frac{2F}{R_m^3} , \qquad F = \int_0^{2\eta} xbc \, d\eta . \qquad (3.100)$$

With this result the relations (3.97) may be reduced to a form similar to the Bredt formulas

$$S = \frac{M_t R_m^2}{2F R^2} , \qquad \tau_m = \frac{M_t 2\pi b}{B_G 4F^2} , \qquad (3.101)$$

where F denotes the area bounded by the middle line (profile) of the cross-section of the tube.

Turning to the beams with *open cross-sections* we have, as mentioned, $1/C_B = 0$. Formulas (3.97) here give (with the profile length $\eta_2 b - \eta_1 b \equiv 2\pi b$)

$$\tau_m = \frac{M_t}{4D_G 2\pi b}, \qquad \tau = \tau_m \frac{R_m^2}{R^2}. \tag{3.102}$$

In the foregoing (present chapter, §§2–5), the stress and deformation of tubes loaded on their ends were determined in the frame of the principle of Saint-Venant. This allows the *resultant* force and moment of the loading to be set arbitrarily. But no conditions on the tube ends prescribing the *distribution* of the load and certainly no conditions of fastening or stiffening (ribs, flanges) of the edges can be considered in this way. On the other hand, these factors may influence the stressed state decisively. Consider now a more general problem of tube deformation, taking account of concrete conditions on the tube ends—prescribed fastening or stiffening and given distribution of load. The more general approach will allow also the problem of bifurcation buckling to be considered.

6. Tubes with given conditions on the edges

Curved tubes loaded at the ends are considered. The basic stress state, slowly varying along the tube, is described by the semi-momentless theory. The excluded edge-effect deformation can be superimposed.

The ordinary Fourier-series solution of the linear problem is discussed in a general form and later specialized to obtain approximate formulas for the cases of moderate curvature. A different solution, more effective for larger values of μ and b/R_m, is developed in present chapter, §7.

6.1. Conditions on tube ends

There are two boundary conditions to be imposed on the edge $\xi = \text{const.}$ for the semi-momentless stress state (Chapter 2, §4.4).

On the edge loaded by given external *forces* the boundary conditions can be imposed on the stress resultants $T_{(1)}$ and $S_{(1)}$ defined in (2.112). The conditions imposed on *deformation* of an edge $\xi = \text{const.}$

must be stated in terms of the strain resultants $\kappa_{(2)}$ and $\tau_{(2)}$ defined in (2.113).

For circular tubes and linear approximation, formulas (2.112) and (2.113) become substantially simpler. We express in (2.112) and (2.113) the resultants Q_1 and λ_2 in terms of H and γ with the aid of the fifth of equations (1.56) and (1.36), respectively. After omitting the terms with M_1 and ε_2 (Chapter 2, §4.4) and expressing S_1 and τ_2 in terms of S, H and τ, γ the four formulas are reduced to

$$T_{(1)} = T_1 , \quad S_{(1)} = S - 2U_2 \frac{H}{b} , \quad U_2 \equiv -\frac{1}{2}\frac{\partial}{\partial\eta}\frac{1}{r^2}\frac{\partial}{\partial\eta}r^2 - \frac{1}{2}\frac{\partial^2}{\partial\eta^2} - 1 , \tag{3.103}$$

$$\kappa_{(2)} = \kappa_2 , \quad \tau_{(2)} = \tau + U_2\gamma/b .$$

The four resultants of edge stress and strain can be expressed in terms of the four dependent variables constituting in what follows the resolving system of equations. Using the elasticity equations (2.101) we eliminate H and γ and write the edge resultants (3.103) in the nondimensional form

$$T , \quad S^\circ - k_1 U_2 \tau^\circ ; \quad \kappa , \quad \tau^\circ + k_2 U_2 S^\circ ; \tag{3.104}$$

where

$$T = \frac{T_1}{Ehh^\circ} , \quad S^\circ = \frac{S}{Ehh^\circ\sqrt{h^\circ}} , \quad \kappa = \kappa_2 b , \quad \tau^\circ = \frac{\tau b}{\sqrt{h^\circ}} , \tag{3.105}$$

$$k_1 = h^\circ 2(1 - \nu) , \quad k_2 = h^\circ 2(1 + \nu) ,$$

or, for more general, orthotropic, elastic properties,

$$T = T_1 B_1' \frac{1}{h^\circ} , \quad S^\circ = SB_1' \frac{1}{h^\circ\sqrt{h^\circ}} , \quad h^\circ = \sqrt{D_2 B_1'}\frac{1}{b} , \tag{3.106}$$

$$k_1 = h^\circ \frac{4D_G}{D_2} , \quad k_2 = h^\circ \frac{B_G'}{B_1'} .$$

Consider three cases of conditions on tube edges, which are of more interest.

(a) An edge is reinforced by a *"thin" flange*. Deformation of the edge in its own plane—the end cross-section of the tube—is prevented. But the flange cannot resist the warping of the edge *out* of its plane. It transmits the external longitudinal forces T_{1B} to the edge. The conditions on the edge are

$$T_1 = T_{1B}, \qquad \kappa_2 = 0.$$

In particular, the case of the simplest distribution of the applied longitudinal forces T_{1B} varying *linearly* with the distance x from the axis of the cross-section, is of practical interest. It represents the situation at a tube end with the flange and elastic washer. These conditions can be written in the form (Fig. 31)

$$T_1 = \frac{N}{2\pi b} + \frac{M}{\pi b^2} \cos \eta, \quad \kappa_2 = 0 \quad \text{or} \quad T = N^\circ + M^\circ \cos \eta, \quad \kappa = 0,$$

$$(3.107)$$

where M and N are the moment and the longitudinal component of the resultant of all forces applied at the edge.

(b) A *"stiff" flange* prevents any deformation of the edge. This is expressed by vanishing of both strain parameters κ_2 and $\tau_{(2)}$. In terms of the dimensionless resultants (3.104) these edge conditions are

$$\kappa = 0, \qquad \tau^\circ + k_2 U_2 S^\circ = 0. \qquad (3.108)$$

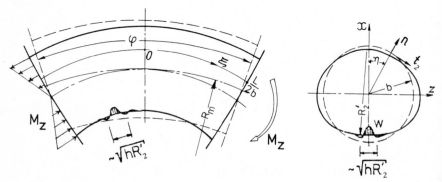

FIG. 31. Bending of tube with (thin) flanges. A local buckling mode.

(c) An edge is connected to that of another tube which may differ in curvature, wall-thickness, etc. The boundary conditions express the *continuity* of the inner forces and strain through equality of the four parameters (3.103) on both sides of the dividing (boundary) line $\xi = \text{const.}$

6.2. Ordinary differential equations for tube bending

The Fourier-series solution discussed in Chapter 2, §8 allows a reduction of the semi-momentless equations of the tube problem to a system of ordinary differential equations. Let the tube be deformed symmetrically to the plane of its axis. Measuring the coordinate η from the symmetry plane as shown in Fig. 31, we have the Fourier-series expansion of the governing functions in the form

$$T \equiv T_1/(Ehh^\circ) = N^\circ + M^\circ \cos \eta + \sum_{2,3,\ldots} g_n(\xi) \cos n\eta ,$$

$$\kappa \equiv \kappa_2 b = \sum_{2,3,\ldots} f_n(\xi) \cos n\eta . \tag{3.109}$$

The coefficients of these expansions—the functions $g_n(\xi)$ and $f_n(\xi)$—are determined by the system of ordinary differential equations (2.194). Consider this system in more detail. Besides the diagonal matrix W, the system is shaped by the matrix L. We write out this matrix with s_+ from (3.23) and c_+, Λ and Λ_+^2 according to (2.148) and (2.156)

$$L = Ed^2 - \frac{\mu}{2}
\begin{bmatrix}
0 & 0 & 0 & 0 & 0 & \cdots \\
2 & 0 & -1 & 0 & 0 & \cdots \\
0 & 2 & 0 & -2 & 0 & \cdots \\
0 & 0 & 3 & 0 & -3 & \cdots \\
\cdots & \cdots & \cdots & \cdots & \cdots & \cdots
\end{bmatrix}
- \frac{\mu}{2}
\begin{bmatrix}
0 & 0 & 0 & 0 & 0 & \cdots \\
2 & 0 & 1 & 0 & 0 & \cdots \\
0 & 4 & 0 & 4 & 0 & \cdots \\
0 & 0 & 9 & 0 & 9 & \cdots \\
\cdots & \cdots & \cdots & \cdots & \cdots & \cdots
\end{bmatrix},$$

$$c_+ = \frac{1}{2}\begin{bmatrix} 0 & 1 & 0 & 0 & \cdots \\ 2 & 0 & 1 & 0 & \cdots \\ 0 & 1 & 0 & 1 & \cdots \\ 0 & 0 & 1 & 0 & \cdots \\ \cdots & \cdots & \cdots & \cdots & \cdots \end{bmatrix}; \qquad d^2 = \frac{d^2}{h^\circ \, d\xi^2}.$$

As indicated in Chapter 2, §8.4, the equations of the system (2.194), which correspond to the rows numbered 0 and 1 of the matrices W and L, are satisfied identically and can be omitted. We replace g, f and L, W, respectively, by the column matrices $\check{g}$ and $\check{f}$—lacking the elements number zero and number one—and by square matrices $\check{W}$ and $\check{L}$ lacking both the rows and columns with these numbers.

With the new notation, the system (2.194) may be rewritten for pure plane bending (when $N = 0$ and $d^2 M^\circ = 0$) in the form

$$\begin{bmatrix} \check{W} & -\check{L} \\ \check{L} & \check{W} \end{bmatrix}\begin{bmatrix} \check{f} \\ \check{g} \end{bmatrix} = \begin{bmatrix} -3\mu M^\circ \\ 0 \\ 0 \\ \cdots \end{bmatrix}, \qquad (3.110)$$

where the matrices and their nonzero elements are

$$[\check{g} \ \check{f}] = \begin{bmatrix} g_2 & f_2 \\ g_3 & f_3 \\ \cdots & \cdots \end{bmatrix}, \qquad \check{W} = \begin{bmatrix} W_{22} & 0 & \cdots \\ 0 & W_{33} & \cdots \\ \cdots & \cdots & \cdots \end{bmatrix},$$

$$\check{L} = \begin{bmatrix} d^2 & L_{23} & \cdots \\ L_{32} & d^2 & \cdots \\ \cdots & \cdots & \cdots \end{bmatrix}, \qquad (3.111)$$

$$W_{jj} = j^4 - j^2, \quad L_{jj} = d^2, \quad L_{j \cdot j \pm 1} = (\pm j - j^2)\mu/2.$$

The diagonal matrix $\check{W}$ is easily inverted and the system (3.110) solved for $\check{g}$. This yields the system

$$\check{g} = -\check{W}^{-1}\check{L}\check{f},$$

$$D\check{f} = -\frac{\mu}{4}M^\circ \begin{bmatrix} 1 \\ 0 \\ \cdots \end{bmatrix}, \quad D = E + (\check{W}^{-1}\check{L})^2. \qquad (3.112)$$

The matrices $\check{W}^{-1}\check{L}$ and D are symmetric.

It is, obviously, not difficult to write out simple formulas for the elements of the matrices of equations (3.112).

Inserting the series for $g(\xi, \eta)$ defined by (3.112) into formula (3.109) we have tne expression for T in terms of $f_n(\xi)$,

$$T = N^\circ + M^\circ \cos \eta$$

$$- \sum_{n=2,3,\ldots} \left[\frac{d^2}{n^3 - n} f_n - \frac{\mu/2}{n-1} f_{n-1} - \frac{\mu/2}{n+1} f_{n+1} \right] \frac{\cos n\eta}{n}. \qquad (3.113)$$

The equations of the boundary problem have, of course, as their special limiting case, the corresponding relations of the Saint-Venant problem. Indeed, in the part of the tube sufficiently distant from the ends the deformation is nearly constant with respect to the longitudinal coordinate ξ. The quantities $f_n(\xi)$ are varying so insignificantly that their derivatives (the terms $d^2 f_n$) may be neglected in equations (3.110) and (3.112). The equations are then algebraic and determine the deformation of the middle part of a long tube to be identical to the solution of present chapter, §2. This becomes apparent if we note that for $f_n = $ const. the terms $D_{i1} f_1, D_{i3} f_3, \ldots$ of the even-numbered ($i = 2, 4, \ldots$) equations (3.112) disappear and the remaining odd-numbered equations ($i = 3, 5, \ldots$) give $f_3, f_5, \ldots = 0$. Furthermore, according to (2.9) and (3.21) we have

$$\kappa = \kappa_2 b = \frac{d\vartheta}{d\eta}, \qquad f_n = na_n, \qquad (3.114)$$

making the quantities f_n/n identical in their geometric meaning to the a_n of present chapter, §2. Finally, we have according to (3.30) and (3.24) the expression of M° in terms of m,

$$M^\circ = b_1 = m + \tfrac{1}{2}\mu a_2. \qquad (3.115)$$

Introducing this and the relation (3.114), in the form $f = \{2a_2 \ 0 \ 4a_4 \ \cdots\}$, into equations (3.112) and omission of the identically satisfied equations transforms (3.112) to a system equivalent to (3.25).

6.3. Equations encompassing γ, H and $R \neq R_m$

For shorter tubes or tubes made of orthotropic material the shear strain may be of substantial influence. Consider application of the corresponding extended theory of Chapter 2, §4.4. The basic equations follow from the linearized equations (1.36) and (1.59) after dropping the ε_2 and M_1 terms. Specialized for circular tubes (Fig. 31) of constant curvature

$$R_2 = b = \text{const.} , \quad a = br , \quad r = \frac{R}{R_m} ;$$

$$\frac{1}{R_1} = \frac{c}{R} , \quad \frac{1}{\rho_1} = \frac{s}{R} ; \quad c = \cos \eta , \quad s = \sin \eta ,$$

the equations become

$$T_{1,1} + US + \frac{1}{cR_m}(c^2 H)_{,2} + \frac{1}{R_1} UH = 0 ,$$

$$S_{,1} + (rT_2)_{,2} + \frac{1}{b} H_{,1} + \frac{b}{R_m} sT_1 + rQ_2 = 0 ,$$

$$rT_2 = \frac{1}{rb} UH_{,1} + (rQ_2)_{,2} - \frac{b}{R_m} cT_1 ,$$

$$brQ_2 = (rM_2)_{,2} + H_{,1} , \quad U = \frac{1}{r} \frac{\partial}{\partial \eta} r^2 ;$$

$$\kappa_{2,1} - U\tau + \frac{1}{cR_m}\left(c^2 \frac{\gamma}{2}\right)_{,2} - \frac{1}{R_1} U \frac{\gamma}{2} = 0 ,$$

$$\tau_{,1} - (r\kappa_1)_{,2} - \frac{1}{2b} \gamma_{,1} - \frac{b}{R_m} s\kappa_2 - r\lambda_1 = 0 ,$$

$$r\kappa_1 = \frac{1}{2rb} U\gamma_{,1} + (r\lambda_1)_{,2} - \frac{b}{R_m} c\kappa_2 ,$$

$$br\lambda_1 = -(r\varepsilon_1)_{,2} + \frac{1}{2} \gamma_{,1} , \quad \left(\lambda_2 = \frac{-1}{2rb} U\gamma\right) .$$

(3.116)

The operator U will be used in the following.

The distributed-load terms are dropped in (3.116). They can easily be recovered.

Together with the elasticity relations (2.98), equations (3.116) constitute a full system. After elimination of all variables, except those four that describe the stress state and deformation of the section $\xi =$ const., the resolving system can be presented in the following non-dimensional form

$$\begin{bmatrix} d & 0 & U & k_1 U_1 \\ 0 & d & k_2 U_1 & -U \\ K & V & d & -k_1 U_2 d \\ V & -K & k_2 U_2 d & d \end{bmatrix} \begin{bmatrix} T \\ \kappa \\ S^\circ \\ \tau^\circ \end{bmatrix} = 0 , \quad d = \frac{1}{\sqrt{h^\circ}} \frac{\partial}{\partial \xi} ; \tag{3.117}$$

where, besides the variables and operators defined in (3.103), (3.105), (3.106) and (3.116) we use the operators

$$U_1 = \frac{b}{2R_m} \left(\frac{c}{r} U + c \partial_2 - 2s \right) ,$$

$$K = -\mu(\partial_2 c - s) , \quad V = (\partial_2^3 + \partial_2)r , \quad \partial_2 = \partial/\partial\eta .$$

The solution will be sought in the trigonometric-series form (3.109), which implies

$$S^\circ = \sum_1^m S_n^\circ \sin n\eta , \quad \tau^\circ = \sum_1^m \tau_n^\circ \sin n\eta . \tag{3.118}$$

Inserting the series of the four variables into the system (3.117) we obtain (as described in Chapter 2, §7) a system of ordinary differential equations for the coefficients $g_n(\xi)$, $f_n(\xi)$, $S_n^\circ(\xi)$ and $\tau_n^\circ(\xi)$. This system can be directly stated in the matrix form. We need only to replace in (3.117) the dependent variables by the column matrices of their Fourier coefficients $T = \{N^\circ \ M^\circ \ g_2 \ g_3 \ \cdots\}$, κ, S, τ and the operators d, U, U_1, U_2, K and V by the corresponding rectangular matrices (Chapter 2, §§7, 8):

$$PX \equiv \begin{bmatrix} Ed & 0 & U & k_1 U_1 \\ 0 & Ed & k_2 U_1 & -U \\ K & V & Ed & -k_1 U_2 d \\ V & -K & k_2 U_2 d & Ed \end{bmatrix} \begin{bmatrix} T \\ \kappa \\ S \\ \tau \end{bmatrix} = 0 ,$$

$$\tag{3.119}$$

$$U = r_+^{-1} \Lambda r_- r_- , \quad U_1 = \frac{b}{2R_m} (c_+ r_+^{-1} U - c_+ \Lambda - 2s_-) ,$$

$$K = \mu(\Lambda_+ c_+ + s_+) , \quad U_2 = \tfrac{1}{2} \Lambda r_+^{-1} U + \tfrac{1}{2} \Lambda^2 - E , \quad V = (\Lambda^2 - E)\Lambda_+ r_+ .$$

For slender tubes, when $r \approx 1$, the equations are drastically simplified: the U_1 elements disappear, $U = \Lambda$, $U_2 = \Lambda^2 - E$ and $V = (\Lambda^2 - E)\Lambda_+$.

The system can be profitably contracted (similarly to (3.110)). Indeed, the quantities $N°$, $S_1°$ and $M°$ represent the normal and transverse resultant forces and respectively the moment of all stresses acting on the cross-section of the tube. These quantities are known (for pure flexure $N°$, $S_1 = 0$ and $M° = $ const.) or can be determined with the help of beam statics. According to the continuity conditions, f_0, f_1 and $\tau_1°$ are for a tube equal to zero (cf. Chapter 2, §6.2). This allows us to simplify the system by freeing the matrices of dependent variables of six known elements. It is achieved by transferring the corresponding terms of the equations to the right-hand sides and dropping those (six) equations (those lines in the matrix P) that correspond to the known elements of X. We denote the so-abridged matrices by the sign $\check{}$ and rewrite the contracted system in the form

$$\check{P}\check{X} = R, \quad \check{X} = \{\check{g} \ \check{f} \ \check{S} \ \check{\tau}\}, \quad R = \{0 \ 0 \ -M°k \ -M°v\}. \quad (3.120)$$

The matrix $\check{P}$ is obtained from P by replacing U, U_1, U_2, V and K with the respective matrices $\check{U}, \ldots, \check{K}$ containing only the lines and columns numbered in the initial matrices $2, 3, \ldots, m$. Matrices k and v are obtained from the columns number one in K and V (i.e. from $K(0: m, 1)$ and $V(0: m, 1)$) by dropping the elements $K(0, 1)$, $K(1, 1)$, $V(0, 1)$ and $V(1, 1)$. Matrices $\check{g}$ and $\check{f}$ are displayed in (3.111), $\check{S} = \{S_2° S_3° \cdots\}$ and $\check{\tau} = \{\tau_2° \tau_3° \cdots\}$. Equations (3.120) are further transformed and solved in present chapter, §7.

6.4. Simplest solution

The linear equations (3.110) and (3.112) and the analogous equations (3.119) have constant coefficients. Such equations have the well-known Euler-form solution: the dependent variables $f_n(\xi), \ldots$ are determined by a linear combination of functions $\exp(\xi\sqrt{h°}d_k)$, where d_k are the appropriate eigenvalues (cf. [104, 176]). The eigenvalues and the relevant eigenvectors, composed of coefficients by $\cos n\eta$, can be effectively computed with the aid of the "LR transformation" (described, for instance, in STOER and BULIRSCH, *Einführung in die Numerische Mathematik II*, Springer, Heidelberg, 1973) of the system matrix.

Alternatively, the (rather simple) equations can be integrated numerically. The computations of the shooting method (present chapter, §8) require in these cases not much more time than those realizing the exponential-form solution. The respective programs run on a personal computer. A more straightforward solution, one using double Fourier-series, is discussed in present chapter, §7.

There is, however, a large class of tube bends allowing a *closed-form solution*. The resulting explicit formulas are useful for routine design and as a framework for parameter studies. This class encompasses tubes designed for high pressure or high reliability. Having comparatively thick walls these tubes satisfy the condition

$$R_m h/b^2 > 0.4 \quad \text{or} \quad \mu < 8 .$$

For $\mu < 8$ we can retain in the series of κ merely one term $f_2 \cos 2\eta$. This means retaining in (3.112) only one unknown and only the terms $g_n \cos n\eta$ with $n \le 3$ in the expansion (3.109) of T. The matrix equation (3.112) reduces to

$$D_{22} f_2 \equiv \left(1 + \frac{d^4}{144} + \frac{\mu^2}{144}\right) f_2(\xi) = -\tfrac{1}{4}\mu M^\circ , \quad d = \frac{1}{\sqrt{h^\circ}} \frac{\mathrm{d}}{\mathrm{d}\xi} .$$

The general integral of this equation has the form

$$f_2(\xi) = 2a_2 + \sum_1^4 C_j \exp(\xi d_2 \sqrt{h^\circ}) , \quad 2a_2 = -\tfrac{1}{4}\mu M^\circ \Big/ \left(1 + \frac{\mu^2}{144}\right) .$$
$$(3.121)$$

The quantity d_2 is the root of the auxiliary equation

$$d_2^4 + 144 + \mu^2 = 0 , \quad d_2 = \pm(1 \pm i)\alpha_2 , \quad \alpha_2 = \left(\frac{144 + \mu^2}{4}\right)^{1/4} .$$
$$(3.122)$$

Applicability of the simplest expression $\kappa = f_2(\xi) \cos 2\eta$ with f_2 of the form (3.121) for somewhat higher values of μ can be achieved with the quantities a_2 and d_2 determined by a system (3.112) with a sufficient number of equations retained. The expression $f_2 \cos 2\eta$ can be interpreted as the dominant term of the eigenfunction corresponding to the lowest eigenvalue d_2.

The sum of the four exponent terms of (3.121) may be expressed as a linear combination of four real functions—Kryloff functions defined in Table 4. Retaining the notation C_j for the (now real) integration constants, we have

$$f_2 = 2a_2 + \sum_1^4 C_j K_j(\alpha_2 \xi \sqrt{h^\circ}) . \tag{3.123}$$

The four constants C_j are determined by conditions on the edges.

The stress state and deformation are explicitly expressed in terms of $\kappa = f_2 \cos 2\eta$ with the aid of the formulas discussed in the foregoing.

TABLE 4.

j	$K_j(x)$	$\frac{d}{dx}K_j$	$\frac{d^2}{dx^2}K_j$	$\frac{d^3}{dx^3}K_j$	$\frac{d^4}{dx^4}K_j$
1	$\cosh x \cos x$	$-4K_4$	$-4K_3$	$-4K_2$	$-4K_1$
2	$\frac{1}{2}(\cosh x \sin x + \sinh x \cos x)$	K_1	$-4K_4$	$-4K_3$	$-4K_2$
3	$\frac{1}{2}\sinh x \sin x$	K_2	K_1	$-4K_4$	$-4K_3$
4	$\frac{1}{4}(\cosh x \sin x - \sinh x \cos x)$	K_3	K_2	K_1	$-4K_4$

6.5. Displacements. Angle of flexure

The *small* elastic displacements will be presented as a sum of two parts. The first is determined by the beam formulas of the Strength of Materials, which disregard the deformation of the cross-sections of the tube and the corresponding Kármán effect. The second part is constituted by elastic displacements connected with the deformation of the cross-sections of a curved tube bent in the symmetry plane (Fig. 31).

The components of this second part of elastic displacement and the corresponding angle of rotation (denoted u, v, w and $\vartheta(\xi, \eta)$) are determined in terms of the resolving function κ by (2.125). With Fourier expansions the formulas yield (for $R_2 = a = b = \text{const.}$),

$$w = \sum_n w_n(\xi) \cos n\eta , \qquad v = \sum_n v_n(\xi) \sin n\eta = - \sum_n \frac{1}{n} w_n \sin n\eta , \tag{3.124}$$

$$u = u_0(\xi) + \sum_n u_n(\xi) \cos n\eta = u_0 - \sum_n \frac{dw_n}{d\xi} \frac{\cos n\eta}{n^2}, \tag{3.125}$$

$$\frac{w_n}{b} = \frac{f_n}{n^2 - 1} \qquad \left(\kappa = \sum_n f_n(\xi) \cos n\eta \right), \tag{3.126}$$

where a natural notation for the Fourier coefficients is introduced.

In (3.124)–(3.126)

$$f_0 = 0, \qquad f_1 = 0, \qquad w_0 = 0. \tag{3.127}$$

The last relation follows from the condition of inextensibility of the profile (2.124).

Taking into account the symmetry of deformation with respect to the plane $\eta = 0$ (Fig. 31), the constant of integration in the expression (2.125) for v is set equal to zero.

The first harmonics $w_1 \cos \eta$, $v_1 \sin \eta$, $u_1 \cos \eta$ describe the rigid-body movement of the cross-section. Indeed, projecting the vector of displacement in the plane of a cross-section $t_2 v_1 \sin \eta + n w_1 \cos \eta$ on the axes x and z (Fig. 31) and taking into account the relation $v_1 = -w_1$ from (3.124) we find $u_x = w_1$ and $u_z = 0$. These displacements of all points of a cross-section are equal. Further, we note that the term $u_1 \cos \eta$ of the longitudinal displacement u represents a rigid-body rotation of a cross-section around the z-axis (Fig. 31). The angle of this rotation is equal to $u_1(\xi)/b$.

The harmonics $u_1 \cos \eta$, $w_1 \cos \eta$ and $v_1 \sin \eta$ thus characterize the deformation of the "tube axis", caused by the deformation of the cross-sections. Equations (3.124)–(3.126) do not suffice to express these harmonics in terms of f_n directly. We obtain these relations in another way: by comparing the expression of the longitudinal stress resultant T_1 in terms of κ, presented in (3.113), with its expression in terms of the elastic displacements. Writing $T_1 = Eh\varepsilon_1$ with the extension ε_1 determined as a sum of a beam deformation and the contribution to ε_1 of the elastic displacements, determined according to (3.124)–(3.126) and (1.40), renders

$$T_1 \frac{1}{Eh} = \frac{N}{2\pi bEh} + \frac{M_z}{EI} b \cos \eta + \frac{du_0}{b\, d\xi}$$

$$- \sum_{n=1,2,\ldots} \left(\frac{1}{bn^2} \frac{d^2 w_n}{d\xi^2} - \frac{w_{n+1}}{2R_m} \frac{n+2}{n+1} - \frac{w_{n-1}}{2R_m} \frac{n-2}{n-1} \right) \cos n\eta .$$

Equating this with T_1 as obtained from (3.113) yields

$$\frac{d^2 w_1}{b\, d\xi^2} - \frac{w_2}{2R_m} \frac{3}{2} = 0 . \tag{3.128}$$

Hence, with $w_2 = bf_2/3$ and $u_1 = -dw_1/d\xi$ from (3.125),

$$\frac{d^2 w_1}{d\xi^2} = \frac{b^2}{4R_m} f_2(\xi), \qquad u_1 = -\frac{dw_1}{d\xi} = -\frac{b^2}{4R_m}\int f_2\, d\xi. \quad (3.129)$$

Now adding to the angle of flexure u_1/b, connected with the deformation of the cross-section, the angle $M_z L/EI$, contributed by the beam-type bending, yields the rotation of one end of the tube ($\xi = \xi_2$) in relation to another ($\xi = \xi_1$)

$$\varphi^* - \varphi = \frac{M_z L}{EI} + \left[\frac{u_1}{b}\right]_{\xi_1}^{\xi_2} = \frac{M_z L}{EI} - \frac{b}{4R_m}\int_{\xi_1}^{\xi_2} f_2\, d\xi. \quad (3.130)$$

Consider now the angle of rotation $\vartheta(\xi, \eta)$ at *any* point of the tube and derive the formula for the flexure angle of the tube elbow avoiding the use of displacements and taking into account the shear strain γ.

The angular displacement $\vartheta(\xi, \eta)$ is expressed in terms of the change of curvature of the reference surface by (1.39) and (1.25),

$$\frac{\partial \vartheta}{a\, \partial \xi} = -t_1 \tau_1 + t_2 \kappa_1 + n\lambda_1. \quad (3.131)$$

Since the variables τ_1, κ_1 and λ_1 can be expressed in terms of the function κ_2, the integration of (3.131) provides the corresponding expression for ϑ. In particular, the relevant component of ϑ, the angle ϑ_z of rotation in the plane of bending, is given by

$$\vartheta_z = \int (\tau_1 \cdot 0 + \kappa_1 \cos \alpha + \lambda_1 \sin \alpha) a\, d\xi,$$

where the relations (2.42) between the unit vectors of the tangents to the coordinate lines have been taken into account. Inserting here the expressions for κ_1 and λ_1 from (3.116) gives

$$\vartheta_z = -h° \int_{\xi_1}^{\xi} (\mu\kappa \cos^2 \eta + T_{,22} \cos \eta + T_{,2} \sin \eta)\, d\xi$$

$$+ \frac{1}{2}\left[\frac{c}{r} U\gamma + \gamma_{,2} \cos \eta + \gamma \sin \eta\right]_{\xi_1}^{\xi}. \quad (3.132)$$

The γ terms have the order of magnitude of the strain γ and can be omitted while determining the angle of flexure of the tube.

When, as usual, the ends of the tube elbow may deform, the angle $\varphi^* - \varphi$ of the flexure of the tube can only represent some *average* (over each of the edges) value of the relative rotation. The corresponding formula follows from (3.132), when solely the constant term of the $\cos n\eta$-series of the relative rotation angle ϑ_z is considered. The formula so derived coincides with (3.130).

Formula (3.130) for the angle of flexure may be presented in a form more convenient for calculations. The starting point for this is provided by the relations (3.114). For long tubes, $f_2 = 2a_2$. Expressing a_2 in terms of the stiffness-reduction factor of a *long* tube, K, using (3.115) and (3.30), we have

$$\tfrac{1}{2} \mu a_2 = M^\circ \left(1 - \frac{1}{K}\right). \qquad (3.133)$$

Now, (3.130) may be rewritten introducing a flexibility factor F of the tube elbow, expressed in terms of K,

$$\varphi^* - \varphi = F\frac{ML}{EI} \qquad (L = \varphi R_m), \qquad (3.134)$$

$$F = 1 + \left(\frac{1}{K} - 1\right)\delta, \quad \delta = \frac{b}{L} \int_{\xi_1}^{\xi_2} \frac{f_2(\xi)}{2a_2} \, d\xi. \qquad (3.135)$$

In the limit case of a very long tube ($b/L = 0$) the deformation of the tube away from the end zones is identical to that in the Saint-Venant problem. For this case, formula (3.135) gives $F = 1/K$.

The ordinary-series solution with only *one* term retained in the expansion of κ ($\kappa = f_2(\xi) \cos 2\eta$) is elementary simple and provides a good approximation for stress and strain in *small-curvature* tubes.

Formula (3.135) with the f_2 from (3.123) gives a simple estimate for the bending stiffness of tubes. When conditions on both ends of a tube are similar, only the two even Krylov functions K_1 and K_3 (cf. Table 4) remain in f_2. With the expression (3.123), formula (3.135) becomes

$$F = 1 + \left(\frac{1}{K} - 1\right)\delta(l^\circ),$$

$$\delta(l^\circ) = 1 + \frac{1}{l^\circ}[C_1 K_2(l^\circ) + C_3 K_4(l^\circ)], \quad l^\circ = \frac{La_2}{2b}\sqrt{h^\circ}. \qquad (3.136)$$

The value of α_2 from (3.122) and the expressions (3.123) and (3.136) remain approximately valid for the analysis taking into account H and γ. The eigenvalues remain very nearly the same. The constants C_1 and C_3 are determined by the conditions on the tube ends. For *thin flanges* we have the conditions (3.107). The series form of the variables T and κ (3.109) and the expression (3.113) reduce (3.107) to

$$f_n = 0, \qquad f_{n,\xi\xi} = 0 \quad (n = 2, 3, \ldots). \tag{3.137}$$

With C_1 and C_3 determined by inserting (3.123) into (3.137), formula (3.136) gives for thin flanges

$$\delta = 1 - \frac{1}{2l^\circ}\frac{\sinh 2l^\circ + \sin 2l^\circ}{\cosh 2l^\circ + \cos 2l^\circ}, \quad l^\circ = \left(\frac{144 + \mu^2}{4}\right)^{1/4} \frac{L}{2b} \sqrt{h^\circ}. \tag{3.138}$$

For *stiff flanges* the edge conditions are less simple. Applied to the series (3.109) and (3.118) the conditions (3.108) mean

$$f = 0, \qquad \check{\tau}^\circ + k_2 U_2 S^\circ = 0.$$

To eliminate τ° and S° we multiply the second condition by U and insert into it the expressions $\check{U}\check{\tau}^\circ = d\check{f}$ and $\check{S}^\circ = -U\,d\check{g}$, which follow from (3.119) and (3.120) (without the U_1 terms). The two conditions become

$$\check{f} = 0, \qquad \frac{d}{d\xi}(\check{f} - k_2\check{U}_3\check{g}) = 0, \qquad \check{U}_3 = \check{U}\check{U}_2\check{U}^{-1}. \tag{3.139}$$

For slender tubes, when $r \cong 1$, one easily finds $\check{U}_3 = \Lambda^2 - E$ and the conditions (3.139) are equivalent to

$$f_n = 0, \qquad \frac{d}{d\xi}[f_n - k_2(n^2 - 1)g_n] = 0 \quad (n = 2, 3, \ldots). \tag{3.140}$$

The *one-term solution* (3.123) is determined by the $n = 2$ conditions (3.140). With $g_2 = -d^2 f_2/(2^4 - 2^2)$, following from (3.109) and (3.113), the relevant conditions (3.140) become $(df_2 = (df_2/d\xi)/\sqrt{h^\circ})$

$$f_2 = 0, \qquad df_2 + k_2\,d^3 f_2/4 = 0. \tag{3.141}$$

In the limit case $h/b \to 0$, or if the theory disregarding γ and H is used, the k_2 term of (3.141) disappears. With the corresponding values of C_1 and C_2, formula (3.136) renders for *stiff flanges*

$$\delta = 1 - \frac{1}{l^\circ} \frac{\cosh 2l^\circ - \cos 2l^\circ}{\sinh 2l^\circ + \sin 2l^\circ}. \tag{3.142}$$

The functions $\delta(l^\circ)$ are presented in Fig. 32.

The factor δ has a specific mechanical meaning. Introduced in (3.134) as $\delta = (F-1)/(K^{-1}-1)$ it indicates the part of the flexibility contribution of the Kármán effect, which is retained for the tube with end stiffening.

For longer tubes, Fig. 32 suggests asymptotic formulas: in the case of stiff flanges $\delta = 1 - 1/l^\circ$, for the thin flanges $\delta = 1 - 1/2l^\circ$. The quantities $1/l^\circ$ or $1/2l^\circ$ can be interpreted as a fraction of the tube's length, adjoining the flanges, where the cross-sections remain undeformed.

It must be borne in mind that (3.138) and (3.142) as well as the solution (3.123) with (3.141) are intended for $\mu < 8$. The accuracy may prove to be satisfactory also for any short tube, when $\delta \ll 1$ and has little influence on F, and for long tubes, when $\delta \cong 1$ and $F \cong 1/K$.

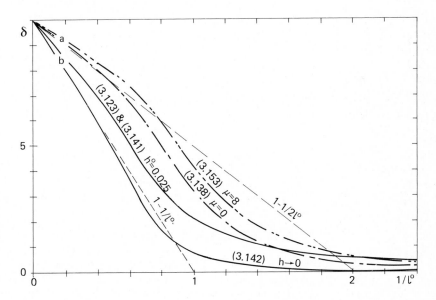

FIG. 32. Stiffening effect of end flanges for tubes of restricted curvature.

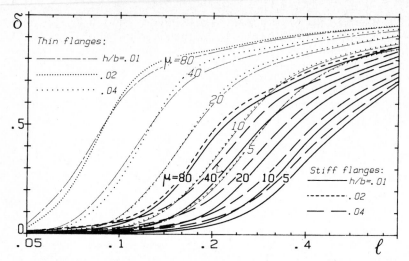

FIG. 33. Stiffening effect of end flanges of two limiting-case types.

The results of a more exact solution of §7, presented in Fig. 33,[14] allow a check of the approximate formulas and of the one-term solution itself.

7. Tubes of unrestricted curvature

The bending problem is considered with the aid of the extended semi-momentless theory. The results are checked against those of the full thin-shell equations. Discussion is restricted to the deformation symmetric with respect to both the plane of curvature of the tube and its midlength plane. The extension to unsymmetrical cases is direct.

7.1. Semi-momentless solution

The resolving system (3.120) can easily be reduced to the form appropriate for numeric integration,

$$d\check{X} = A\check{X} + B .$$
(3.143)

Indeed, with all the elements of the system matrix $\check{P}$, which include the

---

[14] The programming, computation and plotting for Figs. 33, 36, 40, 41, 42 and 44 are due to V.E. Axelrad.

differentiation, written explicitly the system (3.120) takes the form

$$(C + Dd)\check{X} = R \, . \tag{3.144}$$

The matrices C and D are directly formulated starting from (3.117) and (3.119),

$$C = \begin{bmatrix} 0 & 0 & \check{U} & k_1\check{U}_1 \\ 0 & 0 & k_2\check{U}_1 & -\check{U} \\ \check{K} & \check{V} & 0 & 0 \\ \check{V} & -\check{K} & 0 & 0 \end{bmatrix}, \qquad D = \begin{bmatrix} E & 0 & 0 & 0 \\ 0 & E & 0 & 0 \\ 0 & 0 & E & -k_1\check{U}_2 \\ 0 & 0 & k_2\check{U}_2 & E \end{bmatrix}. \tag{3.145}$$

The matrices $\check{U}, \check{U}_1, \dots, E$ have dimensions $(m-2) \times (m-2)$ with m defined in (3.109) and (3.118). Elements of these matrices and thus of C and D are constants. The system (3.143) is obtained by multiplying (3.144) by the matrix D^{-1},

$$A = -D^{-1}C \, , \qquad B = D^{-1}R \, . \tag{3.146}$$

The numerical integration of the system (3.143), e.g. with the aid of shooting method, is (in contrast with the full shell equations) a very effective operation. Owing to the basic features of the semi-momentless theory (Chapter 2, §4), computation of the eigenvalues and eigenvectors of the system (3.143) is also comparatively simple.

A still more transparent solution effectively realized even by personal computers is provided by applying to the system (3.143) the Fourier-series method. We first eliminate the unknown $\check{S}$ and $\check{\tau}$ from the system (3.120) or (3.144). This leads to

$$M\begin{bmatrix} \check{f} \\ \check{g} \end{bmatrix} = \hat{R} \, , \qquad M = \begin{bmatrix} L & -W_2 \\ W_1 & L \end{bmatrix}, \qquad \hat{R} = M^\circ \begin{bmatrix} \check{U}v \\ -\check{U}k \end{bmatrix},$$
$$L = -Ed^2 + \check{U}\check{K} \, , \quad W_n = \check{U}\check{V} - k_n U_3 d^2 \quad (n = 1, 2) \, . \tag{3.147}$$

The matrices v, k and U_3 have been defined at the end of §6.3, present chapter, and in (3.139). The matrix $\hat{R}$ is written here in a form corresponding to pure bending. Matrix M is in (3.147) simplified by neglecting the $\check{U}_1$ terms in the original system (3.144) and (3.145). These terms stem from the H and γ additives in the first and the fifth of equations (3.116), respectively. Calculations show the influence of these terms to be insignificant.

Solution of the system (3.147) will be sought in the form

$$\begin{bmatrix} \check{f} \\ \check{g} \end{bmatrix} = \sum_{1,3,\ldots} \begin{bmatrix} f^j \\ g^j \end{bmatrix} \sin j\hat{\xi} + \begin{bmatrix} f^B \\ g^B \end{bmatrix}, \quad \hat{\xi} \equiv \frac{\pi b}{L}\xi . \tag{3.148}$$

The coordinate $\hat{\xi}$ has values 0 and π on the tube ends. The absence of the terms $j = 2, 4, \ldots$ reflects the symmetry with respect to the mid-length plane $\hat{\xi} = \pi/2$.

The additives f^B and g^B are necessary in (3.148) as the nonzero values of f and g on the edges, representing the boundary values $f^B(\eta)$ and $g^B(\eta)$ of $f(\xi, \eta)$ and $g(\xi, \eta)$.

$$[f^B \; g^B] = \sum_{n=2}^{m} [f_n^B \; g_n^B] \cos n\eta , \quad f^B = \{f_2^B \; f_3^B \; \cdots\} ,$$

$$g^B = \{g_2^B \; g_3^B \; \cdots\} .$$

The set of constants f_n^j and g_n^j determines the double Fourier series of the respective variables, constituted by (3.109) and (3.148),

$$\begin{bmatrix} T \\ \kappa \end{bmatrix} = \begin{bmatrix} N^\circ + M^\circ \cos\eta \\ 0 \end{bmatrix} + \sum_{n=2,3,\ldots} \left(\sum_j \begin{bmatrix} g_n^j \\ f_n^j \end{bmatrix} \sin j\hat{\xi} + \begin{bmatrix} g_n^B \\ f_n^B \end{bmatrix} \right) \cos n\eta .$$

$$\tag{3.149}$$

We introduce the expansions (3.148) together with the $\sin j\hat{\xi}$-expansions of the constant matrices f^B, g^B and R, obtained as product of each of them with

$$1 = \frac{4}{\pi} \sum_{1,3} \frac{1}{j} \sin j\hat{\xi} , \tag{3.150}$$

into equations (3.147). Equating the Fourier coefficients of both sides of each of equations (3.147) we obtain algebraic systems determining f^j and g^j separately for every j-value of $j = 1, 3, \ldots,$

$$M_j\begin{bmatrix} f^j \\ g^j \end{bmatrix} = \frac{4}{\pi j}\left(\hat{R} - M_0\begin{bmatrix} f^B \\ g^B \end{bmatrix}\right). \tag{3.151}$$

The matrix M_j is obtained from M by replacing the operator d^2 by $-j^2/l^2$, where l represents the length of the tube $(L = \varphi R_m)$,

$$l = \frac{L}{\pi b}\sqrt{h^\circ}. \tag{3.152}$$

Consider two boundary problems:

(a) Bending of a tube with "*thin*" *flanges* on the ends. The boundary conditions of this case (3.107) are obviously satisfied by the series (3.148) with f^B, $g^B = 0$. This solution is rendered directly by (3.151); denoting it with the subscript a we have

$$\begin{bmatrix} f_a^j \\ g_a^j \end{bmatrix} = \frac{4}{\pi j}M_j^{-1}\hat{R}. \tag{3.153}$$

(b) Bending of a tube with "*stiff*" *flanges*, defined by the edge conditions (3.139). The first condition, $f = 0$ by $\hat{\xi} = 0, \pi$, is satisfied by the solutions (3.109) and (3.148) with $f^B = 0$. The second condition, concerning the warping of the edge, determines the reactive forces (g^B) on the edge. To calculate g^B we insert into this condition the expressions of $f_{,\xi}$ and $g_{,\xi}$, following from (3.148) after differentiating and setting $\cos j\hat{\xi} = 1, -1$ (for $\hat{\xi} = 0, \pi$),

$$\sum_{1,3,\dots} (jf_j - k_2 U_3 jg_j) = 0. \tag{3.154}$$

The expressions of f_j and g_j in terms of g^B follow from (3.151)

$$\frac{\pi}{4}\sum_{1,3,\dots} j\begin{bmatrix} f^j \\ g^j \end{bmatrix} = \sum_{1,3,\dots} M_j^{-1}\hat{R} - \sum_{1,3,\dots} M_j^{-1}M_0\begin{bmatrix} 0 \\ g^B \end{bmatrix} \equiv \begin{bmatrix} b_1 \\ b_2 \end{bmatrix} - \begin{bmatrix} N_{12}g^B \\ N_{22}g^B \end{bmatrix} \tag{3.155}$$

with b_1, b_2 and N_{12}, N_{22} denoting the (matrix) elements of the respective matrices.

Substituting the expressions of $\sum jf^j$ and $\sum jg^j$ into (3.154) we obtain the equation for g^B,

$$b_1 - k_2 U_3 b_2 - (N_{12} - k_2 U_3 N_{22})g^B = 0.$$ (3.156)

With $f^B = 0$ and the edge tractions g^B known, (3.151) and (3.148) yield the Fourier coefficients of the resolving functions.

Results of the solutions just described are illustrated in Figs. 33 and 34, discussed in present chapter, §§7.3, 7.4.

The effectiveness of the solution and the clarity of the overall picture of the stress state provided by it are determined by the *convergence* of the Fourier series involved. The sin $j\hat{\xi}$-series (3.148) converges particularly favourably for the middle-length tubes, i.e. for the very class of tubes that requires the two-dimensional solution. For instance, for a tube with *thin* flanges and length $l = 0.5$ the sin $\hat{\xi}$ terms determine 95% of the stress and deformation in the midlength cross-section. For tubes with *stiff* flanges the sin $j\hat{\xi}$-series converge less intensively but also cause no difficulties.

The convergence of the cos $n\eta$-series for middle-length tubes is not much different from that of the long tubes. The number of cos $n\eta$ terms required can be estimated from (3.33).

The angle of flexure of a tube bend can be determined with the formula (3.130) also for the double Fourier-series solution. We need only to substitute into formula (3.130) the expression (3.148) $f_2 = \Sigma f_2^j \cos j\hat{\xi} + f_2^B$ and to set as limits of integration those coordinates

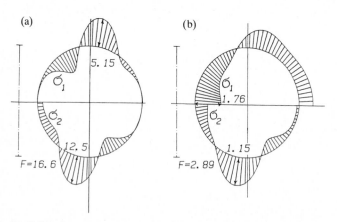

FIG. 34. Flexibility factor and stresses in the midlength cross-section of tubes with $h/b = 0.02$, $b/R_m = 1/3$, $\mu = 55.07$, $\varphi = 90°$. (a) Thin flanges. (b) Stiff flanges. Notation: $\sigma_1 = T_1/(h\sigma_B)$, $\sigma_2 = M_2 6/(h^2\sigma_B)$, $\sigma_B = M_2/(\pi b^2 h)$.

$\xi = 0$, L/b associated with the tube edges in the $\sin j\hat{\xi}$-series solution. This yields

$$\varphi^* - \varphi = F \frac{M_z L}{EI}, \quad F = 1 - \frac{\mu}{4M^\circ}\left(f_2^B + \frac{2}{\pi}\sum_{j=1,3,\ldots} \frac{1}{j}f_2^j\right). \quad (3.157)$$

Pure bending, investigated in this section, is a central but, of course not the single interesting case. The $\sin j\xi$-series solution can be directly extended to any conditions on the edges. The main difference to the foregoing consists in the $N^\circ(\xi)$ and $M^\circ(\xi)$: the two parameters must be determined by the relevant stress resultants in the section $\xi = \text{const.}$, to be found as for a curved beam. The solution can be carried out without the transformation of (3.119) to the form of (3.120) (as in present chapter, §8.2).

7.2. Solution of the complex shell equations

Consider now the problem of bending of a curved tube on the basis of the general thin-shell equations. This will provide the opportunity of testing the accuracy of the semi-momentless analysis of the problem. The check is essential for tube bends of middle and large curvature, which do not satisfy the (sufficient) condition (2.120) of applicability of the semi-momentless theory—that of $b \ll R_m$.

The discussion is concerned with circular tubes. With the characteristics of the local geometry of these shells (Fig. 31),

$$a = rb, \quad r = 1 + \lambda \cos \eta, \quad \lambda = \frac{b}{R_m}, \quad \frac{1}{R_1} = \frac{\cos \eta}{R_m}, \quad R_2 = b = \text{const.},$$
$$(3.158)$$

the complex shell equations (1.122) assume the form

$$\begin{bmatrix} (r + ih^\circ\lambda c)\dfrac{\partial}{\partial\xi} & ih^\circ\lambda c\dfrac{\partial}{\partial\xi} & \dfrac{\partial}{\partial\eta}r^2 \\[2ex] \lambda s + ih^\circ r\dfrac{\partial}{\partial\eta} & \dfrac{\partial}{\partial\eta}r + ih^\circ r\dfrac{\partial}{\partial\eta} & \dfrac{\partial}{\partial\xi} \\[2ex] \lambda c - ih^\circ\Delta & r - ih^\circ\Delta & 0 \end{bmatrix}\begin{bmatrix} \tilde{T}_1 \\[2ex] \tilde{T}_2 \\[2ex] \tilde{S} \end{bmatrix} = \begin{bmatrix} -r^2bq_1 \\[2ex] -rbq_2 \\[2ex] rbq \end{bmatrix},$$
$$(3.159)$$

$$\Delta = \frac{1}{r}\frac{\partial^2}{\partial\xi^2} + \frac{\partial}{\partial\eta}r\frac{\partial}{\partial\eta}.$$

This system can be written in the form of equations of first order,

$$\frac{\partial}{\partial \xi}\{\tilde{T}_1 \ \tilde{T}_2 \ \tilde{S} \ \tilde{U}\} = A\{\tilde{T}_1 \ \tilde{T}_2 \ \tilde{S} \ \tilde{U}\}, \quad \tilde{U} \equiv \frac{\partial}{\partial \xi}(\tilde{T}_1 + \tilde{T}_2).$$

Introducing the Fourier series $\tilde{T}_1 = \Sigma \ \tilde{T}_{1n}(\xi) \cos n\eta, \ldots$ we have for $\tilde{T}_{1n}, \ldots, \tilde{U}_n$ a system of ordinary differential equations. This system is easily written out in the matrix form (Chapter 2, §7) and can be numerically integrated in the context of the shooting method (present chapter, §8.3).

Consider another solution, one using the two-dimensional Fourier series. As in the preceding section the deformation is symmetric with respect to two planes (Fig. 31): the plane $\eta = 0, \pi$ of the axis of the tube and the midlength cross-section of the tube, $\xi = L/2b$.

This deformation can be described by the double Fourier series of the form similar to (3.149)

$$[\tilde{T}_1 \ \tilde{T}_2] = \sum_j [\tilde{T}_1^j(\eta) \ \tilde{T}_2^j(\eta)] \sin j\hat{\xi} + [\tilde{T}_1^B(\eta) \ \tilde{T}_2^B(\eta)],$$

$$\tilde{S} = \sum_j \tilde{S}^j(\eta) \cos j\hat{\xi} \quad (j = 1, 3, \ldots);$$

(3.160)

$$[\tilde{T}_1^j \ \tilde{T}_2^j \ \tilde{T}_1^B \ \tilde{T}_2^B] = \sum_n [\tilde{T}_{1n}^j \ \tilde{T}_{2n}^j \ \tilde{T}_{1n}^B \ \tilde{T}_{2n}^B] \cos n\eta,$$

$$\tilde{S}^j = \sum_n \tilde{S}_n^j \sin n\eta \quad (n = 0, 1, \ldots).$$

(3.161)

Introduction of the series (3.160) into (3.159) results in a separate system of ordinary differential equations for the three complex functions $\tilde{T}_1^j$, $\tilde{T}_2^j$ and $\tilde{S}^j$, for each integer $j = 1, 3, \ldots$.

Solution of the three simultaneous equations in the form (3.161) leads to a system of algebraic equations for the coefficients $\tilde{T}_{1n}^j$, $\tilde{T}_{2n}^j$ and $\tilde{S}_n^j$ of the double Fourier series. This system can be written directly in the matrix form (Chapter 2, §7)

$$A^j \begin{bmatrix} \tilde{T}_1^j \\ \tilde{T}_2^j \\ \tilde{S}^j \end{bmatrix} = -\frac{4}{\pi j} A^0 \begin{bmatrix} \tilde{T}_1^B \\ \tilde{T}_2^B \\ 0 \end{bmatrix},$$

(3.162)

$$A^j = \begin{bmatrix} (r_+ + i\lambda h^\circ c_+)m_j & i\lambda h^\circ c_+ m_j & -\Lambda r_-^2 \\ \lambda s_+ - ih^\circ \Lambda_+ & -\Lambda_+ r_+ - ih^\circ r_- \Lambda_+ & Em_j \\ \lambda c_+ - ih^\circ \Delta^j & r_+ - ih^\circ \Delta^j & 0 \end{bmatrix},$$

$$\Delta^j = -m_j^2 r_+^{-1} - \Lambda r_- \Lambda_+ ,$$

where $m_j = j\pi b / L$. The terms reflecting the distributed load (q_1, q_2, q) are not included in (3.162). The matrix A^0 represents the operator A^j with $j = 0$, it takes into account that the derivatives of the quantities $\tilde{T}_1^B(\eta)$ and $\tilde{T}_2^B(\eta)$ with respect to the coordinate ξ are equal to zero. The right-hand side in (3.162) reflects the expansions (3.150).

Investigate now the pure bending of a tube with the "thin" flanges on both edges. The boundary conditions (3.107) of the semi-moment-less theory remain valid. But they must be (for the full equations considered now) supplemented by two conditions on each of the edges. We choose the additional conditions that are as close as possible to those of the semi-momentless solution discussed in the preceding section.

On the edges we thus require

$$T_1 = Ehh^\circ M^\circ \cos \eta , \qquad \kappa_2 = 0 ,$$
$$M_1 = 0 , \qquad \varepsilon_2 = -\nu\varepsilon_1 , \tag{3.163}$$

Recalling the definitions (1.114) of the complex variables $\tilde{T}_1$ and $\tilde{T}_2$ and using the elasticity relations (1.83) for T_2 and κ_1 we rewrite the conditions (3.163) in the form

$$\hat{\xi} = 0, \pi: \qquad \tilde{T}_1 = \tilde{T}_1^B = Ehh^\circ M^\circ \cos \eta , \qquad \tilde{T}_2 = \tilde{T}_2^B = 0 . \tag{3.164}$$

Besides this, the solution must satisfy the conditions of continuity of the closed cross-section (Chapter 2, §6), which require the $\cos 0\eta$ terms in the series of T_1 and κ_2 (constituting the $\tilde{T}_1$) and the $\cos \eta$ term in κ_2 to vanish.

The algebraic equations (3.162) may be solved directly or after separating the real and imaginary parts. A more convenient solution is obtained by first eliminating the unknowns $\tilde{T}_2^j$ and $\tilde{S}^j$. The resulting equation may be transformed to determine the nondimensional function $T^j - \mathrm{i}\kappa^j = \tilde{T}_1^j/Ehh°$. Upon separation of the real and imaginary parts this gives two simultaneous matrix equations for T^j and κ^j, which correspond to the system (3.143). These equations demonstrate the relation between the full and the semi-momentless equations, briefly discussed in Chapter 2, §5.2.

For the boundary conditions (3.163), the series (3.160) and (3.161) converge as effectively as in the semi-momentless solution presented in this chapter, §7.2. But the convergence of the sin $j\hat{\xi}$-series may be less favourable when the boundary conditions introduce a considerable edge-effect contribution. This can be avoided by calculating the edge effect separately, e.g. in a way similar to that in (2.183)–(2.185).

7.3. Effects of tube geometry

Results of the foregoing analysis of tube flexure allow an insight into some properties of tube bends as determined by their geometry. They also supplement the *a priori* estimates (Chapter 2, §§4, 5) of accuracy of the employed theory and provide an opportunity to check the accuracy.

The basic assumption (2.5)—$\varepsilon \ll 1$ in $|F_{,11}/a^2| \sim \varepsilon |F_{,22}/b^2|$—can be checked directly, once the solution is obtained. A more convincing test provides a comparison with calculations based on the general thin-shell theory and with experiments.

For tubes with the *thin flanges* the comparison was carried out elsewhere [176] using the data obtained by calculations based on the complex equations equivalent to (3.159). It has been found that, when these tubes are not too short ($l > 0.1$) and when $b/R_m < 0.25$ (notation of Fig. 17), their properties do not depend on the value of b/R_m and on γ and H. Such tubes are adequately described by the simplest semi-momentless theory based on setting $r = 1$ and disregarding the γ and H terms. This conclusion is supported by the graphs of Fig. 33. The flexural stiffness and maximum stress of tubes with thin flanges depend on the parameter b/R_m only weakly. Indeed, for a fixed μ-value, tubes with $h/b = 0.01$ and 0.04, which for $\mu = 40$ means $b/R_m = \mu h° = 0.12$ and 0.48, respectively, have nearly the same values of δ and thus of F.

For tubes with *stiff flanges* the influence of the slenderness parameter on maximum stress and flexural stiffness remains immaterial even for the radius b as large as $R_m/2$. That is, the theory simplified by setting $r = 1$ remains applicable for $b/R_m < 1/2$. This is demonstrated

TABLE 5. Different solutions for tubes with stiff flanges, values of $F(h/b, b/R_m, \varphi)$.

h/b	0.01			0.02			0.05		
R_m/b	2	4	8	2	4	8	2	4	8
μ	165.2	82.6	41.3	82.6	41.3	20.65	33.04	16.52	8.26
$\varphi = \dfrac{\pi}{2}$ 1	1.346	2.64	6.376	1.379	2.516	4.450	1.341	2.20	2.547
2	2.55	3.89	7.19	2.49	3.55	4.81	2.40	2.86	2.65
3	2.51	3.74	7.22	2.48	3.49	4.93	2.37	2.88	2.69
4	2.63	3.88	7.32	2.56	3.56	4.96	2.34	2.86	2.69
$\varphi = \pi$ 1	10.82	19.39	13.95	8.625	11.17	7.549	5.75	5.355	3.282
2	16.73	20.83	14.20	12.48	11.94	7.69	7.55	5.66	3.33
3	14.50	21.45	14.32	11.5	12.21	7.74	7.45	5.75	3.35
4	15.1	21.8	14.4	11.9	12.3	7.79	7.41	5.77	3.36

1. Semi-momentless theory (SMT) with $r = 1$ and γ, $H = 0$.
2. SMT with $r = 1$ but γ, $H \neq 0$.
3. SMT taking into account both $r = r(\eta)$ and γ, $H \neq 0$.
4. WHATHAM solution [214] of full thin-shell equations.

in Table 5, which exploits in the lines 4 the exact-solution results obtained by WHATHAM [214].

Coincidence of the semi-momentless solution with that [192] of the full shell-theory equations is demonstrated, besides Table 5, by the graph representing [192] in Fig. 35 for the range of tubes with $b/R_m = 1/3$ and $\varphi = 90$.

The error estimates, stated in (2.120) for the simplest version of the semi-momentless theory, find confirmation in Table 5. The rows "1" show deviations of just the order of b/R_m, compared to the quantities in the rows 2, 3 and 4. Taking into account the γ and H terms (rows 2) leads to an improvement in accuracy, which corresponds very closely to that estimated at the end of §4.4, Chapter 2.

The difference between the δ-values (cf. Figs. 33 and 32) for different h/b by equal μ-values represents for stiff flanges the effect of γ and H. This influence accounts for a contribution of merely the order of $2(1 \mp \nu)h/b \sim 2h/b$ to the deformation determined by the edge forces. This amount is negligible in the thin-shell theory. However, the actual deformation is a *difference* between the deformation of an infinite tube and that caused by the self-equilibrating reactive forces of the flanges. Thus, the error in the quantity δ has the order of $(2h/b)/\delta$. It can be considerable for short tubes having $\delta \sim h/b$. In such cases the theory encompassing γ and H becomes necessary. We note that tubes with such edge stiffening retain little flexibility. They do not in fact belong to the class of flexible shells. For the deformation of shells that are actually flexible, the γ and H terms *are* negligible and the canonical semi-momentless theory of Chapter 2, §§4.2, 4.3, 5 is adequate.

7.4. Influence of warping stiffness of flanges

The spectrum of possible conditions on tube ends is wide. The models of thin flanges and of stiff flanges (defined in present chapter, §6.1) belong to the variety of limiting cases. They represent the case of extreme low warping stiffness of the flanges and that of infinitely high resistance to warping. The data of Figs. 32–35 show the flexibility (F) of some tubes with thin flanges reaching many times that calculated for the similar tube with the stiff flanges. The influence of the warping constraint can be spectacular also for the stresses; this is illustrated in Fig. 34.

The results of various experimental investigations, presented in Fig.

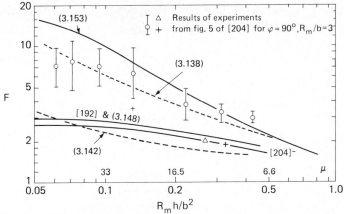

FIG. 35. Flexibility of tubes with freely warping and with stiff flanges, respectively, constitutes the upper and the lower limits for the experimental values which are mostly nearer to F of thin flanges.

35 (which reproduces Fig. 5 of [204] with the four curves, indicating the results of the foregoing, being added) fully confirm this role of the warping deformation. Nearly all the experimental values of flexibility lie between those calculated for the two types of flanges. For stiff flanges the solution (3.148) coincides[15] with the graph of the exact analysis [192]. The experimental points lie (with two exceptions) above or far above this curve. They are somewhat nearer to the graphs representing the case of *thin* flanges, even to that of the elementary formula (3.138).

Apparently, the flanges of the specimen could not, contrary to the model of stiff flanges, prevent the warping deformation of the edges. This is to be expected also in most of real cases.

The adequacy of the edge conditions imposed, not the approximations of the theory, is at present critical for the accuracy of the tube analysis.

8. Nonlinear bending of tubes with given conditions on edges

The nonlinearity of flexure is more pronounced for tubes that are initially straight or have a comparatively small initial curvature. The

[15] The difference between the flexibility computed in [192] and the F-values of [204], to be observed in Fig. 35, can be explained by "the $(1 - \nu^2)$ term . . . included for consistency" in [204].

following discussion concerns these tubes. It uses the semi-momentless theory in its simplest form disregarding the γ and H terms and setting $r = 1$.

8.1. Influence of normal pressure

A strictly linear model used in §§6, 7, present chapter, does not detect the influence of the constant external or internal pressure on the flexure of circular tube. The load term q_Σ representing the $q = \text{const.}$ in (2.194) is equal to zero for $r = 1$ and determines only a small secondary effect when $r = r(\eta)$ is taken into account (Chapter 5, §5).

The constant-pressure influence on the flexure of a tube becomes substantial only as a consequence of the noncircularity of the cross-sections (initial or caused by the elastic deformation of a bent tube).

Consider in this section the tube flexure with a change of curvature restricted by the relation between the bending moment and the initial curvature of the tube,

$$M^\circ \ll \mu \quad \text{or} \quad M_z/EI \ll 1/R_m . \tag{3.165}$$

This condition allows us to use (2.140) and (2.197) omitting those terms comprising products of $T^{(0)}$ by the unknowns. Thus, the governing equations contain, besides the linear-deformation terms, only terms dependent on the normal pressure q. Equations (2.197) rewritten in a form similar to (3.110) become

$$\begin{bmatrix} \check{W} + q^\circ \Lambda^2 & -\check{L} \\ \check{L} & \check{W} \end{bmatrix} \begin{bmatrix} \check{f} \\ \check{g} \end{bmatrix} = \begin{bmatrix} -3\mu M^\circ \\ 0 \\ 0 \\ \cdots \end{bmatrix} . \tag{3.166}$$

For $q = 0$ this matrix equation coincides with (3.110). (Of course, an equation similar to (3.166) can be derived in an analogous way for $r = r(\eta)$. The implications of setting $r = 1$ are discussed in present chapter, §7.3.)

The further solution in the form of ordinary Fourier series does not differ in any substantial respect from that in present chapter, §6; the evaluation with the aid of the double Fourier series coincides with that described in §7.1, present chapter, for $q = 0$.

In particular, formulas (3.134), (3.136) and (3.157) for the angle of flexure of a tube bend remain valid. Naturally, the values of F and α_k depend not only on the curvature (μ), but also on the pressure (parameter $q°$). The influence of the normal pressure is immaterial as long as $|q|$ is small compared to the critical external pressure (discussed in present chapter, §10).

8.2. Second-order-theory solution

The ordinary trigonometric-series solution of the "second-order-theory" equations leads to the differential equations (2.197). For a circular cross-section the equations may be presented in the form

$$\begin{bmatrix} W^* & -L^* \\ L' & W \end{bmatrix} \begin{bmatrix} f \\ g \end{bmatrix} = \begin{bmatrix} 0 \\ -2\mu'N° \\ -3\mu'M° \\ 0 \\ \dots \end{bmatrix} . \tag{3.167}$$

The matrices W^*, L^*, L' and W are determined by the following formulas for the elements of the nth rows and $(n \pm 1)$th, nth and $(n \pm 2)$th columns:

$$W_{nn} = n^4 - n^2 , \quad W_{nn}^* = W_{nn} - \frac{n^2}{2}(1 + \delta_{n-1})\mu'M° + q°n^2 ,$$

$$W_{n\beta}^* = -n^2\mu'N°\delta_\beta , \quad \beta = n \pm 1 , \quad W_{n\cdot n \pm 2} = -\frac{n^2}{2}\mu'M°\delta_{n\pm 2} ,$$

$$L_{nn}' = d^2 = \frac{1}{h°}\frac{d^2}{d\xi^2}, \quad L_{nn}^* = L_{nn}' - (n^4 + n^2)N° ,$$

$$L_{n\cdot n \pm 1}' = -\frac{n^2 \mp n}{2}\mu'\delta_{n\pm 1} , \tag{3.168}$$

$$L_{n\beta}^* = L_{n\beta}' - \tfrac{1}{2}n\beta(n\beta + 1)M°$$

$$(n, n \pm 1, n \pm 2 = 0, 1, \dots ; \delta_0 = 2, \delta_j = 1 \text{ for } j \neq 0) .$$

All the remaining elements of these matrices are equal to zero. The columns f and g are written out in (2.197); the parameters $N°$, $M°$ and μ' are *in these equations* constants.

The solution of equations (3.167) can be obtained with the aid of numeric integration or along the lines of present chapter, §§6, 7. In most cases the double Fourier-series solution is the more convenient of the methods discussed.

Equations (3.167) are linear with respect to the unknowns f and g. But they determine the stresses and the deformation of the shell, taking into account the geometric nonlinearity. The coefficients of the system and its right-hand side depend on the loading.

The nonlinearity terms are of two kinds. First, the products of the longitudinal or the circumferential forces, represented by $N°$, $M°$ and $q°$ without any reflexion of the deformation of the cross-sections and of the changes of curvatures (κ_1, κ_2). The second group of these terms consists of products of the (full) longitudinal stress resultants T_1 and the change of curvature of the longitudinal fibres, determined without taking account of the deformation of the cross-sections.

The approximations mentioned restrict, of course, the applicability of the "second-order theory" (cf. Chapter 2, §6.3). It may be useful only in so far as the stress resultants and curvatures do not differ too much from their approximate values assumed in the nonlinear terms. These conditions are fulfilled for tubes only when the curvature is small enough, i.e. remains within the bounds

$$\mu' = \mu + M° < 2 , \quad M° < 2 . \tag{3.169}$$

To ascertain the bounds of applicability of the second-order-theory solution and to investigate the nonlinear deformation outside these bounds we employ the full system (2.96)–(2.98).

8.3. Large deformation

The basic equations for numeric integration will be derived by applying to the system (2.96)–(2.98) the Fourier-series solution. These equations will be used in the form

$$\frac{1}{\sqrt{h°}} \frac{\mathrm{d}}{\mathrm{d}\xi} X = AX + B . \tag{3.170}$$

The first step of the derivation of (3.170) consists in eliminating from (2.96)–(2.98) all the variables except T_1, κ_2, S and τ, which determine the stresses in the sections ξ = const. as well as deformations of these sections and tube edges. The resulting system of four equations is reduced to the following nondimensional form (the distributed-load terms—B in (3.170)—are not considered for the sake of brevity, they can be easily reintroduced):

$$
\frac{1}{\sqrt{h^\circ}}\frac{\partial}{\partial\xi}
\begin{bmatrix} T \\ \kappa \\ S^\circ \\ \tau^\circ \end{bmatrix}
=
\begin{bmatrix}
0 & 0 & -\dfrac{1}{r}\dfrac{\partial}{\partial\eta}r^2 & 0 \\
0 & 0 & 0 & \dfrac{1}{r}\dfrac{\partial}{\partial\eta}r^2 \\
-U' & -V & 2\dfrac{\partial}{\partial\eta}r\rho'\tau^\circ & 0 \\
-V & U & 0 & \dfrac{\partial}{\partial\eta}r\rho'\tau^\circ
\end{bmatrix}
\begin{bmatrix} T \\ \kappa \\ S^\circ \\ \tau^\circ \end{bmatrix}.
$$

$$(3.171)$$

We use here the dimensionless variables

$$
S^\circ = \frac{S}{Eh\sqrt{h^{\circ 3}}}, \qquad \tau^\circ = \frac{\tau b}{\sqrt{h^\circ}}, \qquad \rho' = \frac{R_2'}{b} \qquad (3.172)
$$

and the operators

$$
U' = \frac{br}{\rho_1'h^\circ} - \frac{\partial}{\partial\eta}\rho'\frac{br}{R_1'h^\circ} \equiv U - (rT)^{\cdot} + \frac{\partial}{\partial\eta}\rho'^2[(rT)^{\cdot\cdot} + \mu c\kappa - r\tau^{\circ 2}],
$$

$$
U = \mu s - \mu\frac{\partial}{\partial\eta}\rho'c, \qquad V = \frac{\partial}{\partial\eta}\rho'\frac{\partial^2}{\partial\eta^2}r + \frac{1}{\rho'}\frac{\partial}{\partial\eta}r, \qquad (3.173)
$$

$$
(\ )^{\cdot} = \frac{\partial(\ )}{\partial\eta}.
$$

For the deformation symmetric with respect to the plane $\eta = 0$, π, we seek the solution in the series form

$$
\begin{bmatrix} T \\ \kappa \end{bmatrix} = \begin{bmatrix} g_0 \\ 0 \end{bmatrix} + \sum_1^m \begin{bmatrix} g_n \\ f_n \end{bmatrix}\cos n\eta, \qquad
\begin{bmatrix} S^\circ \\ \tau^\circ \end{bmatrix} = \sum_1^m \begin{bmatrix} S_n^\circ \\ \tau_n^\circ \end{bmatrix}\sin n\eta. \qquad (3.174)
$$

Inserting these expansions into (3.171) we obtain for the coefficients

$g_0(\xi)$, $g_1(\xi)$, ..., $\tau_m^\circ(\xi)$ a set of ordinary differential equations. The equations can be written out using the procedure of Chapter 2, §7 (i.e. similarly to (2.197)) in the matrix form (3.170). The column matrix $X = \{T, \kappa, S^\circ, \tau^\circ\}$ is composed of the four column matrices encompassing the Fourier coefficients of the respective dependent variables. The system (3.170) contains $4m + 2$ equations, with m being the number of the last coefficient retained in each series.

Equations (3.170) are nonlinear: the system matrix A depends as defined by (3.171) on the actual shape and stress state. It must be "updated" for a current level of loading.

The integration of the system (3.170) was carried out[16] for each step of loading with the aid of multiple-shooting method (as described by BULIRSCH and STOER, *Einführung in die Numerische Mathematik* II, Springer, Berlin–New York, 1973).

The longitudinal coordinate is in (3.171) represented only with the quantity $\xi\sqrt{h^\circ}$. Correspondingly, the actual length of the tube L is represented by the dimensionless parameter l defined in (3.152).

For a relatively unfavourable dimension and loading—a cylinder of length $l = 0.65$ under nearly critical bending moment $M^\circ = 0.9$—it is sufficient to retain $m = 5$ terms of each Fourier series. The Pascal-written program required on the Burroughs 7800 the machine time 500–700 sec. (The second-order-theory calculation lasts 10^3 times less.)

This solution was used to investigate the flexure of tubes with the two limiting cases of end constraints, described by the boundary conditions formulated in (3.107), (3.108), (3.137) and (3.140) with $k_2 = 0$. Some results of the large-deformation analysis as well as those of the second-order theory are presented in Fig. 36 and in Fig. 42, discussed in present chapter, §9. In contrast to the linear approximation, different levels of loading, in our case different values of the bending moment, generate qualitatively different stress distributions and patterns of deformation. The examples of Fig. 36 concern tubes with the "thin" flanges, loaded by the bending moments as large as those causing buckling of the respective tube (cf. present chapter, §9). The graphs of Fig. 36 display three characteristic types of deformation of the midlength cross-sections. $\sigma_{cl} \equiv 2Eh^\circ \cong 0.605Eh/b$.

---

[16] The algorithm and program can be found in the report: Valery AXELRAD, Forschungsbericht 03/84, Institut für Mechanik, FB LRT, Hochschule der Bundeswehr München (1984).

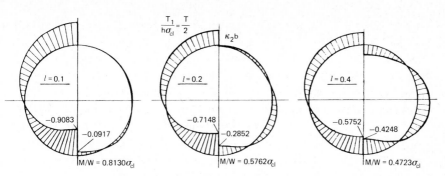

FIG. 36. Cylinders of different lengths with thin flanges on ends under critical bending moments. Longitudinal stress and $\kappa_2 b$ in the middle cross-section. $\sigma_{cl} = 2Eh°$.

The tube with $l = 0.4$ deforms in the middle part in a way resembling the behaviour of an infinitely long tube (Fig. 17). The stress state here strays not very much from being double symmetric. This example represents a transition case between middle-length and long tubes.

The graphs for $l = 0.1$ represent, in Fig. 36, the other limit of the middle-length range of tubes. Despite the considerably larger bending load, the deformation of the cross-section is here small and the longitudinal stress does not differ very much from that of a straight beam. The reduction of curvature $1/R'_2$ (which is shown in present chapter, §9 to be responsible for the reduction in the buckling load) observed here is substantial. But in contrast to the longer tubes this deformation is a local one. It has the character of a shallow and narrow zone of flattening. For this example the curvature $1/R'_2$ is at a point reduced by $0.09/b$ to $0.91/b$ while $|w| < 0.009b$.

The case of tube with $l = 0.2$ (Fig. 36) illustrates the behaviour of the *middle-length* cylinders, which is decisively dependent on the length l.

The "stiff" flanges at the ends reduce the deformation of the cross-sections of the tubes more strongly than do "thin" flanges (cf. present chapter, §6.4).

Tubes with initial curvature, $\mu \neq 0$, have larger lateral distortion. The external pressure on the tube wall increases the deformation of the cross-section also. However, as long as q is less than a third of its critical value (indicated by relations of present chapter, §10), the influence remains moderate. The *internal* pressure q has, naturally, the opposite effect.

Note that the deformation just discussed develops gradually with the growth of the bending load. It has nothing in common with instability until the critical load is reached. Buckling is discussed in present chapter, §9.

The calculations show the influence of the edge constraints to fade away at a distance from the edge of about $2b/\sqrt{h^\circ}$. That is, the constraints practically do not influence the deformation in the middle part of a tube with the dimensionless length $l \geq 1$. For such a long tube, in its middle part, the system (3.171) gives results identical to those of the extended Reissner–Meissner equations (2.30), i.e., of axisymmetric (Saint-Venant) flexure problem.

The reader has not failed to note: the solution presented contains no use of displacements. It contradicts in this respect the tastes prominent in the shell-theory literature. Thus, contrary to the often expressed opinion, there is no need to evaluate displacements even when (as in the cases under discussion) *kinematic* boundary conditions are specified. The shell shape, resulting from *any* displacements and rotations, is computed directly by way of the strain resultants κ_2, τ and ε_1. These parameters determine the angles of bending and torsion of an element of the shell as well as its extensions. It remains to sum up—to integrate.

How much more complicated an analysis of a nonlinear problem becomes by using the displacements as dependent variables can be observed in such a solution of the tube-flexure problem in [95]. It involved merely the second-order theory. For unrestricted deformations a solution in terms of displacements would be incomparably more complicated.

9. Collapse of bent cylinder shells and curved tubes

The tube bending is one of the problems where the buckling starts in some part of the shell, locally. Focusing the stability analysis on the zone of initial buckling allows radical simplifications. First, the relevant hypothesis is formulated. Its corollary leads to stability criteria asymptotically exact for very thin shells. The criteria render closed-form conditions for the local character of buckling and for the adequacy of the simplest asymptotic analysis itself (§9.2). Next, the analysis is extended to shells of finite wall-thickness (§§9.3, 9.4). The technique is employed in §9.5 for the investigation of failure of tubes under bending.

9.1. Local-stability hypothesis

The classic stability analysis studies bifurcation buckling which can start globally—over the entire shell including its edges. (This approach appears to stem from the L. Euler work on compressed columns.)

However, for thin shells *local* buckling is typical, particularly so by nonuniform shape and stress, resulting from the prebuckling deformation (cf. Fig. 31).

Of course, local buckling does not necessarily lead to an immediate failure of the structure. But the initial-buckling load should give a conservative estimate of the collapse load.

Several intrinsic features of shell structures make it worthwhile or even necessary to employ the additional, specialized, instrument of local-buckling analysis.

The effect of imperfections can be drastic. But the buckling analysis of an *imperfect* shell is not a simple issue. Even for the canonical case of a cylinder shell under axial compression the very problem of describing the problem that arises has proved to be a formidable one.

The prebuckling deformation introduces severe complications for flexible shells.

The global stability analysis involves an eigenvalue boundary problem for a nonlinear high-order system of partial differential equations (e.g. of the Donnell type). A rigorous solution of such a problem is at present hardly attainable. Experience of computational mechanics is expressed in warnings (e.g. [186]) against overconfidence in numerics.

But just those factors complicating the *global* stability analysis make the buckling *local* and the respective local-stability investigation effective. Any variable with respect to surface coordinates stress state and local shape present in some part of the shell a conjunction more favourable for buckling. The initial buckling is confined to this zone. The corresponding mode is determined by a particular solution of the stability equations tending to zero away from the initial-buckling zone. The boundary conditions as well as the stresses and geometry of the most part of the shell need not be involved.

The ultimate intent of the concept of local stability becomes more lucid in the limiting case: when the initial-buckling zone is very small, we can speak of a stability condition "at a point",

$$f(T_{\alpha\beta}, R_{\alpha\beta}, h, E, \nu) \geqslant 0 . \tag{3.175}$$

The asymptotic condition of this type stands in the same sort of relation to the stability analysis encompassing the entire shell as the condition of allowable stress (e.g. with the R. von Mises plasticity condition) to the failure analysis of a structure as a whole.

The further discussion is based on the following statement interpreting the fundamental feature of the local-buckling phenomenon in the form of *hypothesis* [193].

The buckling instability is determined by the stress state and the shape of shell inside the zone of the initial buckle(s).

Conditions outside the zone of initial buckling have a negligible effect. Functions describing the stress and shell shape may be set (analytically extended) outside the buckling zone in any way helping to simplify the analysis. The only restriction: the analytical extension must not amount to assuming zones of the shell, which are more susceptible to buckling than the zone under consideration.

The hypothesis directly leads to the *corollary*:

In so far as in the domain of initial buckle(s) the stress resultants and the curvature parameters of the shell are approximately constant, they may be assumed constant in the stability analysis.

A concept amounting to the last statement has been the nucleus of the local-stability approach. It can be traced back to the work in 1936 of STAERMAN [18] on conical shell under axial compression. The buckling is supposed in [18] to start at a point where the compressive stress T_1/h first reaches the value T_1^*/h,

$$\frac{T_1^*}{h} = \frac{Eh}{R_2\sqrt{3(1 - \nu^2)}}, \qquad (3.176)$$

where $1/R_2$ denotes the normal-section curvature at the "buckling point".

This concept of local stability was employed in 1965 [95] for the analysis of cylinders and initially curved tubes under bending. In consequence of nonlinear elastic deformation the stresses and the local shape vary in these cases with respect to both surface coordinates. The stability condition was stated in the form similar to $T_1 \leq T_1^*$ with the curvature $1/R_2'$ of the *deformed* shell. The ground for this has been rendered by the result of SEIDE and WEINGARTEN [83] for linear bending of a circular cylinder of radius R_2: the critical stress at the centre of any initial buckles (Fig. 37(a)) is nearly equal to that (σ_{cl}) of uniform axial compression

$$T_{1cr}(0) = \lambda \sigma_{cl} h , \quad \sigma_{cl} = Eh/[R_2\sqrt{3(1 - \nu^2)}] . \quad (3.177)$$

The values of λ are presented in Table 6.

Leaving the quantitative test of the hypothesis to §9.4, present chapter, we consider two outstanding results providing a direct confirmation of the hypothesis [17].

The first is rendered by the rigorous investigation of the stability of a circular cylinder, compressed inside a longitudinal strip of breadth c_σ (Fig. 37(b)), due to HOFF, CHAO and MADSEN [93]: When the breadth c_σ is not less than $3.5\sqrt{hR_2}$ the critical stress is very nearly equal to that of uniform compression (to σ_{cl}). The situation outside the buckling zone is of no influence.

The second confirmation comes from experiment. Obviously, the conditions most unfavourable for *local* character of buckling are those of *uniform* over the shell stress and curvature. Cases of just this nature have been investigated by ESSLINGER and GEIER [168] and by TENNYSON [129]. These experiments show beyond any doubt that even the utmost carefully produced circular cylinders under as uniform axial compression as ever possible buckle *locally*. The buckling starts with a single small buckle with both breadth and length having (for $R_2/h = 100/0.251$) approximate dimensions

$$2\pi R_2/18 \cong 7(hR_2)^{1/2} . \quad (3.178)$$

TABLE 6. Buckling stress $T_{1cr}/h = \lambda \sigma_{cl} \cos N\eta$ of circular cylinders.

N	$\dfrac{R_2}{h}$	[83]		[183]	(3.194)	
		λ	m	λ	λ	m
1	100	1.015	0.979	1.0138	1.0150	0.963
1	200	1.009	0.972	1.0084	1.0095	0.970
1	1000	1.003	0.992	1.0027	1.0033	0.982
2	100	—	—	1.0368	1.0379	0.943
3	100	—	—	1.0652	1.0649	0.927
2	1000	—	—	1.0072	1.0082	0.972

---

[17] An important indirect confirmation can be seen in the recent results of W.T. KOITER in [202] and in *Behaviour of Thin-Walled Structures*, J. Rhodes and J. Spence, Eds., Elsevier Appl. Sci. Pbls., London (1984) 35–46. Here the effect of local imperfections is shown to be nearly as strong as that of double-periodic ones.

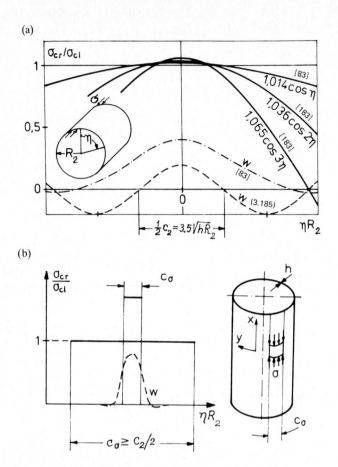

FIG. 37. Global solutions: buckling is determined by the stress inside a zone of breadth $3.5\sqrt{hR_2}$, equal to $c_2/2$ from (3.185).

Consider first the simplest, limiting, case of local buckling: when the basic issue of delimiting the initial-buckling zone and taking into account variations of the stress and shell shape inside it does not arise.

9.2. Asymptotic approximation

Consider stability of a stress state in an area around some point of the middle-surface, assuming the hypothesis in the most restrictive form, that of the *corollary*:

The stress resultants T_i and the curvatures $1/R_i'$ will be treated in the analysis as constants—equal to their respective values at the point (ξ_0, η_0).

Consider the stress state and the shape of the buckling zone to be symmetric to the lines $\xi = \xi_0$ and $\eta = \eta_0$. In particular, $S = 0$ and $\tau = 0$.

Following the hypothesis, we may set outside the zone of initial buckling any appropriate functions T_i and R_i'. We set them equal to their values at the point (ξ_0, η_0). The uniform, with respect to the surface coordinates, stress and curvature suggest a *periodic* buckling mode. A simple mode of this kind is defined by

$$W = A \cos mx \cos ny; \quad A, m, n = \text{const.}, \quad (3.179)$$

where x and y are the dimensionless distances from the centre of the buckling zone, defined in (1.133).

Substitution of (3.179) into the stability equation (1.134) with $S, \tau = 0$ and with the dimensional parameter R set equal to $R_2'(\xi_0, \eta_0)$ (making, $k_2 = 1$) leads to

$$A\left[(m^2 + n^2)^4 - \frac{2}{T_*}(T_1 m^2 + T_2 n^2)(m^2 + n^2)^2 + (m^2 + k_1 n^2)^2\right] = 0. \quad (3.180)$$

The buckling occurs $(A \neq 0)$, when the expression in the brackets is equal to zero. This renders for the critical values of the resultants (T_i^*) the equation

$$T_1^* + T_2^* H^2 = T_* \frac{1}{2}\left[m^2(1 + H^2)^2 + \frac{(1 + k_1 H^2)^2}{m^2(1 + H^2)^2}\right],$$

$$H = \frac{n}{m}, \quad T_* = \frac{Eh^2}{R\sqrt{3(1 - \nu^2)}}. \quad (3.181)$$

(This equation was obtained in [95] and [176, p. 228]; an independent derivation is given in [202]. For axial compression of a cylinder T_2, $k_1 = 0$, (3.181) reduces to (2.9.19) of [215].)

The buckling load, determined by (3.181) through the T_i^*, must be minimized with respect to m and n. Thus, the length $2\pi c/m = c_1$ of a buckling-mode wave and its breadth $c_2 = 2\pi c/n$ are calculated simultaneously with the critical stress resultants T_i^*. For any fixed ratios T_2/T_1

and H, the extremum condition $\partial T_1^*/\partial m = 0$ and (3.181) give [176, 205]

$$c_1 = 2\pi c/m = 2\pi c(1 + H^2)/(1 + k_1 H^2)^{1/2}, \qquad c_2 = c_1/H ; \quad (3.182)$$

$$T_1^* + T_2^* H^2 = T_*(1 + k_1 H^2) \quad (c = \sqrt{hR}/[12(1 - \nu^2)]^{1/4}). \quad (3.183)$$

Equation (3.183) determines the value of H—one corresponding to the minimum critical load—with an indirect but essential restriction. The ratio H renders through (3.182) the dimensions of a buckle. But the basic assumption of T_1 and R_i' being nearly constant inside the initial-buckling area holds only when this area is small enough. This requires, as illustrated in the following, that c_i be small compared to R_1' and R_2' as well as to the overall dimensions of the shell.

Relations (3.182) and (3.183) provide a commentary on some well-known results of the theory and experiment. These, in turn, throw light on the applicability of this simplest solution.

A *spherical shell* $(R_1 = R_2)$, loaded by a uniform normal pressure, is uniformly compressed (outside the edge zone) to the radius $R_1' = R_2' = R$ and $k_1 = 1$. For the compressive stress resultants $T_1 = T_2 = qR/2$, formula (3.183) renders $T_i^* = T_*$ and thus the critical pressure equal to the classical value

$$q_{cr} = 2Eh^2/(R^2\sqrt{3(1 - \nu^2)}). \quad (3.184)$$

The length-to-breadth ratio H of a buckle has no influence on this q_{cr}. The smallest buckles correspond to $H = 1$: $c_1 = c_2 = \sqrt{2}\, 2\pi c$.

For a *cylinder shell* $(k_1 = 0)$ under the longitudinal stress T_1/h, the stability condition $T_1 \leq T_1^*$ contains T_1^* from (3.183), coinciding with (3.176). For circular shell, $T_1^* = h\sigma_{cl}$. This critical stress corresponds to various modes (3.179) including, in particular, the axisymmetric mode (with $H = 0$, $c_1 = 2\pi c$) and the "square" buckles,

$$m = n = 0.5, \qquad c_1 = c_2 = 4\pi c \approx 7(hR)^{1/2}, \qquad R = R_2. \quad (3.185)$$

The dimensions (3.185) of one wave of a periodic mode coincide with those (3.178) of a single buckle, found experimentally.

Equation (3.183) gives for a circular cylinder shell *the same* critical stress $T_1^*/h = \sigma_{cl}$ for *any* variation of T_1 over the shell. The global solutions [83, 183] of the stability problem for $T_1 = h\sigma_m \cos N\eta$ $N = 1, 2, \ldots$ confirm this: the critical value of $h\sigma_m$ is nearly equal to $h\sigma_{cl}$, provided the variation of T_1 remains moderate inside the area of the breadth not less than $c_2/2 = 2\pi c = 3.5\sqrt{hR_2}$ (Table 6, Fig. 37(a)).

For the stress $T_1/h = \sigma_m(1 - a\xi/L)$, varying *along* a circular cylinder with hinged edges, the global analysis [180] gave the critical value $\sigma_m = 1.048\sigma_{cl}$. The length L of the shell was 10 times the $c_1 = 2\pi c$ of a "free" axisymmetric buckle. Such a buckle occurring in the zone of maximum stress (adjoining the edge $\xi = 0$) would have its centre at $a\xi = 0.05\,L$. At this point $(T_1/h)_{cr} = 1.048\sigma_{cl} \cdot 0.95 = \sigma_{cl}$—as predicted by (3.183).

Experiments [189] confirm (3.183) for T_1 varying with *both* coordinates.

The simplest solution (3.183) proves quite useful also when not the stress but the curvature varies over the shell surface. This is displayed by the rigorous results of VOLPE, CHEN and KEMPNER [196, p. 576] on uniform axial compression of oval cylinders defined by the relation

$$\frac{b}{R_2'} = 1 - e \cos N\eta = 1 - e \cos ny , \qquad (3.186)$$

where $b = \text{const.}$, $2\pi b$ is equal to the perimeter of the cross-section. For all cylinders, with $e \le 0.3$ and $N = 2$, the T_1^* according to (3.183) is nearly equal to T_{1cr} of [196] (cf. Table 7, line 2).

The examples discussed and the calculations of present chapter, §§9.3, 9.4 indicate the order of magnitude of the initial-buckling-zone dimensions: it is not larger than one wave of an "unrestrained" mode having the length and breadth $c_i \sim \sqrt{hR_2'}$, defined by (3.182) (in terms of the curvature $1/R_2'$ of the section in which the predominant compression stress T_1/h acts).

This entails a perception of equation (3.183) (of the type (3.175)) as *asymptotically exact* for very thin shells. Indeed, the parameters of stress and shape T_i and R_i' or k_i do not directly depend on h (except in the areas of edge effect or local loading). Thus, inside a very small $\sim\sqrt{hR_2'}$ buckling zone they are very nearly constant. With the same (in actual cases, better) accuracy the critical stress is determined by the relations (3.182) and (3.183).

The asymptotic analysis gives a lower-limit estimate of the buckling load. It investigates the stability by way of sampling those zones of the shell, where the conjunction of the stresses and local shape is the *most* susceptible of buckling. The neighbouring parts of the shell have a *supporting* influence. It is this support that the asymptotic approximation neglects by treating T_i and R'_i as constants. The quantities T^*_i are determined by (3.183) as characteristics of *resistance* to buckling *at a point* of the shell.

Consider a more general investigation of stability, allowing for variability of stress and shape inside the initial-buckling zone, i.e. for finite wall-thickness.

9.3. Local buckling by finite wall-thickness

The stress state and shape of a shell, in particular of an imperfect and deformed one, can be complicated even locally. But for *local* approximation of the relevant functions $T_i(x, y)$ and $k_i = R/R'_i(x, y)$ a well-tested and simple device is provided by the power-series (Taylor) expansions:
$T_i(x, y) = T_i(0, 0) + xT_{i,x}(0, 0) + \cdots; \cdots$.

In this context the asymptotic solution of the previous section is based on the representation of the stress state and shape merely by the constant, zeroth, terms of the expansions of $T_i(x, y)$ and $k_i(x, y)$. Retaining also the next nonvanishing terms of the expansion for each of the stress and shape parameters should provide the estimate of the accuracy of the simplest approximation and the possibility of a direct and unified analysis for situations substantially variable inside the initial-buckling zone. Consider such a solution. To avoid complications we confine the analysis to problems symmetric with respect to lines $x = 0$ and $y = 0$ ($\xi = \xi_0, \eta = \eta_0$) and to $S = 0$ and $\tau = 0$.

As in the foregoing, we choose R equal to R'_2 at $x = 0$, $y = 0$. The stress state and shape are approximated by the expressions defining the constant parameters k_{ij} and n_{ij},

$$\begin{bmatrix} k_2 \\ k_1 \\ T_1/T_1(0,0) \\ T_2/T_1(0,0) \end{bmatrix} = \begin{bmatrix} 1 & k_{21} & k_{22} \\ k_{10} & k_{11} & k_{12} \\ 1 & n_{11} & n_{12} \\ n_{20} & n_{21} & n_{22} \end{bmatrix} \begin{bmatrix} 1 \\ x^2 \\ y^2 \end{bmatrix}. \tag{3.187}$$

We formulate now the local-stability condition generalizing the relations (3.175) and (3.183) to ($h° = h/(b\sqrt{12(1 - \nu^2)})$)

$$-T_1 \le T_{1cr} = \lambda T_*, \quad T_* = 2Ehh°\frac{b}{R'_2}. \tag{3.188}$$

The factor λ depends on the parameters k_{ij} and n_{ij} defined in (3.187). It takes into account the situation in the *area* around the point, for which the condition (3.188) is stated, and also the influence of k_1 and T_2. When k_1 and T_2 vanish, the factor λ has the meaning of a correction to the asymptotic solution (3.183) and $\lambda \to 1$ when $h \to 0$.

Consider first the stress and shape varying substantially only in the y-direction and the factors T_2 and $1/R_1'$ to be of secondary importance (as in the tube-stability problem), specializing (3.187) to

$$k_2 = 1 + k_{22}y^2 , \qquad T_1 = (1 + n_{12}y^2)T_1(0,0) ,$$
$$k_1 = k_{10} , \qquad T_2/T_1(0,0) = n_{20} . \tag{3.189}$$

This situation allows a buckling-mode which is periodical with respect to x. The mode must be local with respect to y. We seek this mode as an expansion in terms of the Bessel functions Λ_j (plotted in Fig. 38),

$$W = \sum C_j \varphi_j ,$$

$$\varphi_j = \cos mx \, \Lambda_j(ny) , \tag{3.190}$$

$$\Lambda_j = \sum_{p=0}^{\infty} \frac{j!}{(j+p)!p!}\left(-\frac{n^2 y^2}{4}\right)^p \quad (j, p = 0, 1, \ldots) .$$

The functions Λ_j have been chosen on the ground of their similitude to the modes found for circular cylinders in [83, 93].

Substitution of (3.190) into (1.134) leads to linear Galerkin equations for C_j,

$$\sum_j D_{ij}C_j = 0 , \qquad D_{ij} = \int\int D(\varphi_j)\varphi_i \, dx \, dy , \tag{3.191}$$

where the integration is to be carried out over the entire shell.

The nonzero solution of the system (3.191) determines the buckling mode. The lowest eigenvalue λ and the corresponding m and n are found by equating the determinant of the system (3.191) to zero,

$$\det(D_{ij}) = 0 , \qquad D_{ij} = D_{ij}(\lambda, m, n) , \qquad \lambda = T_{1cr}(0,0)/T_* . \tag{3.192}$$

By calculating the D_{ij} the integration with respect to x need not go

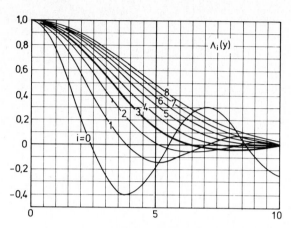

FIG. 38. The Bessel functions.

further than one period of cos mx. The functions $\Lambda_j(ny)$ vary symmetrically with respect to $y = 0$ and diminish with the growth of y. An increase of the limit of integration over $|ny| = 12$ has little influence on the values of D_{ij} for $i, j \geqslant 1$. Formulas (3.191) with D from (1.134) determine the following expressions of D_{ij} (written here briefly, without the indices ij by the constant factors A_p and B_p),

$$D_{ij} = m^8(A_0 - 4A_2 + 6A_4 - 4A_6 + A_8)$$
$$+ m^4(A_0 + 2k_{22}B_0 - 2k_{10}A_2) - 2\lambda[A_0 - 2A_2 + A_4$$
$$+ n_{12}(B_0 - 2B_2 + B_4) - n_{20}(A_2 - 2A_4 + A_6)m^{-2}]m^6. \quad (3.193)$$

Consistent with the accuracy assumed in (3.187) the terms with products of the parameters k_{ij}, n_{ij} are neglected. The integrals in (3.191) can be computed conveniently with the help of the appropriately truncated series (3.190). The solution was programmed on the personal computer HP-45 (by V.E. Axelrad).

For the range of local shapes and stress distributions, presented in Fig. 39, the solution with only the Λ_3 term retained in the mode (3.190) renders practically the same value of λ as the solution with up to eight terms $C_j \cos mx \, \Lambda_j(ny)$. Thus, the simplest version of equation (3.192), $D_{33}(\lambda, m, H) = 0$, may be used, providing the formula

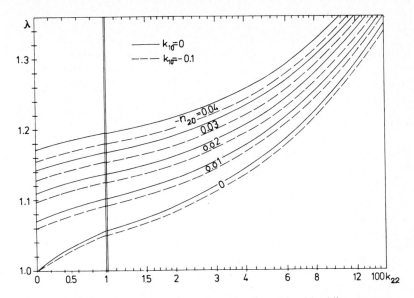

Fig. 39. Correction factor to the asymptotic value of local-buckling stress.

$$\lambda = \frac{m^2 F + m^{-2}(1 + 6.869k_{22}n^{-2} + 0.1538k_{10}H^2)}{2 + 0.3077H^2 + 0.0308H^4 + n_{12}Un^{-2} + n_{20}VH^2} ,$$

$$F = 1 + 0.3077H^2 + 0.0923H^4 + 0.0181H^6 + 0.0017H^8 , \qquad (3.194)$$

$$U = 6.869 + 0.2606H^2 - 0.2048H^4 ,$$

$$V = 0.1538 + 0.0614H^2 + 0.0091H^4 .$$

This relation determines also the values of m and $n = Hm$, those rendering a minimum for λ.

The results of the global analysis, available for cylinder shells [83, 183, 196], confirm those of local-buckling investigation. In particular, formula (3.194) proves adequate even for profiles (3.186) with $e = 0.9$, i.e. for the curvature $1/R_2' = 1/R_2$ varying as much as from $0.1b$ to $1.9b$ (Tables 6 and 7).

The solution presented provides an explanation for the (mentioned above) *local* character of buckling of uniformly compressed circular cylinder shells [129, 168]. The mode $W = C \cos mx \, \Lambda_3(ny)$ corresponds

TABLE 7. Critical uniform stress for shells with variable local shape. $h = 0.01b$.

| No. | In (3.200) | | | | | $\lambda = T_{1cr}R_2(0,0)\sqrt{3(1-\nu^2)}/Eh^2$ | | | |
	m	N	e	k_{22}	n	$\dfrac{T_2}{T_1}$	[196]	(3.194)	(3.199)
1	0	—	0	0	—	0	—	1.0001	1.0
2	0	2	0.3	0.0037	0.1315	0	1.0357	1.029	1.028
3	0	6	0.0455	0.0037	0.3945	0	—	1.029	1.026
4	0	2	0.6	0.0227	0.1740	0	1.1125	1.097	1.096
5	0	4	0.2720	0.0227	0.3479	0	—	1.097	1.092
6	0	2	0.9733	0.5448	0.1739	0	—	1.708	1.700
7	0	2	0.9	0.5448	0.3479	0	1.720	1.708	1.707
8	0	4	0.8	0.5448	0.5205	0	—	1.708	1.705
9	0	8	0.6412	0.5448	0.7808	0	—	1.708	1.646
10	0	10	0.5	0.3027	0.778	0	—	—	1.467
11	0	10	0.5	0.3027	0.778	0.2	—	—	1.215
12	0.15	10	0.5	0.3027	0.778	0	—	—	1.517
13	0.15	10	0.5	0.3027	0.778	0.2	—	—	1.211
14	0.3	10	0.5	0.3027	0.778	0	—	—	1.579
15	0.3	10	0.5	0.3027	0.778	0.2	—	—	1.259
16	0.6	10	0.5	0.3027	0.778	0	—	—	1.673
17	0.6	10	0.5	0.3027	0.778	0.2	—	—	1.394

to virtually the same critical stress (Table 7, line 1) as the double-periodic mode (3.179) does. Moreover, even a single-buckle mode $W = C\Lambda_0(mx)\Lambda_3(ny)$ leads, for the shells identical to the specimen of [168] to $T_{1cr}/h = 1.015\sigma_{cl} \approx \sigma_{cl}$.

9.4. A more general algorithm

Consider a local-stability investigation free of the restrictions connected with the use of (1.134) and with the simplest description (3.187) of the local stress and shape.

The buckling mode will be sought in the form of a double Fourier series,

$$\begin{bmatrix} W \\ F \end{bmatrix} = \sum_{i=0}^{I} \begin{bmatrix} C_i \\ F_i \end{bmatrix} \varphi_i(x, y) = \sum_{k=1}^{M+1} \sum_{l=0}^{L} \begin{bmatrix} w_{kl} \\ f_{kl} \end{bmatrix} \cos mkx \cdot \cos nly ,$$

$$I = ML + M + L , \quad k = 1 + \text{i.p.}[i/(L+1)] ,$$

$$l = i - (L+1)(k-1) , \tag{3.195}$$

where i.p.[ ] denotes the integer part of the quantity in the brackets; m and n are (constant) parameters of buckling mode.

Substitute the expressions (3.195) into (1.132). Applying the Galerkin method or equating to zero the Fourier coefficients of the left-hand side of each of equations (1.132) one obtains algebraic equations for C_i and F_i. This system can be composed in an easily programmable matrix form extending that of Chapter 2, §7,

$$
\begin{bmatrix} (D_1 + D_2)^2 - \lambda 2T & K \\ -K & (D_1 + D_2)^2 \end{bmatrix} \begin{bmatrix} W \\ f \end{bmatrix} = \begin{bmatrix} 0 \\ 0 \end{bmatrix},
$$

$$
\begin{bmatrix} K \\ T \end{bmatrix} = \begin{bmatrix} K_2 & K_1 \\ T_1 & T_2 \end{bmatrix} \begin{bmatrix} D_1 \\ D_2 \end{bmatrix}, \tag{3.196}
$$

$$
W = \{C_0 \ C_1 \ \cdots \ C_I\}, \qquad f = \{F_0 \ F_1 \ \cdots \ F_I\}.
$$

The matrices D_1 and D_2 represent the operators ∂_1^2 and ∂_2^2, respectively, in such a way that D_1W, D_1f, D_2W and D_2f are column matrices of the Fourier coefficients of the functions $-\partial_1^2 W$, $-\partial_1^2 F$, $-\partial_2^2 W$ and $-\partial_2^2 F$, respectively. Specifically D_1 and D_2 are diagonal matrices determined by formulas for the elements of the line i,

$$
D_1(i, i) = m^2 k^2(i), \qquad D_2(i, i) = n^2 l^2(i),
$$

with $k = 1, 2, \ldots, M + 1$; $l = 0, 1, \ldots, L$ determined by (3.195).

The matrices K_1, K_2, T_1 and T_2 are composed of the coefficients of double Fourier series of the functions k_1, k_2, T_1 and T_2, respectively. For instance, the composition of the matrix K_1 assures that the column matrix $K_1 D_2 W$ determines in (3.196) the Fourier coefficients of the function $-k_1 \partial_2^2 w$. For the expansion, where $k_1(k, l)$ denotes Fourier coefficients of $k_1(x, y)$,

$$
k_1(x, y) = \sum_{k=0}^{2M} \sum_{l=0}^{2L} \frac{1}{\delta_k \delta_l} k_1(k, l) \cos mkx \cos nly \, ;
$$

$$
\delta_0 = 2, \, \delta_\beta = 1, \, \beta \neq 0 \, ; \tag{3.197}
$$

it can be ascertained by means of the relation $2 \cos \beta \cos \gamma = \cos(\beta + \gamma) + \cos(\beta - \gamma)$ that the elements of the matrix K_1 are given by the formula

$$K_1(i, j) = \frac{1}{4\delta_1}[k_1(|k - k'|, |l - l'|) + k_1(|k - k'|, l + l')$$
$$+ k_1(k + k', |l - l'|) + k_1(k + k', l + l')].$$

(3.198)

Here k and l are rendered by the expressions (3.195), the integers k' and l' are defined by the same expressions with j instead of i. Formulas similar to (3.198) determine the remaining matrices of the equations (3.196): K_2, T_1 and T_2.

Eliminating f we obtain out of (3.196) the system

$$\left\{ [(D_1 + D_2)^2 + K(D_1 + D_2)^{-2}K]^{-1}T - \frac{1}{2\lambda} \begin{bmatrix} 1 & 0 & \cdots \\ 0 & 1 & \cdots \\ \cdots & \cdots & \cdots \end{bmatrix} \right\} W = 0.$$

(3.199)

The largest eigenvalue $1/(2\lambda)$ of (3.199) and the respective eigenvector W render the critical load and the buckling mode $W(x, y)$. They are effectively computed by means of the R. von Mises matrix-iteration method (cf. M. ESSLINGER et al., *Der Stahlbau* (1975) 71–74).

This solution does not assume the buckling to be local. It can be exact as far as the Donnell-type equations (1.132) are applicable, the series (3.195) satisfy the boundary conditions and the expansion (3.197), as well as those for k_2, T_1 and T_2, describe the actual prebuckling state. The algorithm was coded on a desk-top computer HP-85. For a 30-term mode (3.195) the calculation requires 7 minutes.

The series solution allows a further check of the local stability concept. The two questions to be clarified are: (1) What are the limits of the zone to which the adequate stability analysis may be restricted? (2) Is the leading-term description (3.187) of the prebuckling state sufficient?

Consider first cylindrical shells—those defined by (3.186). The results of calculations, presented in Table 7 and Fig. 40(a), display for a range of substantially different shapes nearly equal values of λ and the modes $W(x, y)$, which nearly coincide within the buckling zone. In all these cases, λ depends solely on the parameter k_{22}. The actual shape is (for a fixed k_{22}-value) of no influence for any profiles (3.186) with $n < 0.5$. Recalling (3.185) we recognize: the buckling is determined by the

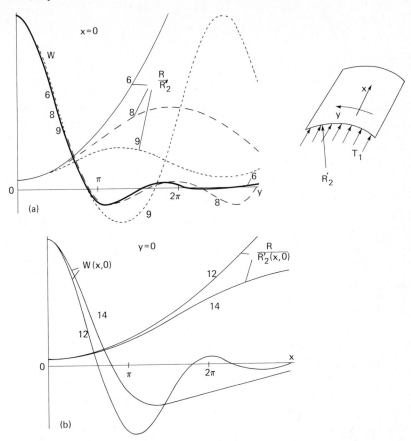

Fig. 40. Local shapes and buckling modes for the examples 6, 8, ... of Table 7.

situation in the zone $|y| < \pi$—one half the wave-breadth c_2 of the unrestrained periodic mode (3.185).

This conclusion applies to variations of stress and to noncylindrical shells. However, the simultaneous involvement of several parameters characterizing stress and shape in a more general case (there are ten in (3.187)) makes the influence of each of them less transparent.

The further discussion is restricted to one type of shells, to T_1, $T_2 =$ const. and to the influence of two factors not discussed in the foregoing. We define these shells by a deviation—distance $W_0 = 0.5h \cos mx \cos 10\eta$—from a circular cylinder of radius b (the Lamé parameter b is, of course, only approximately equal to this radius). For

$h = 0.01b$ we find using (1.128) with $W = W_0$,

$$\frac{b}{R_2'} = \left(\frac{1}{b} - \frac{\partial^2 W_0}{b^2 \partial^2 \eta} \right) b = 1 - e \cos mx \cos ny , \quad e = 0.5 . \quad (3.200)$$

Setting $R = R_2'(0, 0) \equiv 2b$, $\nu = 0.3$ and using (1.133) we have $c = 0.0778b$, $n = Nc/b = 0.778$ ($N = 10$), and according to the definition (3.187), $k_{22} = en^2/(2 - 2e) = 0.3027$.

The influence of the length $2\pi c/m$ of the imperfection wave (3.200) and of the compression stress T_2/h in the circumferential direction (represented by T_2/T_1) on the buckling stress T_1/h is illustrated by Table 7 and Fig. 40(b). The $m = 0$ lines of Table 7 concern a cylindrical shell. The other lines of the table indicate the critical stress to be the higher the shorter the wavelength $2\pi c/m$ of imperfection (3.200). But for any wavelengths not shorter than $2\pi c/0.3$ the critical stress differs only insignificantly from that of a cylindrical shell.

The reduction of the longitudinal critical stress in consequence of the lateral compression is displayed by the lines 11–17 of Table 7. (However, the influence of the lateral compression caused by normal pressure q is not restricted to the *local* stability. It can lead to global buckling and a check is needed whether the shell can withstand this.)

9.5. Failure of bent tubes

The character of collapse of tubes under bending is well known from experiments[18]. Thin tubes invariably collapse by way of buckling inside a small zone.

In what follows, the analysis of failure of tubes consists in calculating the shape of the tube and the stress resultants for successive values of the applied bending moment and checking the stability of each of the equilibrium states. The lowest value of the applied moment, leading to bifurcation or to snap-through, determines the collapse load. The prebuckling deformation is determined in the present chapter, §§2, 3, 8. To bypass the difficulty, caused by the two-dimensionally varying curvatures and stress resultants of the prebuckling configuration, the

---

[18] Reported already in 1927 by BRAZIER [9] (cf. the discussion in present chapter, §1.4).

stability will be investigated with the technique of local analysis set forth in present chapter, §9.3.

In terms of the dimensionless dependent variables $T = T_1/Ehh°$ and $\kappa = \kappa_2 b$, the stability condition (3.188) takes the simple form

$$\frac{-T}{1 + \kappa} \leqslant 2\lambda . \tag{3.201}$$

The check of stability was carried out for each stage of loading—each value of the bending moment—by computing the quantity (3.201) over the shell. The buckling is indicated by the quantity $-T/(1 + \kappa)$ reaching at some point the value of 2λ.

To illustrate the calculations *consider an example*: the check of stability of a *long, initially cylindrical and circular, tube* loaded by a nearly maximum (snap-through) bending moment. This state of the tube is described by Table 2 (present chapter, §2.4) in terms of the Fourier coefficients of ψ and ϑ. Using the definitions (2.9), (2.15) and (2.19) we have $T = \dot{\psi}$ and $\kappa = \dot{\vartheta}$.

To determine the λ we need the y^2 terms of the power expansions of T and κ in the vicinity of the point susceptible of instability. In this case, it suggests itself to check the stability at $\eta = \eta_0 = \pi$ (Fig. 31). With $\cos n\eta = \cos n\pi + (\eta - \pi)^2(n^2/2)(-\cos n\pi)$ we have

$$T(\eta) = \sum jb_j \cos j\eta = T(\pi) - \sum [(\eta - \pi)^2 b_j j^3/2] \cos j\pi ,$$

$$T(\pi) = \sum jb_j \cos j\pi \equiv T .$$

We express now $\eta - \pi$ in terms of the coordinate y defined in (1.133): $\eta - \pi = cy/b = (h°R/b)^{1/2}y$ and $h° = h/[b\sqrt{12(1 - \nu^2)}]$. As stipulated in the foregoing $R = R_2'(\pi)$, we find $b/R_2' = b/R_2 + \kappa_2 b = 1 + \kappa$. Thus, $\eta - \pi = y\sqrt{h°/(1 + \kappa)}$.

Substitution of $\eta - \pi$ into the above expression of $T(\eta)$ determines in terms of y the quantity $T_1/T_1(\pi) \equiv T/T(\pi)$. According to (3.189) with $T_1(x = 0, y = 0) \equiv T_1(\pi)$ this determines n_{12},

$$T(\eta)/T(\pi) = 1 + n_{12}y^2 , \quad n_{12} = -\frac{h°}{2T(\pi)}\frac{1}{1 + \kappa} \sum j^3 b_j \cos j\pi . \tag{3.202}$$

The quantities $\kappa \equiv \kappa(\pi)$ and k_{22} (which is defined in (3.189)) are obtained similarly

$$k_2(\eta) \equiv \frac{R_2'(\pi)}{R_2'(\eta)} = \frac{1 + \kappa(\eta)}{1 + \kappa(\pi)} \cong 1 + k_{22} y^2 \,,$$

$$k_{22} = \frac{-h°/2}{(1 + \kappa)^2} \sum n^3 a_n \cos n\pi \,.$$

(3.203)

The relative curvature $k_1 = R_2'/R_1'$ is determined, according to the definition (2.20) in terms of μ^*. Using[19] $R/kc^* = R_1'$ from (2.9) and $c^* = \cos \alpha^*(\pi) = -1$ and $R \cong R_m$ (Fig. 17), we have $R_m/k = -R_1'(\pi)$. Thus, (2.20) gives $\mu^* = -b/[R_1'(\pi)h°]$ and $b/R_1'(\pi) = -h°\mu^*$. With $b/R_2' = 1 + \kappa$ we find $k_{10} \equiv k_1(\pi)$

$$k_{10} \equiv \frac{R_2'(\pi)}{R_1'(\pi)} = -\frac{h°\mu^*}{1 + \kappa} \,.$$

(3.204)

It remains to substitute the coefficients b_j and a_n from Table 2 to find for the example $T \equiv T(\pi) = -1.2133$ and $\kappa = \kappa(\pi) = -0.60844$, and in the stability condition (3.201) there is $-T/(1 + \kappa) = 6.198$. This quantity does not depend on $h°$. On the other hand, (3.202)–(3.204) give by $h° \rightarrow 0$: $n_{12} \rightarrow 0$, $k_{ij} \rightarrow 0$ and thus $\lambda \rightarrow 1$.

The stability condition (3.201) takes with $\lambda = 1$ its asymptotic form. It shows that the buckling moment is exceeded, by far $(6.198 > 2)$.

This conclusion remains valid for *any* realistic wall-thickness. Indeed, consider the case $h/b = 0.1$—as high as can be consistent with the use of the thin-shell theory. Inserting $h° = h/[b\sqrt{12(1 - \nu^2)}]$, and the just calculated quantities T and κ into (3.202)–(3.204) we find using the b_j and a_n from Table 2: $n_{12} = -0.0175$, $k_{22} = 0.01486$ and $k_{10} = -0.1238$. With this equation (3.194) gives $\lambda = 1.13$. The critical stress is exceeded by far $(6.198 > 2 \times 1.13)$. The maximum (snap-through) bending moment *cannot* be attained. Long tubes collapse by local buckling.

For *finite tubes* the application of the stability condition (3.201) is similar. The prebuckling state continues to be represented by the expressions (3.189).

As mentioned, local buckling is sensitive to *imperfections* of the shell. The case of tube bending is no exception.

The great effort invested in the investigation of the effect of initial

---

[19] In this sentence R has the meaning corresponding to (2.9) and (2.20).

imperfections since the work of DONNELL [15] has clarified many sides of the problem (cf. $[160, 167, 194, 202, 215]^{20}$). But when no reliable data on possible imperfections are available, their eventual influence is taken into account by reducing the theoretical buckling stress by an experimentally assessed factor. For the axial compression of isotropic circular cylinders, the critical stress is assumed to equal $\chi\sigma_{cl}$, with the factor χ recommended in design codes. This experience suggests the introduction into the relations describing local buckling of an imperfection factor $\chi < 1$. Condition (3.201) then becomes

$$\frac{-T}{1 + \kappa} \leqslant 2\lambda\chi . \tag{3.205}$$

The value of χ for the case of flexure must be somewhat higher than that for the axial compression of the same cylinder. The variation of the compressive stress can be expected to *reduce* the unfavourable effect of the imperfections. Indeed, deviations from the cylindrical form of the shell lead to a nonlinearity in the relation between a shortening of a longitudinal element of the shell and the stress. This decreases the maximum compressive stress, shifting the forces in the cross-section to the less stressed parts of it.

Some results of the analysis are shown in Figs. 41 and 42 and in detail in [225]. The decisive factor is the length of a tube, represented by the parameter l, defined in (3.152).

When the length is small enough the end flanges practically eliminate the effects of the deformation of the cross-sections. *Short tubes* bend linearly (Fig. 41) and buckle when the maximum stress reaches the critical value, which is only slightly more than that for the uniform axial compression. This class includes tubes with $l \leqslant 0.03$ for "thin" flanges or those with $l \leqslant 0.05$, when the end flanges also exclude the warping deformation (Fig. 42).

The bending of longer tubes is influenced by the prebuckling deformation of the cross-sections. The influence grows with the length l. This leads to lower values of the buckling load.

The relation $M°(m)$ between the bending moment applied to the tube and the angle of flexure displays a maximum (Fig. 41). The maximum moment *could* constitute the limit-point *collapse load*. How-

---

[20] A comprehensive numerical investigation of the stability of cylinder shells with measured imperfections is given by J. ARBOCZ in Report LR-419, Dept. of Aerosp. Eng., TH Delft, 1984.

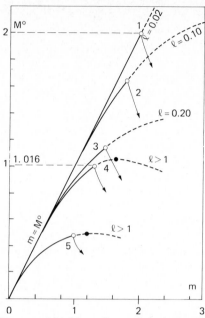

FIG. 41. Nonlinear flexure and buckling. Curves 1, 2, 3, 4: cylinder tubes under pure bending. Curve 5: tube with initial curvature $\mu = 0.5$ under bending and external pressure $q° = -2$. Thin end flanges. $m = (\varphi^* - \varphi)b/(Lh°)$, L—length of a tube.

ever, for all circular tubes and edge conditions investigated, the buckling occurs under a lesser [21] load. This is demonstrated in Figs. 21, 25, 26 and 41 and Table 8 which concerns tubes of unlimited length.

The following remarks may be helpful in putting the theory in proper perspective.

The limitation of the accuracy of the "second-order theory", indicated in present chapter, §8.2, can be observed in Fig. 42. Tubes

TABLE 8.

μ	$q°$	m_{cr}		$M°_{cr}$		$M°_{max}$	$m(M_{max})$
		$h/b=0$	$h/b=0.05$	$h/b=0$	$h/b=0.05$		
0	0	1.34	1.45	1.016	1.040	1.057	1.65
0	−2	0.971	1.09	0.635	0.641	0.642	1.10
0.5	0	1.317	1.459	0.862	0.879	0.885	1.60
0.5	−2	0.963	1.136	0.485	0.492	0.492	1.20

[21] A different case can be found in Fig. 25. It concerns an infinitely long tube with elliptical profile $A/B = 0.7$.

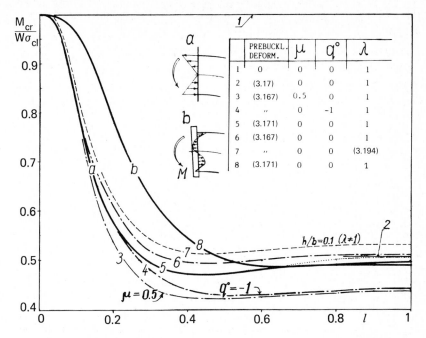

FIG. 42. The buckling moment depends on tube length (l), initial curvature (μ) and on the edge stiffening (cases a and b are defined in present chapter, §6.1).

longer than $l = 0.20$ with "thin" flanges, or $l > 0.35$ by the "stiff" flanges, have a prebuckling deformation, which is too large to be described by the second-order theory accurately. For such tubes the theory underestimates the precritical deformation. This leads to overestimated buckling moments. It is displayed in the right-hand sides of the graphs 6 and 7 of Fig. 42.

Rather unexpectedly, the M_{cr}-values are for some tubes *lower* than those calculated for longer or even for infinitely long tubes (line 2 in Fig. 42). This phenomenon is a consequence of an inherent feature of the influence of end flanges. Namely, the end flanges not merely resist the prebuckling flattening of the cross-sections; they also cause a redistribution of the longitudinal stress resultants T_1. Owing to this, the buckling starts by long enough tubes not in the middle of the tube length. The criterium (3.201) applied only for the midlength cross-section, where the deformation tends with growing length l to that of infinitely long tubes, gives M_{cr}-values (of the dotted line in Fig. 42) approaching for large l-values the M_{cr} of infinite tubes.

The published *experimental data* on stability of tubes in bending mainly concern short tubes. Comparison with the theory is also made difficult by omission in many cases of the information on the length of the specimen. The test series of [118] carried out with the aim of checking the theory just set forth are an exception. These experiments included measurements of the prebuckling deformation.

The experimentally found buckling moments are lower than the theoretical values. The difference can be explained by the effect of initial imperfections.

Initial curvature of a tube enhances the precritical deformation. Even for the value $\mu = 0.5$, which is merely one third of the elastic curvature $\mu^* = m$ corresponding to the buckling of a long cylinder tube, the buckling moment is considerably lower than for a perfect cylinder (Fig. 42).

The last remark concerns the influence of the variability of the curvatures and stress resultants *inside* the zone of eventual local buckling, manifested by $\lambda > 1$ for finite h/b. It is represented in Fig. 42 by the difference between the values of the critical moment for $h/b \rightarrow 0$ and those indicated by the graph for $h/b = 0.10$ (curve 7). The difference is obviously not a substantial one, particularly so when we recall that to buckle in the *elastic* range a metallic tube must be at least 20 times thinner than $h/b = 0.10$. On the other hand, the wall-thickness can hardly be larger than $h = 0.10b$ (for *any* material), if the theory of *thin* shells is to be applied. The data of Fig. 42 and Table 8 show the effect under discussion to be insignificant. Its influence on M_{cr} is merely of the order of magnitude of $h/b = h/R_2$—negligible for the thin-shell theory (cf. Chapter 1, §3.1).

10. Buckling of tubes and torus shells under external pressure

The buckling of tubes with various conditions on edges and of closed torus shells is considered as a single problem interdependent with that of tube bending.

Starting with the formulation of the problem in terms of the semi-momentless equations and boundary conditions the critical pressure is obtained by way of the Fourier-series solution discussed in the foregoing. For tubes of small curvature this leads to simple explicit relations, which can be recognized as an extension of the classical solution for cylinders. In the other limit case (long tubes and closed torus shells), the results of the series solution converge to an asymptotic formula.

10.1. Stability conditions

A uniform external pressure causes in tubes and torus shells of circular cross-section only a negligible wall-bending. For the "slender tubes" the corresponding Reissner equations as well as the semi-momentless theory determine zero wall-bending. The insignificance of this deformation for *any* value of b/R_m is shown in Chapter 5, §5.2.

Let us consider ideally circular tubes and assume the condition of "slenderness" $r = 1$. This reduces the analysis of collapse under external pressure to a linear bifurcation-buckling problem. The canonical version of the semi-momentless theory, disregarding γ and H terms (Chapter 2, §4.3), will be applied.

Tubes with the two limiting-case types of flanges will be considered. In contrast to flexure problem, discussed in the foregoing, conditions on the edges require an additional specification. It concerns the "rigid-body displacement" of the tube edges. Besides the case of one of the edge flanges being free to displace (Fig. 43, *af*, *bf*), tubes with the "rigid-body displacement" being prevented for *both* edges will be considered (Fig. 43, *ar*, *br*).

The stability problem under discussion is described by (3.167). But for the problem of uniform external pressure the equations can be simplified. Indeed, for $\kappa_2 b \ll 1$ we have f_0, $f_1 = 0$ (Chapter 2, §6.2) and the $\cos 0\eta$ and $\cos 1\eta$ coefficients of $T(\xi, \eta)$ are known

$$N^\circ = \frac{q\pi b^2}{2\pi b} \frac{1}{Ehh^\circ} \equiv q^\circ h^\circ /2 , \qquad M^\circ$$

and thus g_0, $g_1 = 0$ in (2.197) and (3.167). This allows us to free (3.167) of the corresponding zeroth and first lines and columns of the matrices W^*, W, L^*, L', f and g, that is, to transform the system similarly to present chapter, §6.2 and represent the buckling mode by the series of the form (3.109).

Further, the matrices W^*, L^* and L' can be simplified by neglecting all the M° and N° terms in formulas (3.168) defining the elements of

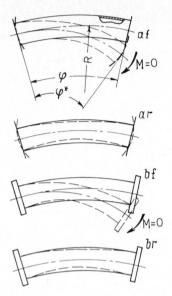

FIG. 43. An end of a tube (stiffened by the thin flange a or by the stiff flange b) may be either free (af, bf) or restricted with respect to the rigid body rotation (ar, br).

these matrices. As follows from the estimates of $M°$ and $N°$, given in (3.208) and present chapter, §10.2, these terms are, at most, of the relative order of magnitude of h/b. Thus, for the present stability problem the system (3.167) reduces to the form

$$
M_j^q \begin{bmatrix} \check{f}^j \\ \check{g}^j \end{bmatrix} = -\frac{4}{\pi j} \begin{bmatrix} 3\mu M° \\ 0 \\ 0 \\ \cdots \end{bmatrix} + M_0^q \begin{bmatrix} f^B \\ g^B \end{bmatrix}; \qquad M_j^q = \begin{bmatrix} \check{W} + q°\Lambda^2 & -\check{L}_j \\ \check{L}_j & \check{W} \end{bmatrix},
$$

$$(3.206)$$

matrices $\check{L}$, $\check{W}$, $\check{f}$ and $\check{g}$ are defined in (3.111), $\check{L}_j$ denotes matrix $\check{L}$ with the operator d^2 replaced by the quantity $-j^2/l^2$.

Consider, first, tubes with one *end-flange free* to displace. In these cases, shown in Fig. 43 by schemes af and bf, the bending moment in any cross-section is equal to zero: $M° = 0$. For *thin flanges* we have, recalling (3.107) and present chapter, §7.1(a): f^B, $g^B = 0$. The critical pressure is indicated by the $j = 1$ system (3.206) with $M° = 0$ naving a nonzero solution. The condition, rendering the lowest value of the critical pressure q_{cr}, is that for the buckling mode $j=1$, extending over the entire length of the tube: $\det(M_1^q)=0$. For the case $M°=0$ and f^B, $g^B=0$, equations (3.206), determining the buckling mode and the critical pressure, can be easily simplified by eliminating g^1 (cf. present chapter, §6.2),

$$D^1\check{f}^1 = 0, \quad D^1 = \check{W} + q°\Lambda^2 + \check{L}_1\check{W}^{-1}\check{L}_1,$$

$$\det[D^1(l, \mu, q_{cr}°)] = 0. \tag{3.207}$$

For *stiff flanges* the reactions g^B must be determined by the edge conditions. As described in present chapter, §7.1(b) this involves deformation described by more than one $\sin j\hat{\xi}$ component.

For the tubes with stiff flanges the critical external pressure will be determined indirectly—as the lowest external pressure causing unboundedly large deformation of the tube for arbitrarily small bending moment. Here we use the solution of the flexure problem presented in (3.151)–(3.155) with the matrices M_j^q, which take account of the pressure q according to formulas (3.206) replacing M_j. The critical pressure is found by repeated trial calculations using (as in the case of equation (3.207) $\det(D^1)=0$) Newtonian interpolation.

The results of the solutions of the two stability problems are presented in Fig. 44.

Turning now to the tubes with *both end-flanges fixed* with respect to rigid-body rotation (Fig. 43) we recognize the bending moment in the cross-sections of the tube to be *not* equal to zero. The moment $M°$ is statically indeterminate. What is known in this case is the angle of flexure of the tube $\varphi^* - \varphi$: as the tube ends cannot rotate, $\varphi^* - \varphi = 0$. This kinematic condition serves to determine the bending moment. With the relation (3.157), where for the flanged tube ends $f_2^B = 0$, we have

$$M° = \frac{\mu}{4}\frac{2}{\pi}\sum_{j=1,3,\dots}\frac{1}{j}f_2^j. \tag{3.208}$$

Being due to the (reactive) forces on the tube ends the moment is constant along the tube. Inserting this expression for $M°$ into the right-hand side of (3.206), we can base the further stability analysis on

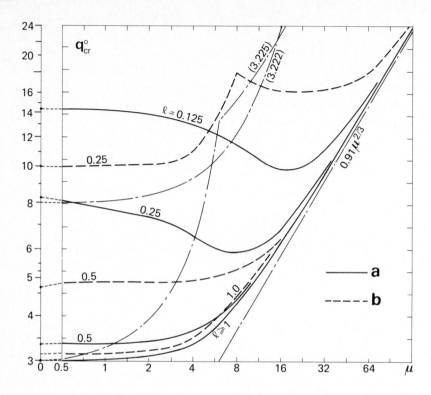

FIG. 44. Buckling pressure for tubes with end conditions indicated in Fig. 43 under *af, bf*.

the pattern just described. Thus, the stability equations for the case of *thin flanges* ($f^B = g^B = 0$) are given by

$$M^q_j \begin{bmatrix} \breve{f}^j \\ \breve{g}^j \end{bmatrix} = -\frac{4}{\pi j} \begin{bmatrix} 3\mu M^\circ \\ 0 \\ 0 \\ \cdots \end{bmatrix}, \qquad (3.209)$$

where M° has the value determined in (3.208). Eliminating g^j we obtain a system determining the buckling mode and critical pressure similar to (3.207). However, in this case (fixed flanges) the buckling mode includes *several $f^k \sin k\hat{\xi}$* terms simultaneously.

The stability system has the form

$$D^k \check{f}^k = -\frac{6\mu^2}{k\pi^2} \begin{bmatrix} 1 \\ 0 \\ \cdots \end{bmatrix} \sum_{j=1,3,\ldots} \frac{f_2^j}{j} \qquad (k = 1, 3, \ldots), \qquad (3.210)$$

the matrix D^k is defined in (3.207).

The calculations indicate some characteristic features of the buckling mode (which apply also to the sin $n\eta$ mode, discussed briefly in present chapter, §10.3).

For a *cylinder* the mode (Fig. 45(a)) includes only *one* cos $n\eta$ harmonic in the circumferential direction.

For a *curved* tube there are *several* harmonics present and the mode includes an overall flexure of the tube (Fig. 43). Tubes with large μ and l have the deformation of the type discussed in present chapter, §2.3. The wall-bending is concentrated inside the zones of breadth $\pi b/n^*$ with n^* estimated by the formula

$$n^* \approx \mu^{1/2}, \qquad n^* \geq 2. \qquad (3.211)$$

The number of terms retained in the cos $n\eta$-series (3.109), representing the buckling mode, and the corresponding order of the matrices D^1, $\check{L}$ and $\check{W}$ should be no less than $\mu^{1/2}$. In addition, this number depends on the length of the tube. As becomes evident in the next section, it cannot be less than $3^{1/8}l^{-1/2}$.

The results of the calculations described, shown in Fig. 44, display possibilities of a considerable simplification of the stability analysis for two wide classes of tubes. These will be examined in the next two sections.

10.2. Small-curvature tubes

A glance at the graphs of Fig. 44 shows a rather weak dependence of the critical pressure on the tube curvature for small values of μ. Thus, q_{cr} may be assumed equal to that of $\mu = 0$ for $\mu \leq 1$, and even for $\mu \leq 2$, when the engineering accuracy is acceptable. The limit value of q_{cr} for $\mu = 0$, i.e. for a *cylinder*, is easily determinable in an explicit way. Let us investigate this possibility.

For a *cylinder*, i.e. $\mu = M^\circ = 0$, the matrices W^*, L_j^*, L_j' and W are determined by formulas (3.168) as *diagonal* matrices. With this the system (3.167) falls apart. Each pair of functions $f_n(\xi)$, $g_n(\xi)$, $n =$

$2, 3, \ldots$ is determined by a separate system. After elimination of g_n the system is reduced to

$$\left\{ [d^4 - N^\circ(n^4 + n^2)d^2]\frac{1}{n^4 - n^2} + n^4 - n^2 + n^2 q^\circ \right\} f_n = 0 ,$$

$$d^2 = \frac{\mathrm{d}^2}{h^\circ \, \mathrm{d}\xi^2} . \tag{3.212}$$

For the case of *thin flanges* the conditions at the edges $\xi = \pm L/2b$, defined in (3.137), are fulfilled by $f_n = C_n \cos \pi b\xi / L$. Inserting f_n into (3.212) renders the condition of buckling ($C_n \neq 0$) as an equation for q_{cr}°,

$$q_{cr}^\circ - \frac{N^\circ}{n^2 l^2} \frac{n^2 + 1}{n^2 - 1} = n^2 - 1 + \frac{1}{n^4 l^4} \frac{1}{n^2 - 1} \quad \left(l = \frac{L}{\pi b} h^{\circ 1/2} \right). \tag{3.213}$$

Here the critical-pressure parameter q_{cr}° is taken as *positive*

$$q_{cr}^\circ = |q_{cr} b^3 / D| .$$

The value of n determining the actual buckling mode is found via the condition of minimum for the right-hand side of (3.213). Of course, n (the number of waves of the buckling mode around the circumference of the cylinder) must be an integer greater than or equal to 2. Thus,

$$n = 2, 3, \ldots \text{ and optimally } n \approx \frac{1}{l^{1/2}} \frac{(3 - 2/n^2)^{1/8}}{(1 - 1/n^2)^{1/4}} \approx 2.74 \left(\frac{b^3}{L^2 h} \right)^{1/4} .$$

$$\tag{3.214}$$

The shorter the tube, and with it the length L of a buckle, the smaller the breadth $2\pi b/n$ of the buckle (Fig. 45(a)).

Before we start the discussion of the critical pressure, consider the longitudinal-force term of (3.213).

In the simplest case of a cylinder shell under *radial pressure* only, the longitudinal force N in a cross-section is zero.

The all-round normal pressure q acts also on the bottoms of a cylinder. As stated in present chapter, §10.1, the pressure determines $N^\circ = q^\circ h^\circ /2$. In this case the N term in the buckling equation (3.213) is

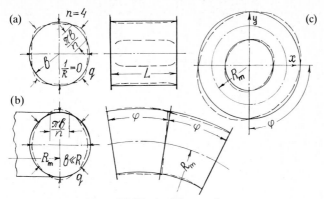

FIG. 45. The buckling modes.

negligible. Indeed, compared to q_{cr}° the N° term has the order of magnitude

$$\frac{h^\circ}{2n^2 l^2} = \frac{1}{2}\left(\frac{\pi b/n}{L}\right)^2 .$$

Furthermore, $(\pi b/Ln)^2 \ll 1$ may be recognized as the condition of applicability of the semi-momentless theory, formulated in Chapter 2, §4.1: $\pi b/n$ is the half-wave length of the deformation in the circumferential direction while L is the same in the longitudinal direction. To check whether the condition is in fact fulfilled we rewrite it with the n from (3.214),

$$\left(\frac{\pi b}{Ln}\right)^2 = 1.313\frac{\sqrt{hb}}{L} \ll 1 . \tag{3.215}$$

It becomes clear that the use of the semi-momentless theory is justified and, simultaneously, the N° term in the buckling equation (3.213) may be dropped for the case of external-pressure load, provided the length of the cylinder L is sufficiently large compared to the length c_1 of one "free" buckle according to (3.185).

A final remark on the case of *uniform axial compression* is made. In this limiting case, equation (3.213) determines the critical compressive longitudinal stress $\sigma_{cr} = -N/2\pi b$,

$$\frac{\sigma_{cr}}{\sigma_{cl}} 2\frac{n^2+1}{n^2-1} = (n^4 - n^2)l^2 + \frac{1}{(n^4 - n^2)l^2},$$

$$\sigma_{cl} \equiv 2Eh^\circ = \frac{Eh}{b[3(1 - \nu^2)]^{1/2}}.$$

Minimizing the right-hand side with respect to $(n^4 - n^2)l^2$ we have

$$\sigma_{cr} = \sigma_{cl}(n^2 - 1)/(n^2 + 1), \qquad n^4 - n^2 \approx 1/l^2, \qquad n = 2, 3, \dots ;$$

where $l = Lh^{\circ 1/2}/\pi b$ represents the *length L* of a *half-wave of buckling*. Thus (3.213) gives a correct account of the axial buckling stress even for $q = 0$ (cf. further discussion in [205], p. 185).

For the case of *thin flanges* the critical pressure (either radial or all-round) is determined by simplified relations[22] (3.213) and (3.214),

$$q^\circ_{cr} = n^2 - 1 + \frac{1}{l^4 n^4 (n^2 - 1)},$$

$$n = 2, 3, \dots \text{ and ideally } \approx \frac{3^{1/8}}{l^{1/2}} \equiv 2.74\left(\frac{b^3}{L^2 h}\right)^{1/4}. \tag{3.216}$$

For *cylindrical* tubes the stability problem presents no substantial difficulty for *other* edge *conditions* either. Equation (3.212) has a general solution of the form

$$f_n = \sum_{j=1}^{4} C_{nj} \exp(\xi h^{\circ 1/2} d_{nj}).$$

The constants C_{nj} and d_{nj} are determined by the boundary conditions and as the roots of the auxiliary equation corresponding to (3.212). This equation, simplified by dropping the small N° term, is (q°_{cr} taken as positive):

$$d_{nj} = [(n^4 - n^2)(q^\circ_{cr} n^2 - n^4 + n^2)]^{1/4} \quad (j = 1, 2, 3, 4). \tag{3.217}$$

---

[22] For the tube lengths (3.215) formulas (3.216) give q°_{cr} near to those calculated according to "European Recommendations for Steel Constr.", ECCS, 1984. For shorter cylinders (3.216) are conservative. The warping stiffness of the flanges leads to q°_{cr} of (3.218), which is substantially higher than according to ECCS.

The quantities C_{nj} and d_{nj} can be complex, determining the final analytical form of the real functions $f_n(\xi)$.

Consider now the case of *stiff flanges*, characterized by the conditions (3.140) with $k_2 = 0$, i.e. $f_n = 0$, $\mathrm{d}f_n/\mathrm{d}\xi = 0$. With the origin of the coordinate ξ in the midlength section constituting the plane of symmetry we have in the f_n only two terms, which are *even* functions of ξ. Calculations show that under these conditions the solution of equation (3.212) has the form

$$f_n = C_{n1} \cos(\xi h^{\circ 1/2} d_n) + C_{n2} \cosh(\xi h^{\circ 1/2} d_n) , \quad d_n = |d_{nj}| .$$

The conditions $f_n = 0$ and $\mathrm{d}f_n/\mathrm{d}\xi = 0$ on the edges result in a system of equations for C_{n1} and C_{n2}. The condition of instability is provided by equating the determinant of the system to zero,

$$\begin{vmatrix} \cos l^* & \cosh l^* \\ -d_n \sin l^* & d_n \sinh l^* \end{vmatrix} = 0 , \quad l^* = d_n \frac{L}{2b} h^{\circ 1/2} \equiv d_n \frac{\pi}{2} l .$$

The smallest root of this equation yields

$$d_n = \frac{1.5056}{l} .$$

Inserting here the expression (3.217) for $d_n = |d_{nj}|$ renders an equation for q°_{cr}, which differs from (3.216) in only one respect: the length parameter l is replaced by the quantity $l/1.5056$. That is,

$$q^\circ_{\mathrm{cr}} = n^2 - 1 + \frac{5.14}{n^4 l^4} \frac{1}{n^2 - 1} , \quad n = 2, 3, \ldots \approx \frac{1.41}{l^{1/2}} \equiv 3.36 \left(\frac{b^3}{L^2 h} \right)^{1/4} . \tag{3.218}$$

Thus, the tube with "stiff flanges" has the same value of the buckling pressure q_{cr} as has a tube 1.5056 times shorter but having thin flanges that do not resist any warping of the edges.

Formulas (3.216) and (3.218) determine the number n of waves of the buckling around the circumference of a cylinder. With n known in terms of l the graphs of Fig. 44 indicate a more specific estimate of the range of tubes, for which q_{cr} is *not* substantially *influenced by* the *curvature*. These are tubes with

$$\mu/n^2 \ll 1 \quad \text{or} \quad \sqrt{12(1 - \nu^2)} \frac{(b/n)^2}{Rh} \ll 1 . \tag{3.219}$$

That is, the simplified analysis of stability is possible when the interval of variation of the buckling mode (b/n) is small compared to $(Rh)^{1/2}$.

Curved tubes with no stiffening ribs do not, as a rule, satisfy the conditions (3.219) of *small curvature or small length*. But they are in most cases eligible for another simplification of the buckling analysis. Namely, the stresses resulting from *variation* of the buckling mode in the longitudinal direction are negligible. This situation is discussed in the next section.

10.3. Long tubes

When the distance $(L = \varphi R_m$, Fig. 45) between two stiffened cross-sections is large enough, there exists some middle part of the tube, where the buckling-mode deformation varies in the longitudinal direction with negligible intensity. In this part of the tube the differential terms of the stability equations following from (3.167) and the corresponding l terms of the algebraic equations of present chapter, §10.1 determining the q_{cr} are insignificant and may be dropped. For the leading elements of the instability equations (3.207) or (3.210)—the D_{nn} with n corresponding to (3.211)—this simplification is evidently possible when *one* of the following two conditions is fulfilled:

$$l > \mu^{-1/2} \quad \text{or} \quad l > 1 . \tag{3.220}$$

The data presented in Fig. 44 support these relations as conditions under which the approximation $q_{cr}(\mu, l) \approx q_{cr}(\mu, \infty)$ is possible. Let us consider the stability of these *long* tubes.

Without the l terms the instability equations (3.206), (3.207) or (3.210) become separable. Each of them spawns two determinantal equations: one containing only the even-numbered elements of any row and any column of the system matrix, the other with the matrix composed of the remaining rows and columns. For instance, the system (3.207) separates into a system for $f_2, f_4, \ldots$ and one for the remaining coefficients of the buckling mode $f_3, f_5, \ldots$. The respective instability equations determining q_{cr} are

$$\begin{vmatrix} C_{22} & C_{24} & \cdots \\ C_{42} & C_{44} & \cdots \\ \cdots & \cdots & \cdots \end{vmatrix} = 0 , \qquad C_{mn} = [D^1_{mn}]_{l=\infty} , \tag{3.221}$$

$$\begin{vmatrix} C_{33} & C_{35} & \cdots \\ C_{53} & C_{55} & \cdots \\ \cdots & \cdots & \cdots \end{vmatrix} = 0 . \tag{3.222}$$

These two equations yield quite *different* values of the critical pressure. The q_{cr} determined by (3.221) is one compelling a tube with *free to rotate* ends and $l = \infty$ to bend under zero bending moment (Fig. 43). It is the *lowest* critical pressure possible for a tube with a given curvature μ. This lower bound is attained by tubes with "thin flanges" and length above $l = 1$ (curve $l \geqslant 1$ in Fig. 44).

Equation (3.222) gives much higher values of q_{cr}. It corresponds to buckling modes (determined by odd-numbered $\cos n\eta$ harmonics) which *do not include* the tube *flexure*, i.e. leave some of the ξ-lines undeformed. Naturally, this sort of buckling (without flexure of the tube) is also possible with the $\cos 2\eta, \cos 4\eta, \ldots$ terms. The corresponding stability equation can be obtained from the system (3.167) by assuming the buckling deformation to be constant in the longitudinal direction and the change of curvature of the tube to be zero. With constant elements f_n, g_n of $\check{f}$, $\check{g}$ the system separates into two. One of the two systems resulting from (3.167) determines $f_2, f_4, \ldots, g_3, g_5, \ldots$. Upon elimination of the g_n this system becomes

$$\begin{bmatrix} C_{22} & C_{24} & \cdots \\ C_{42} & C_{44} & \cdots \\ \cdots & \cdots & \cdots \end{bmatrix} \begin{bmatrix} f_2 \\ f_4 \\ \cdots \end{bmatrix} = \begin{bmatrix} -M°3\mu \\ 0 \\ \cdots \end{bmatrix} . \tag{3.223}$$

It differs from the system leading to the stability equation (3.221) by the presence of the nonzero element $M°3\mu$ on the right-hand side. The condition of zero change of curvature determines $M°$. Setting $m = 0$ in (3.115) we obtain with (3.114),

$$M° = \frac{\mu}{4} f_2 . \tag{3.224}$$

The instability equation (3.223) assumes the form

$$\begin{vmatrix} C_{22} + \dfrac{3}{4}\mu^2 & C_{24} & \cdots \\ C_{42} & C_{44} & \cdots \\ \cdots & \cdots & \cdots \end{vmatrix} = 0 . \tag{3.225}$$

An exactly equivalent equation for the q_{cr} is obtained by equating the determinant of the matrix A of the system (3.25) to zero. Equations (3.225) or (3.222) (the one that yields the lower value of q_{cr}°) provide the *upper bound* of the critical pressure of unboundedly long tubes; the graphs are presented in Fig. 44. This is what occurs in tubes that are stiffened in some way, for instance, by longitudinal ribs, preventing any flexure in the longitudinal direction. (Cf. [226].)

For tubes stiffened merely in certain cross-sections, the critical pressure $q_{cr}(\mu, l)$ for sufficiently large values of the curvature μ approaches the lower-bound q_{cr} of *long* tubes. As is seen for the examples shown in Fig. 44, the curves of $q_{cr}^{\circ}(\mu, l)$ converge to that for tubes $l \geqslant 1$. And the curve $q_{cr}^{\circ}(\mu, 1)$ approaches for larger curvatures μ the straight-line asymptotic representing the formula

$$q_{cr}^{\circ} = 0.91 \mu^{2/3} . \tag{3.226}$$

The results of calculations, plotted in Fig. 44, allow a *rough estimate* of the critical pressure for tubes of any curvature and length between the stiffened cross-sections. It is the higher of the two limiting values: $q_{cr}^{\circ}(\mu, \infty)$ of (3.226) and $q_{cr}^{\circ}(0, l)$ provided by the cylinder formulas (3.216) or (3.218).

In the foregoing, only the buckling modes *symmetric* with respect to the plane of curvature were accounted for. A solution having, instead of (3.149), the form of a sine series and in the equations (2.140) the resultants of the *out-of-plane bending*,

$$T^{(0)} = T_M = N^{\circ} + M^{\circ} \sin \eta , \quad M^{\circ} = M_x b/(EIh^{\circ}) ,$$

leads to equations differing from those in (3.167) in the meaning attributable to M°, f and g. This yields stability equations identical in form to (3.206)–(3.211) and similar results for q_{cr}. (The relevant similarity of the effect of normal pressure for the *in- and out-of-plane bending* has been discussed in present chapter, §3.3.)

10.4. Closed toroidal shells

The buckling modes of a toroidal shell closed along the parallel circles must be consistent with the continuity of these ξ-lines. As shown in Fig. 45(c), such a mode includes not less than two full waves of deformation in the circumference of the ξ-line. Hence, the half-wave length L of the buckling mode along the ξ-line of a toroidal shell does not exceed $\pi R/2$. Inside each of the four half-waves the buckling mode is analogous to that in a tube with free-to-rotate thin flanges. The dimensionless length of this half-wave is

$$l = \frac{L}{\pi b} h^{o1/2} = \mu^{-1/2}\left(\frac{R_m}{4b}\right)^{1/2}. \qquad (3.227)$$

A check of this length with the conditions (3.220) shows that for all *closed* toroidal shells with

$$R_m \geqslant 4b$$

the critical pressure is nearly equal to that of a *long* tube which is free to bend. This critical pressure is determined as a function of μ by the curve $l \geqslant 1$ in Fig. 44.

The shells with $R_m < 4b$ are not "long" enough to have the lower-bound critical pressure. But for these cases also the curve $q_{cr}(\mu, l \geqslant 1)$ gives a *conservative* estimate for the critical pressure. A comparison with published experimental data, presented in [185], indicates the actual critical pressure to be sufficiently close to the q_{cr} calculated for $l = \infty$. There is good agreement not only with the experiments of SOBEL and FLÜGGE [113] on relatively slender shells with $R_m \approx 8b$ but also for specimens with $R_m \sim 2b$ as investigated by ALMROTH, SOBEL and HUNTER [121], NORDELL and CRAWFORD [156] and JORDAN [154]; the deviations remained very moderate even for $b = 0.85 R_m$. Moreover, no correlation between the deviation and the value of b/R_m of the specimens could be discerned.

This is significant since $b/R_m = 0$ was in fact assumed in the basic equations (3.167). The negligible influence of the assumption $b \ll R_m$ even when $b = 0.85 R_m$ is not accidental. It is a result of the concentration of the buckling deformation inside the zone of small breadth $\pi b/n \sim \pi b/\sqrt{\mu}$ (Fig. 45), where the assumption $R = R_m + b \cos \eta \approx R_m$ *is* accurate.

11. Bourdon tubes

For noncircular curved tubes, used as pressure-measuring instruments, stresses and displacements are investigated with the aid of the series solution of the Reissner–Meissner equations. Graphs are presented for practical calculations.

11.1. Basic problems

The Bourdon tube shown schematically in Fig. 46 is a pressure-measuring element of widespread application. An increase in the internal pressure q causes a decrease of the curvature $1/R_m$. The free end of the tube is thereby displaced. The displacement U indicates the change of pressure.

The unbending of the tube can be explained as follows: The internal pressure deforms the tube displacing the linear elements on the convex side of the tube (for instance, 11 in Fig. 46) away from the centre of curvature. The elements of the other side of the tube, such as 22, are moved nearer to the centre of curvature. It is clear on inspection of Fig.

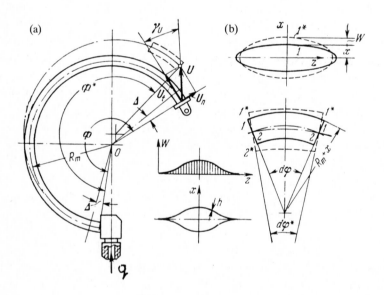

FIG. 46. Bourdon tube.

46 that with the deformation of the cross-sections just described the tube unbends, we have $d\varphi^* < d\varphi$.

Assuming all cross-sections to remain plane and their deformation to be constant along the tube, it is easy to express the longitudinal extension ε_1 in terms of the change-of-curvature parameter k' and of the radial displacement W (Fig. 46),

$$\varepsilon_1 = \frac{(R_m^* + x + W)\,d\varphi^* - (R_m + x)\,d\varphi}{(R_m + x)\,d\varphi}$$

$$\approx \frac{W + k'x}{R_m + x} \qquad \left(k' = \frac{d\varphi^* - d\varphi}{d\varphi} \right).$$

The highest sensitivity, i.e. the largest value of the ratio of k' to the pressure q causing the deformation, would be realized by the deformation without membrane strain—when $\varepsilon_1 = 0$. In this case the change of curvature of the tube brought about by the pressure q would require energy only for the *bending* of the thin tube wall, accompanying the deformation of the cross-sections in their planes. This ideal tube must have a profile $x = x(z)$ of the cross-section making $W + k'x = 0$. The graphs illustrating the displacement W and the corresponding profile of the tube of maximum sensitivity are presented in Fig. 46. The profile is determined by the relation $x = -W/k'$.

Six different shapes realized for Bourdon tubes are shown in Fig. 47. The sensitivity-optimal profile is approximated by that labelled (e). Other profiles have their good qualities and are preferable for particular applications. For instance, the cross-section form (a) is very far from being sensitivity-optimal. Under identical conditions, the displacement U (Fig. 46) of a tube with profile (a) is much smaller than could be obtained with the profile (e). But the profile (a) is better suited to withstand large pressure. The sensitivity-optimal profile (e) has a definite stress concentration on the ends of the longer axis, where the radius of curvature of the profile is small.

The flat-oval profile (b) is extensively used because of good sensitivity and comparatively easy manufacturing.

The cross-section of a Bourdon tube used for high pressure is presented in Fig. 47(c). It is a case where the thin-shell theory is clearly insufficient, particularly for the determination of the stresses.

The thin-walled tubes with profiles similar to an ellipse (Fig. 47(d))

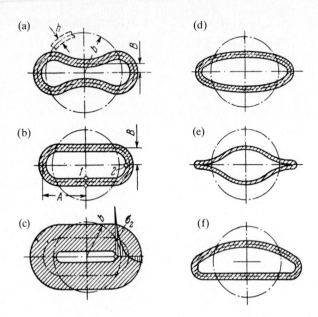

FIG. 47. Cross-sections of Bourdon tubes.

provide better sensitivity than the profiles (a) or (b). The same effect is achieved with the profiles of the type (f) (which is somewhat easier to manufacture).

Another basic scheme of the Bourdon pressure instrument is presented in Fig. 48. In contrast to the scheme of Fig. 46, both ends of the tube are fixed and the pressure is measured via the forces or moments

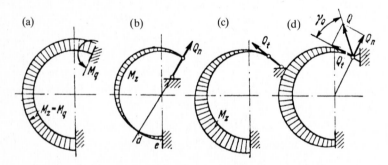

FIG. 48. Components of end moment or force.

exerted by an end of the pressure element. The four versions of this scheme shown in Fig. 48 correspond to the different components of the end force or the end moment, chosen to indicate the pressure.

To finish this brief review, pressure elements in the form of *twisted* tubes should be mentioned. (The twisted element was patented by Bourdon simultaneously with the curved pressure element in 1851.) The pressure applied to the twisted tube from within or from outside deforms its noncircular cross-sections (Fig. 49). This compels the tube either to unwind or to increase the twist. The elastic deformation results in rotation of the free end of the tube around its axis. First investigation of this pressure element is due to CHEN CHU [42]. A more precise formulation of the problem provides the dissertation of WÖBBECKE [158]. As a case of the theory of curved and twisted beams with deformable cross-sections [151], the twisted tube was investigated by BEGUN [161].

The first practically applicable method of evaluation for Bourdon tubes (of elliptic profile) was obtained in the student work of FEODOS'EV, in 1940. Later (1949), FEODOS'EV [32] considered the flat-oval tubes. These solutions were obtained by the Ritz method. The displacements were assumed proportional to those of a cylinder of the same profile under the normal pressure. FEODOS'EV [32] and WUEST [58, 87] have applied to the evaluation of Bourdon tubes the equations of cylinder shells. A wealth of experimental data has been presented by KARDOS [68] and MASON [57]. A review of earlier investigations on Bourdon

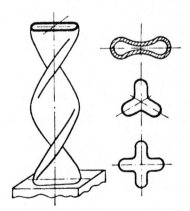

FIG. 49. Twisted Bourdon tube and three versions of profile.

tubes has been given by JENNINGS [55]. The problem is comprehensively treated in the monograph of WUEST [87].

In what follows, the analysis of Bourdon tubes is discussed on the basis of the Fourier-series solution of the Reissner–Meissner equations and of some results of numerical solution obtained by HAMADA, KATSUHISA and KIYOSHI [108]. Additional information on the series solution and a wide comparison of its results with experimental data may be found in the articles of VASIL'EV [98, 103].

All the investigations referred to so far consider the Bourdon tube without taking into account the influence of the end fixtures. This simplification is used also in what follows. However, it may be recommended to introduce a correction based on an evaluation of the sensitivity (present chapter, §11.2): the angular length Φ of the tube (Fig. 46) should be diminished by an angle Δ for each of the parts of the tube adjoining an end fixture and stiffened by it. (A possibility of theoretical evaluation of the end effect is provided by the semi-momentless theory.)

It is proved by experiments that the elastic displacements of Bourdon tubes depend on the load q linearly. The geometric nonlinearity is negligible even for those displacements of the tube end (Fig. 46) that exceed by many times the wall-thickness. Therefore, the analysis is restricted to the linear approximation. The nonlinearity can be investigated in the rather direct way indicated in present chapter, §4.

11.2. Sensitivity and displacements

The characteristics describing the Bourdon tube from the standpoint of practical applications are in fact determined by the coefficients of the linear relation between the bending moment M_z in the cross-section, the relative flexure parameter k' and the normal pressure q. This basic relation is provided by the formulas (3.52). It remains to reduce it to a form more convenient for practical calculations and to do the parametric evaluation of the entire class of tubes including reference stresses.

We now write (3.52) in the form

$$M_z = k' EI° Bh^2 + qk_q I° b^2 R_m , \qquad (3.228)$$

where $I°$ and k_q denote dimensionless factors of flexural stiffness and of

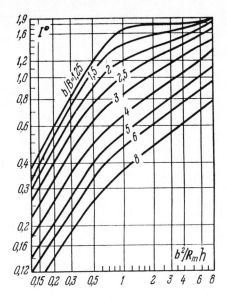

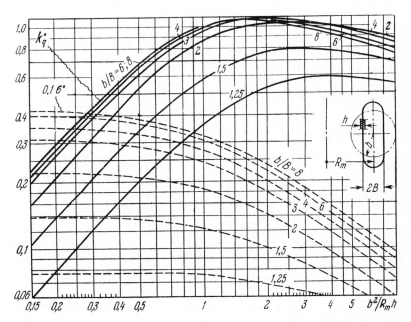

Fig. 50. Factors of flexural stiffness, sensitivity and stress of flat-oval tubes. (b/B from p. 190.)

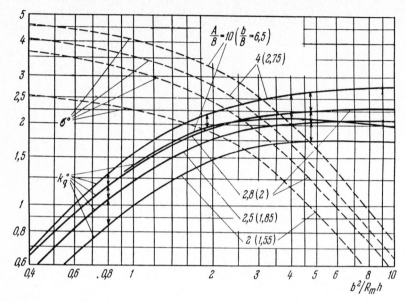

Fig. 51. Stress and sensitivity factors of elliptic tubes.

sensitivity of the tube. The factors represent the dependence of the two characteristics of the Bourdon tube on the shape of its cross-section (in terms of the Fourier coefficients s_j of $\sin \alpha$) and on the curvature of the tube μ (in terms of the quantities b_j determined by solving (3.56)),

$$I^\circ = \frac{b^2}{R_m h} \frac{b}{B} \sum_j \pi b_j^m s_j, \qquad k_q = \sqrt{12(1-\nu^2)} \frac{B}{b} \sum_j b_j^q s_j \Big/ \sum_j b_j^m s_j.$$

$$(3.229)$$

For tubes with flat-oval and elliptic cross-sections the values of the factors I° and k_q, calculated with (3.229) and §3.2 of Ch. 3, are plotted in Figs. 50 and 51.

For the two basic cases of application of the pressure-responsive element, shown in Fig. 46 and Fig. 48, respectively, special cases of (3.228) apply.

In a tube with one free end, loaded by normal pressure (Fig. 46), the bending moment M_z is equal to zero. With $M_z = 0$, formula (3.228) gives the relative change of curvature caused by the pressure q

$$k' = -k_q \frac{q}{E} \frac{b^2}{h^2} \frac{R_m}{B} \, . \tag{3.230}$$

In a tube loaded by normal pressure and *both* ends fastened with respect to rotation (Fig. 50, *a*) there arises a constant bending moment M_z equal to the reactive moments M_q on tube ends. Setting in (3.228) $k' = 0$ (the condition of zero flexure angle) we obtain the reactive moment

$$M_z = M_q = qk_q I^\circ b^2 R_m \, . \tag{3.231}$$

The displacement of the end of the tube is easily determined in terms of $k'(\varphi) = (d\varphi^* - d\varphi)/d\varphi$, where φ denotes the angle shown in Fig. 52.

Indeed, the displacement dU of the tube end, caused by flexure of an element of the tube $R_m \, d\varphi$—long (Fig. 52), is determined by the angle $d\varphi^* - d\varphi = k' \, d\varphi$ of rotation of the part of tube between the element and the tube end,

$$dU = -ak' \, d\varphi \, .$$

Projections of the displacement dU on the directions of the normal and the tangent to the tube axis at the end (determined according to the Fig. 52 with $a \cos \beta = R_m \sin \varphi$) are

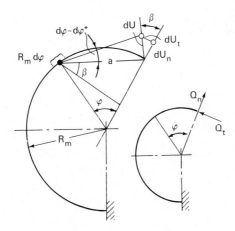

FIG. 52. Displacement of tube end.

$$dU_n = \cos \beta \, dU = -k'R_m \sin \varphi \, d\varphi \,,$$

$$dU_t = \sin \beta \, dU = -(R_m - R_m \cos \varphi)k' \, d\varphi \,.$$

Integration renders the two projections of the displacement of the tube end as (Φ denoted in Fig. 46)

$$U_n = -\int_0^\Phi R_m k' \sin \varphi \, d\varphi \,,$$

$$U_t = -\int_0^\Phi (R_m - R_m \cos \varphi)k' \, d\varphi \,. \tag{3.232}$$

For a relative change of curvature k', constant along the tube, formulas (3.232) yield

$$U_n = -k'\delta_n R_m \,, \qquad U_t = -k'\delta_t R_m \,,$$

$$\delta_n = 1 - \cos \Phi \,, \qquad \delta_t = \Phi - \sin \Phi \,, \qquad \gamma_u = \arctan \frac{\delta_n}{\delta_t} \,. \tag{3.233}$$

Under the internal pressure the tube unbends as shown in Fig. 46. That is, for $q > 0$ and profiles similar to those in Fig. 47, there is $k' < 0$ and $U_n, U_t > 0$.

Consider now the calculations for those tubes represented by the schemes of Fig. 48(b), (c) and (d) when the end of the tube is loaded due to a constraint. Using also for this case the relative change of curvature k' we must determine it as a *variable* along the tube $k' = k'(\varphi)$.

Of course, (3.228) is derived for deformation that is constant along the tube. However, recalling the analysis of present chapter, §5, we shall use (3.228) also to determine the change of curvature k' for the bending moment M_z varying as $\cos \varphi$ or $\sin \varphi$.

The conditions of equilibrium of a part of tube with the opening angle φ (Fig. 52) yield for the bending moment in a cross-section of the tube the expression

$$M_z = -Q_n R_m \sin \varphi - Q_t R_m (1 - \cos \varphi) \,.$$

Inserting the M_z into (3.228) we obtain $k'(\varphi)$. The corresponding displacements of the tube end are determined by (3.232). For $q = 0$ this results in

$$\begin{bmatrix} U_n \\ U_t \end{bmatrix} = \begin{bmatrix} \delta_{nn} & \delta_{nt} \\ \delta_{tn} & \delta_{tt} \end{bmatrix} \begin{bmatrix} Q_n \\ Q_t \end{bmatrix} \frac{R_m^2}{EI^\circ Bh^2} , \tag{3.234}$$

where we have introduced the notation

$$\delta_{nn} = \tfrac{1}{2}\Phi - \tfrac{1}{4}\sin 2\Phi , \qquad \delta_{tt} = \tfrac{3}{2}\Phi - 2\sin\Phi + \tfrac{1}{4}\sin 2\Phi ,$$
$$\delta_{tn} = \delta_{nt} = 1 - \cos\Phi - \tfrac{1}{2}\sin^2\Phi . \tag{3.235}$$

Let us determine the reaction forces Q_n, Q_t and Q (Fig. 48), which are measured in the instruments as indicating the normal pressure q.

We begin with the force Q_n (Fig. 48(b)). It is the reaction eliminating the displacement of the tube end, denoted in Fig. 46 by U_n. Equating the displacement U_n of the free end, caused by the force Q_n, to the U_n caused by pressure q we find through (3.234), (3.233) and (3.230)

$$Q_n = \Gamma_n M_q / R_m , \qquad \Gamma_n = \delta_n / \delta_{nn} \quad (Q_t = 0) . \tag{3.236}$$

In a similar way we determine the reactive force in the tangential direction (Fig. 48(c))

$$Q_t = \Gamma_t M_q / R_m , \qquad \Gamma_t = \delta_t / \delta_{tt} \quad (Q_n = 0) . \tag{3.237}$$

The reactive force Q eliminating the displacement of the tube end in *any* direction (Fig. 48(d)) is found by equating the displacement caused by the components Q_n, Q_t to that produced by the pressure q. For each of the two projections of displacement (U_n, U_t), this results in an equation. With (3.234) and (3.233) the equations are

$$\begin{bmatrix} \delta_{nn} & \delta_{nt} \\ \delta_{tn} & \delta_{tt} \end{bmatrix} \begin{bmatrix} Q_n \\ Q_t \end{bmatrix} \frac{R_m^2}{EI^\circ Bh^2} = -\begin{bmatrix} \delta_n \\ \delta_t \end{bmatrix} k' R_m .$$

Solution of these equations yields, using k' from (3.230),

$$[Q_n \ \ Q_t] = [\Gamma_n' \ \ \Gamma_t'] M_q / R_m . \tag{3.238}$$

The values of the functions $\Gamma_n, \ldots, \Gamma'_t$ can be determined with the aid of the graphs in Fig. 53.

11.3. Change of the inner volume

Some applications of the Bourdon tube use the relation between the change of the inner volume of the tube (filled in this case by a liquid) and the elastic displacement of its free end.

Let us determine the change of volume attendant upon the deformation of the tube, caused by the normal pressure q.

The change of volume V is a result of the change of the area F inside the middle line of the cross-section of the tube. For a tube with the effective length $\Phi_e R_m = (\Phi - 2\Delta)R_m$ (Fig. 46) we have

$$V = (F^* - F)R_m \Phi_e . \qquad (3.239)$$

The area is expressed in terms of the coordinates x and z of the profile (denoted in Fig. 46) by the well-known formula

$$F = \oint x \, dz = -\oint z \, dx = -\int_{-\pi}^{\pi} z(\eta) \frac{dx}{d\eta} d\eta , \qquad (3.240)$$

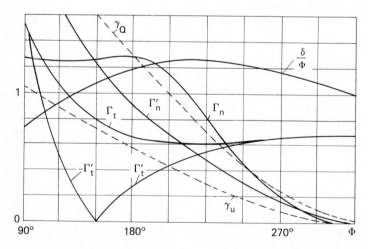

FIG. 53. Factors of formulas (3.233), (3.236)–(3.238). $\delta = (\delta_n^2 + \delta_t^2)^{1/2}$.

which gives the area F^* *after* the deformation, when x and z are replaced by their values x^* and z^* for the *deformed* tube.

Inserting the series expressions of x and z from (3.38) into (3.240) and integrating with the help of (2.154) we have

$$F^* - F = \pi b^2 \{(c^*)^T \Lambda^{-1} s^* - c^T \Lambda^{-1} s\} . \tag{3.241}$$

With this and (3.239) the change of volume is determined for unlimited displacements. The corresponding linear expression in terms of the function $\vartheta(\eta)$ is obtained from (3.241) with the expressions (2.172) for c^* and s^*. Dropping the nonlinear terms we have

$$F^* - F = \pi b^2 \{c^T \Lambda^{-1} c_- \, \vartheta - (s_- \, \vartheta)^T \Lambda^{-1} s\} .$$

The function ϑ is determined by the change of curvature of the tube and, independently, by the inner pressure according to (3.47). After expressing the pressure q with the help of (3.230) in terms of the change of curvature k', we obtain the relation between a change of the volume and a change of curvature

$$V = -vk' AB\Phi_e R_m .$$

The factor v is determined, as a function of the ratio of the dimensions A and B of elliptic (Fig. 23) and flat-oval (Fig. 47(b)) profiles, in Table 9 [85].

11.4. Stresses

The two cases of the application of Bourdon tubes, presented in Fig. 48(a) and Fig. 46 are basically different in their stress states. This is illustrated in Fig. 54 by the results of calculations with formulas (3.48) and (3.49).

TABLE 9.

A/B	1.5	2	3	4	6	8	∞
v, Flat-oval profile	1.88	2.73	3.38	3.60	3.79	3.88	4.00
v, Elliptic profile	1.695	2.26	2.62	2.76	2.86	2.89	2.90

The tube with both ends fixed with respect to rotation (Fig. 48(a)) does not bend and has predominantly membrane stresses for the usual curvature range $\mu < 5$. The maximum stress value is T_1/h at the point denoted by 1 in Fig. 54(a).

The Bourdon tubes with a free end (Fig. 46) have, in addition to the stresses of the case $k' = 0$ just described, considerably larger stresses caused by tube flexure. The superimposed system consists of predominantly wall-bending stresses σ_2. The lateral stresses σ_2 determine the maximum stress occurring at the point denoted by 2 in Fig. 54.

In both cases the stress state at the points of maximum stress can be assumed uniaxial for the evaluation of the strength of the tube.

The calculation of the design stress may be substantially simplified by graphs or tables of stress-ratios. For the two points of the tube, where the maximum stresses are expected, the values of $\dot{\vartheta}$, ϑ, ψ and $\dot{\psi}$, determining the stresses in (3.48) and (3.49), may be computed and presented in a compact form for a range of geometry parameters, encompassing most of the tubes.

For tubes with both ends restricted with respect to rotation we obtain (in the way just described) formulas for the maximum stress T_1/h at the point 1 (Fig. 54) and for the maximum wall-bending stress at the point 2,

$$\sigma_{T1} = \sigma_q^\circ q \frac{R_m + B}{h}, \qquad (3.242)$$

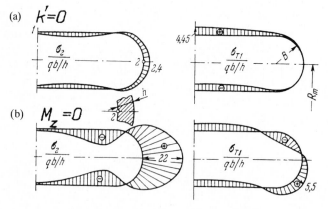

FIG. 54. Two cases of stress state. (Cf. Fig. 22.)

$$\sigma_2 = \sigma_q^\circ q \frac{R_m}{h}\frac{1 - h/6R_2}{1 - h/2R_2} + \frac{qb^2}{hA}. \qquad (3.243)$$

The factor $(1 - h/6R_2)/(1 - h/2R_2)$ is introduced in accordance with the theory of nonthin shells (Chapter 2, §2.4). It takes into account the large curvature of the tube wall at the point 2, where in many cases $R_2 \sim h$.

For the *flat-oval profile* at the point 2 (Fig. 54) $R_2 = B$. The stress-ratio σ_q° is determined by the graphs of Fig. 55: the broken-line curves give σ_q° for (3.242), while the remaining solid lines give that for (3.243). Formula for b can be found on p. 190.

For the tubes with unrestrained ends and thus $M_z = 0$ (Fig. 46) the maximum stress, occurring at the point 2, can be evaluated with the formula

$$\sigma_2 = \sigma^\circ q \frac{b^2}{h^2}\frac{1 - h/6R_2}{1 - h/2R_2} + \frac{qb^2}{hA}. \qquad (3.244)$$

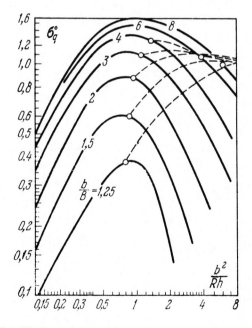

FIG. 55. Stress factor of flat-oval tube with built-in ends.

The stress-ratio σ° may be determined with the help of the graphs of Fig. 50 or 51—for the flat-oval and for the elliptic profiles, respectively.

The foregoing analysis of the two limiting cases $k' = 0$ and $M_z = 0$ provides a basis for evaluation of the stresses in other cases.

The restriction of the displacement of the end of a Bourdon tube in one or two directions (Fig. 48(b)–(d)) produces, as stated in the foregoing, a bending moment M_z which *varies* along the tube. With $M_z = M_q$ the change of curvature is equal to zero. In the cross-sections where $M_z = 0$, for instance at d in Fig. 48(b), the stresses are as high as in the unconstrained tube of Fig. 46. The stresses are still higher in the sections where the bending moment M_z acts in the direction opposite to that of M_q. In such parts of the tube, for instance at $d-e$ in Fig. 48(b), the curvature change k' is larger than in the unrestrained tube. The design stress is approximately proportional to k'. It can be estimated by multiplying the σ_2 of (3.244) by $1 - M_z/M_q > 1$.

12. Layered tubes

All the formulas and relations presented for homogeneous tubes can be extended to a wide class of nonhomogeneous and even orthotropic elastic properties. In what follows, this will be done for the representative case of a tube wall consisting of two layers of constant thickness h_1 and h_2 with different moduli of elasticity E_1 and E_2.

12.1. Parameters of the elasticity relations

As set forth in Chapter 1, §6.3, a shell with a modulus of elasticity E variable with respect to the transverse coordinate (ζ in Fig. 1) can be evaluated as a homogeneous shell with special elastic parameters. The only requirement[23] for this is that the Poisson coefficient ν must be nearly constant through the shell-thickness. For many homogeneous materials (in particular, metals) the values of ν are not much different and seldom known with an accuracy justifying taking this difference into account. Therefore, we take the Poisson coefficients of the two layers of the tube wall to be equal.

For a double-layered shell with $\nu = $ const. equation (1.70) determin-

---

[23] The simplification is in no way dependent on the relation $E_1 h_1^2 = E_2 h_2^2$ or any other relation between E_i and h_i.

ing the optimal position of the reference surface becomes

$$E_1 \int_{\zeta_-}^{\zeta_- + h_1} \zeta \, d\zeta + E_2 \int_{\zeta_- + h_1}^{\zeta_+} \zeta \, d\zeta = 0 . \tag{3.245}$$

This equation gives the coordinates ζ_- and ζ_+ of the side surfaces of the shell (of the inner and outer surfaces of the tube),

$$\frac{\zeta_-}{h_1} = \frac{1}{2} \frac{1 - en^2}{1 + en} - 1 , \qquad \zeta_+ = \zeta_- + h ;$$

$$e = E_2/E_1 , \quad n = h_2/h_1 \quad (h_1 + h_2 = h) . \tag{3.246}$$

Inserting these values of ζ_- and ζ_+ into (1.88) yields the characteristics of the stiffness of the tube wall,

$$E'h = B(1 - \nu^2) = E_1 h_1 + E_2 h_2 ,$$

$$\frac{E''h^3}{12(1 - \nu^2)} = D = \frac{E_1 h_1^3}{1 - \nu^2} \left[\frac{1 + en^3}{3} + \frac{(1 - en^2)^2}{4 + 4en} \right] . \tag{3.247}$$

12.2. Formulas

Consider now the corrections extending the results of analysis of the bending of tubes to the cases of the tube wall made of a two-layer material.

The solutions of the Reissner equations (the functions ψ and ϑ), apart from their dependence on the loading and the tube profile, depend only on the parameter μ. This parameter can be defined for a nonhomogeneous tube wall through formulas (2.31). In particular, with $B_1' = 1/(E'h)$ following from (1.87), (1.88) and (3.247) we have

$$\mu = \left(\frac{E'h}{D} \right)^{1/2} \frac{b^2}{R_m} = \sqrt{12(1 - \nu^2)} \frac{b^2}{R_m h} \left(\frac{E'}{E''} \right)^{1/2} , \qquad m = k'\mu , \tag{3.248}$$

where E' and E'' correspond to the definitions (3.247).

The elasticity constant of the relation (2.31) expressing V in terms of ψ is determined, similarly to (3.248) by

$$V = \psi(DE'h)^{1/2}. \tag{3.249}$$

With the relations (3.248) and (3.249) the results of the evaluation of the functions ψ and ϑ as well as of the stress resultants T_1, T_2, M_1 and M_2 are extended to the double-layered shells and, in particular, tubes. This applies to all formulas and equations of this chapter.

The exception is presented by the formulas for stresses, which must be generalized to allow for the position of the reference surface not being in the middle of the wall-thickness and for the variability of the modulus of elasticity.

With formulas (1.49) and (1.61), as well as the elasticity relations of nonhomogeneous shells (1.88) and (3.247), we obtain the formulas

$$\sigma_1 = \frac{E}{E'}\frac{T_1}{h} + \zeta \frac{E}{E''}\frac{M_1}{h^3/12}, \qquad \sigma_2 = \frac{E}{E'}\frac{T_2}{h} + \zeta \frac{E}{E''}\frac{M_2}{h^3/12}, \tag{3.250}$$

which for $E' = E'' = E$ become identical with (1.90).

Of course, E denotes in (3.250) the value of the modulus *at the point*, where the stress is calculated. Maximum stress occurs at $\zeta = \zeta_+$ or ζ_-.

To illustrate the implications of formulas (3.248)–(3.250) for the extension of the analysis to nonhomogeneous tubes, consider the relation between the change of curvature of a curved tube and the bending moment. For a homogeneous material the relation has the form (3.30). To derive the corresponding formula for nonhomogeneous tubes we start with the (static) relation (2.32). Inserting the (static) relation (2.15) $T_1 = bV^{\cdot}$ and expression (3.249) for V we obtain the bending-moment formulas differing from (3.29) and (3.30) only in one detail—the modulus of elasticity of a homogeneous tube (E) is replaced by the quantity $(E'E'')^{1/2}$. Thus, we have instead of (3.30) a more general relation

$$M_z = K(E'E'')^{1/2}I\frac{k'}{R_m}, \qquad K = \frac{b_1}{m} \left(k' = \frac{\varphi^* - \varphi}{\varphi}\right). \tag{3.251}$$

The stiffness factor K is determined in exactly the same way as for homogeneous tubes. It is a function of the curvature parameter μ, which for nonhomogeneous tubes is defined by (3.248).

More insight into the ways of adapting various relations for the more general case of nonhomogeneous tube wall provides the solution for concrete problems.

12.3. Examples

(1) Let us determine the maximum stress and the bending moment M_z in a cross-section of a bimetallic tube, when the elastic change of curvature is given: $k'/R_m = 10^{-5}\,\text{mm}^{-1}$. The tube has a round cross-section with the middle radius $b = 50\,\text{mm}$, wall-thickness $h = 4\,\text{mm}$ and the curvature radius $R_m = 200\,\text{mm}$. The tube wall is bimetallic with two layers of equal thickness $h_1 = h_2 = 2\,\text{mm}$. The moduli of elasticity of the layers are given as $E_2/E_1 = 1.333$. The Poisson coefficient is $\nu = 0.3$.

The solution starts with formulas (3.246)–(3.248) for the parameters of the tube: $e = 1.33$, $n = 1$, $E' = 1.166E_1$ and $E'' = 1.185E_1$,

$$\zeta_+ = 1.857\,\text{mm}\,, \qquad \zeta_- = -2.142\,\text{mm}\,, \qquad \mu = 10.33\,.$$

For this value of μ the graph of Fig. 19 gives $K\mu/2 = 1.02$. With this we find $K = 0.1975$ and the bending moment according to (3.251): $M_z/E_1 = 3.647\,\text{mm}^3$.

The maximum stress occurs in this case ($\mu = 10.33$) at $\eta = \pi/2$ (Fig. 20) as a circumferential bending stress. Reconsidering the derivation of the formula (3.32) for $\sigma_{2\,\text{max}}$ we obtain with (3.248), (1.89) and (3.27) a formula taking into account the variation of $E = E(\zeta)$,

$$\sigma_{2\,\text{max}} = Ek'\frac{b}{R_m}\sigma_2^\circ\frac{\zeta_\pm}{h/2}\left(\frac{E'}{E''}\right)^{1/2}, \tag{3.252}$$

where E and $\zeta_\pm$ must be the values at the point ζ_+ or ζ_-.

For the example under consideration the larger stress occurs at $\zeta = \zeta_+$ where $E = E_2$ and according to (3.252), with $\sigma_2^\circ = 0.74$ following for $\mu = 10.33$ from Fig. 19, we obtain $\sigma_{2\,\text{max}} = 45 \times 10^{-5}E_1$.

The nonhomogeneous elastic properties—the difference between the two layers of the tube wall—complicate the evaluation only insignificantly.

(2) The tube of the preceding example is uniformly heated, the temperature rise is $t = 50°C$, compared to the unstressed state. The linear expansion coefficients of the two layers are $\beta_1 = 2 \times 10^{-5}$ and $\beta_2 = 3.5 \times 10^{-5}$. Let us evaluate the thermal stresses in the tube.

The thermal expansion varying only through the wall-thickness causes, according to Chapter 1, §7.4, a uniform elongation in all directions. All over the tube (excepting a narrow edge zone) the elongation is determined by (1.111) and (1.108) with B from (3.247):

$$e_1 = e_2 = \varepsilon_1 = \varepsilon_2 = \frac{1-\nu}{E'h}\left(\frac{E_1\beta_1 t}{1-\nu}h_1 + \frac{E_2\beta_2 t}{1-\nu}h_2\right) = 14.29 \times 10^{-4}.$$

Inserting these values of e_i together with E_1, $\beta_1 t$ and ν into the relations (1.107) yields the stresses in the layer 1,

$$\sigma_1 = \sigma_2 = \frac{E_1}{1-\nu^2}(e_1 + \nu e_1) - \frac{E_1\beta_1 t}{1-\nu} = 613 \times 10^{-6}E_1.$$

Replacing here E_1, β_1 by E_2, β_2 we obtain the stresses in the layer 2. They differ only in the sign.

CHAPTER 4

OPEN-SECTION BEAMS

CHAPTER 4

OPEN-SECTION BEAMS

1. Open circular section

Linear and nonlinear flexure of thin-walled beams in the plane of curvature is considered as a Saint-Venant problem. The loading consists of forces distributed at the beam ends and causing a pure-bending stress state, constant along the beam. The Fourier-series solution of the nonlinear Reissner equations (3.17) is applied to the case of a circular section with a narrow slit.

1.1. Solution of the flexure problem

Consider thin-walled beams of not too large a curvature, taking the basic equations in the form (3.17). Without the terms representing the load distributed along the beam (usually irrelevant for open-section beams) the equations are [176]

$$\ddot{\psi} + \mu^* c' \vartheta = -ms , \quad c' \vartheta = s^* - s ,$$
$$\ddot{\vartheta} - \mu^* c^* \psi = 0 . \tag{4.1}$$

Let the plane of curvature of the beam be its plane of symmetry. Then it is also the plane of symmetry of the stress state. Set the line $\eta = 0$ in this plane (as in Fig. 56(c))—and choose the constant b so as to make $\eta = \pm\pi$ at the longitudinal edges. The edges $\eta = \pm\pi$ are free of loading, i.e. on these edges the bending moment M_2 and the stress resultants Q_2 and T_2 are equal to zero. Expressed in terms of the functions ϑ and ψ

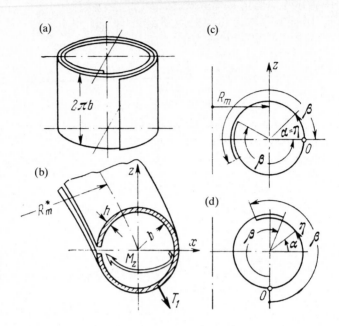

FIG. 56. Slit tubes.

according to formulas (2.23), (2.24) and (2.9), the conditions on the free edges are

$$\eta = \pm\pi: \qquad \frac{M_2}{D} = \frac{\dot{\vartheta}}{b} + \nu\kappa_1 = 0, \quad \psi = 0 \quad \left(\kappa_1 = \frac{kc^* - c}{R}\right). \quad (4.2)$$

The term $\nu\kappa_1$ of the condition $M_2/D = 0$ is negligible for the problem considered in this section: $\nu|\kappa_1| \ll |\dot{\vartheta}|/b$. It is so in all those cases in which the dimensions of the cross-section, measured parallel to the plane of flexure, are large compared to the wall-thickness h.

The solution will be sought in trigonometric-series form, as discussed for boundary problems in Chapter 2, §8.2. From the assumed symmetry of deformation we may write

$$\vartheta = C\eta + \sum a_n \sin n\eta, \qquad \psi = \sum b_n \sin n\eta. \quad (4.3)$$

The term $C\eta$ is essential for the fulfillment of the condition (4.2) for $\dot{\vartheta}$. It has a simple geometric meaning: $\pm C\pi$ is the angle of rotation of the respective edges $\eta = \pm\pi$. Conditions (4.2) without the term $\nu\kappa_1$ are satisfied by expressions (4.3) when

$$C = -\sum_n n a_n \cos n\pi .$$ (4.4)

The equations determining the Fourier coefficients a_n and b_n are obtained by substituting the expressions (4.3) into equations (4.1). They can be easily written in the matrix form starting from the matrix representation of series (4.3) with the columns of their Fourier coefficients given by

$$\vartheta(\eta) \overset{.}{\to} C\boldsymbol{\eta}_s + \boldsymbol{\vartheta}^0 , \qquad \psi \overset{.}{\to} \boldsymbol{\psi} = \{b_1 \ b_2 \ \cdots\} ,$$

$$\boldsymbol{\vartheta}^0 = \{a_1 \ a_2 \ \cdots\} , \qquad C = \tfrac{1}{2}\boldsymbol{\eta}_s^{\mathrm{T}} \Lambda^2 \boldsymbol{\vartheta}^0 .$$ (4.5)

The matrix $\boldsymbol{\eta}_s$ used here to represent the *sine* series of η is defined in (2.159). (Of course, $\ddot{\eta} \overset{.}{\to} \mathbf{0}$, not $\Lambda^2 \boldsymbol{\eta}_s$.)

With (4.5) the matrix form of (4.1) becomes a system for $\boldsymbol{\vartheta}^0$ and $\boldsymbol{\psi}$,

$$\begin{bmatrix} \Lambda^2 & -\mu^* c'_- - \dfrac{\mu^*}{2} c'_- \boldsymbol{\eta}_s \boldsymbol{\eta}_s^{\mathrm{T}} \Lambda^2 \\ \mu^* c^*_- & \Lambda^2 \end{bmatrix} \begin{bmatrix} \boldsymbol{\psi} \\ \boldsymbol{\vartheta}^0 \end{bmatrix} = \begin{bmatrix} ms \\ 0 \end{bmatrix} .$$ (4.6)

Since Λ^2 is a diagonal matrix, the system (4.6) may be easily rewritten (similar to (3.25), (3.26) or (3.37) for tubes) with $\boldsymbol{\psi}$ or $\boldsymbol{\vartheta}^0$ eliminated. Thus,

$$\boldsymbol{\vartheta}^0 = -\mu^* \Lambda^{-2} c^*_- \boldsymbol{\psi} ,$$

$$[\Lambda^2 + \mu^{*2} c'_- \Lambda^{-2} c^*_- + \tfrac{1}{2}\mu^{*2} c'_- \boldsymbol{\eta}_s \boldsymbol{\eta}_s^{\mathrm{T}} c^*_-]\boldsymbol{\psi} = ms .$$ (4.7)

Equations (4.6) or (4.7) can be solved by successive iteration as is indicated in Chapter 3, §2.4. In evaluating c' and c^* the matrix $\boldsymbol{\vartheta}$ and not just a part $\boldsymbol{\vartheta}^0$ of it, must be taken into account. In some cases, particularly when better convergence of the series (4.3) near the edges is important, the solution should include, instead of $\boldsymbol{\eta}_s$, the matrix $\boldsymbol{\eta}_\sigma$ defined in (2.160).

With ψ and ϑ found, the stresses and displacements are easily calculated using formulas in Chapter 2, §2. More explicitly the bending moment M_z can be calculated by using (3.39) and (3.30). (For most cases, when $\psi(\pm\pi) = 0$, (3.30) or (3.39) may be applied to find M_z also for the open sections.)

Let us investigate at a closer range the beams of circular profile.

1.2. Profile slit in the flexure plane

Consider the beams shown in Fig. 56. These slit tubes can be stored in a minimum of space by rolling them up as shown in Figs. 12 and 56(a). They have thus found extensive use, particularly in spacecraft, as storable tubular extendible members, in short STEM. The pioneering work on the analysis of STEM is due to MACNAUGHTON [89], RIMROTT and JAIN [102, 120, 126, 140, 141, 172]. The STEM can have a small initial longitudinal curvature. The similar BI-STEMS are treated in [212].

The extreme flexibility of STEM is taken into account in the following by a nonlinear analysis of bending, including the limit-point Brazier instability.

Other applications concern beams of open cross-section with considerable initial curvature. For these beams the *linear* bending problem will be discussed briefly.

Let us consider first the beams with a slit in the plane of curvature, which are bent in that plane (Fig. 56(c)). Starting with the *linear* approximation we have a circular cross-section, i.e. $\alpha = \eta$. With the matrices $c'_- = c^*_- = c_-$ and s given by (3.23) and η_s from (2.159), equation (4.7) becomes (cf. (3.26))

$$
\left\{ \begin{bmatrix} 1+\mu_2^2 & 0 & \mu_2^2 & 0 & \cdots \\ 0 & 4+\mu_1^2+\mu_3^2 & 0 & \mu_3^2 & \cdots \\ \mu_2^2 & 0 & 9+\mu_2^2+\mu_4^2 & 0 & \cdots \\ 0 & \mu_3^2 & 0 & 16+\mu_3^2+\mu_5^2 & \cdots \\ \cdots & \cdots & \cdots & \cdots & \cdots \end{bmatrix} \right.
$$

$$
\left. + \frac{\mu^2}{2}\, c_- \boldsymbol{\eta}_s (c_- \boldsymbol{\eta}_s)^{\mathrm{T}} \right\}
\begin{bmatrix} b_1 \\ b_2 \\ b_3 \\ b_4 \\ \cdots \end{bmatrix}
=
\begin{bmatrix} m \\ 0 \\ 0 \\ 0 \\ \cdots \end{bmatrix},
\tag{4.8}
$$

$$
(c_- \boldsymbol{\eta}_s)^{\mathrm{T}} = [\, 0 \;\; -\tfrac{1}{2} \;\; \tfrac{1}{1}+\tfrac{1}{3} \;\; -\tfrac{1}{2}-\tfrac{1}{4} \;\; \tfrac{1}{3}+\tfrac{1}{5} \;\; \cdots \,], \qquad \mu_n^2 = \mu^2/4n^2.
$$

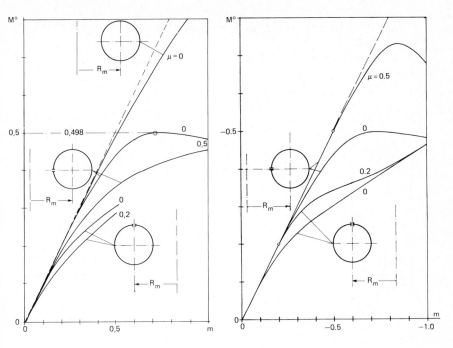

FIG. 57. Bending of slit beams.

Retaining only the first three equations we have

$$\begin{bmatrix} 1 + 3\mu^2/16 & -\mu^2/3 & \mu^2/4 \\ -\mu^2/3 & 4 + 7\mu^2/6 & -\mu^2/2 \\ \mu^2/4 & -\mu^2/2 & 9 + 23\mu^2/64 \end{bmatrix} \begin{bmatrix} b_1 \\ b_2 \\ b_3 \end{bmatrix} = \begin{bmatrix} m \\ 0 \\ 0 \end{bmatrix}. \qquad (4.9)$$

As mentioned before, (3.30) remains valid for the slit section. The influence of the *slit* on the bending stiffness *KEI* is substantial for $\mu \geq 2$. According to (3.31) or Fig. 30, for $\mu = 2$ the stiffness-reduction factor is $K = 0.804$; for a *slit* tube of Fig. 56(c) with $\mu = 2$, equation (4.9) gives $K = 0.67$. The difference is still larger for the wall-bending stresses (σ_2).

Results of the *nonlinear* analysis (4.1)–(4.7) of the flexure of initially straight or slightly curved beams are presented in Fig. 57 (following the work of EMMERLING [200]).

1.3. Profile slit along the neutral layer

For the beams with profiles of the type shown in Fig. 56(d), particularly when $\beta = \pi$, the influence of the Kármán effect is much stronger than for the symmetric profiles considered in the preceding section. Indeed, here the part of the cross-section under extension stresses is connected with the other part, where the stresses are compressive, by only *one thin wall*, which has to transmit a *bending* moment.

The series solution of the form (4.3) is not applicable here; the stress state is not symmetric. A numerical solution of (4.1) was obtained for beams with a slit along the line $\alpha = \pi/2$ and $\beta = \pi$ by EMMERLING [200]. This allowed the investigation of the nonlinear flexure of both initially straight and slightly curved beams.[1] The results shown in Fig. 57 illustrate the features of the beams with a slit along the neutral layer. The flexibility is much higher compared to a slit in the plane of symmetry, both in the linear and, particularly, in the nonlinear range.

From the standpoint of practical applications the foregoing considerations present a first step.

The open profiles are mostly (the STEM, as a rule) overlapped—with $\beta > \pi$. This increases the stiffness drastically.

A further stiffening factor is particularly important for the slit tubes. It is the influence of end fastenings. Investigation of this effect involves, of course, certain boundary conditions and thus a two-dimensional shell problem. A suitable tool for the solution of this problem is provided by the semi-momentless theory (Chapter 2, §§4–6).

This theory should be effective in investigating the related problem of the "ploy region" where the sheet of the STEM passes from the storage roll into the unstrained tubular shape (Fig. 12). The setting of this problem and its engineering theory is provided by the work of JAIN and RIMROTT [141].

Flexure of long beams with various circular open sections is treated by BENSON in [219] and in *J. Appl. Mech.* **51** (1984) 141–145. (The linear distribution of normal stress (T_1/h) in the cross-sections, used in the second work, restricts the analysis to initially straight beams.)

The torsion of the STEM beams was investigated by RIMROTT [126, 172].

Last but not least, the collapse of the open sections in bending may be caused by local *buckling* (cf. Chapter 3, §10). The bifurcation *buckling of*

---

[1] For tubes of small curvature ($\mu^* < 0.5$) the perturbation solution of [176] allows a simple description of the nonlinear flexure. This method may be useful for complicated profiles.

the longitudinal *edges* under longitudinal compressive stress (T_1/h) should be the real cause of collapse in many cases.

2. Shallow profiles

A corrugated plate (Fig. 58(a)) can have a flexural instability of the Brazier type. With the growing elastic curvature, $1/R^*$, the Kármán effect increases, the stresses in the cross-sections become redistributed and the height of the corrugations is diminished by the elastic deformation. For some value of the elastic curvature $1/R^*$ the bending moment reaches its maximum. It is the limit-point snap-through load; the plate collapses. This problem was set and solved for an infinite plate with shallow sinusoidal corrugations by ASHWELL [42]. In the following, the problem is considered as a particular case of a more general one of the flexure of a beam of a *finite breadth* and finite *initial curvature* (investigated in [64]).

Only *shallow* profiles, those satisfying the conditions

$$\cos \alpha^* \approx 1 , \qquad (\sin \alpha^*)^2 \ll 1 , \qquad z \approx \eta b , \qquad (4.10)$$

are discussed here. This assures a substantial simplification compared to the solution of nonshallow beams set forth in present chapter, §1.1. At the same time the solution of equations simplified by the assumption of shallowness (4.10) proves to be of satisfactory accuracy for a wide range of the α^*-values.

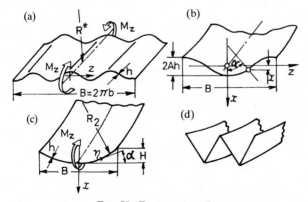

FIG. 58. Four open profiles.

We begin in the next §§2.1, 2.2 with two ways of finding the solution of the Reissner equations (each of the two has its advantages).

2.1. Trigonometric-series solution

The solution in the trigonometric-series form is convenient when the profile of the cross-section (Fig. 58) is set in the form

$$x = h \sum A_n \cos n\eta \,, \qquad \sin \alpha = \frac{h}{b} \sum A_n \sin n\eta \,, \quad b = B/2\pi \,. \quad (4.11)$$

The value of b is chosen to fix the coordinates of the longitudinal edges $\eta = \pm\pi$. Profiles symmetric with respect to the plane of curvature are considered.

For the shallow profiles conforming to condition (4.10), equations (4.1) become extremely simple. With $c' = c^* = 1$ the equations are *linear* with *constant* coefficients:

$$\ddot{\psi} + \mu^* \vartheta = -ms \,, \quad \vartheta = s^* - s \,,$$
$$\ddot{\vartheta} - \mu^* \psi = 0 \,. \qquad (4.12)$$

The nonlinearity of the problem is represented in (4.12) only by the value of $\mu^* = \mu + m$.

The conditions on the longitudinal edges of the beam, indicating the absence of any load there, have the form (4.2). The solution may be conveniently presented in the form of a Fourier series (4.3).

Contrary to the case of nonshallow profiles, the contribution of the longitudinal change of curvature (the term $\nu\kappa_1$) cannot be neglected in (4.2). Boundary conditions (4.2) with (recall $k' = k - 1$ and (2.20))

$$\kappa_1 = \frac{k'}{R} = mh^\circ \frac{1}{b} \qquad (4.13)$$

yield for the coefficient C, appearing in the expression (4.3), for ϑ,

$$C = -\sum_j ja_j \cos j\pi - \nu mh^\circ \,. \qquad (4.14)$$

A system of equations for the Fourier coefficients of the solution (4.3) is obtained by substitution of (4.3) and $s = \sin \alpha$ from (4.11) into (4.12),

$$-n^2 a_n - \mu^* b_n = 0, \quad n = 1, 2, \ldots,$$

$$-n^2 b_n + \mu^* a_n - C\mu^* \frac{2}{n} (-1)^n = -m \frac{h}{b} n A_n. \tag{4.15}$$

The term $C\eta$ of (4.3) is taken into account by the Fourier series (2.159) of η_s.

The system (4.15) and (4.14) without the ν term is equivalent to the system (4.6) with $c' = c^* = 1$.

With the coefficients a_n, b_n and C calculated we can proceed to evaluation of stresses and displacements by using the formulas of Chapter 2, §2.

The relation between the change of curvature and the bending moment in the cross-section of the beam is determined by (2.32), where in the present problem the term with M_1 *cannot* be neglected. We determine the M_1 in terms of $\kappa_1 b = mh^\circ$ and $\kappa_2 b = \dot{\vartheta}$ through the elasticity relation (1.83). Adding the corresponding M_1-contribution to the M_z-value from (3.39) with $s^* = (\ddot{\psi} + \mu s)/\mu^*$ from (4.12) leads to the formula

$$M^\circ = \left[m + \left(\mu \sum_n b_n s_n + \sum_n n^2 b_n^2 \right) \frac{6}{\mu^*} \frac{b^2}{h^2} \right.$$

$$\left. + \frac{\nu}{1-\nu^2} \frac{\mu^*}{h^\circ} \sum_n (-1)^n \frac{b_n}{n} \right] \left(1 + 6 \sum_n A_n^2 \right)^{-1}. \tag{4.16}$$

The dimensionless bending moment M° is defined as

$$M^\circ = \frac{b}{h^\circ} \frac{M_z}{EI_b}, \qquad I_b = \frac{2\pi b h^3}{12} \left(1 + 6 \sum_n A_n^2 \right), \tag{4.17}$$

where I_b is the moment of inertia of the cross-section of the beam.

2.2. Kryloff-functions solution

Equations (4.12) have a closed-form complementary function. A particular integral of the system (4.12) can be very simply formulated in series form, which reduces to an explicit expression for a sinusoidally corrugated profile. Consider a solution combining these two possibilities.

Eliminating ψ from the first of equations (4.12) we have an equation for the angle of rotation $\vartheta = \alpha^* - \alpha$,

$$\psi = \frac{\ddot{\vartheta}}{\mu^*}, \qquad \ddddot{\vartheta} + \mu^{*2}\vartheta = -m\mu^*s .\qquad (4.18)$$

This equation can be transformed by expressing ϑ in terms of the radial displacement $W = x^* - x$ in accordance with Fig. 58,

$$\sin \alpha^* = -\dot{x}^*/b, \qquad \vartheta \approx \sin \alpha^* - \sin \alpha = -\dot{W}/b .\qquad (4.19)$$

Upon integration, the system (4.18) may be presented in the form

$$\ddddot{W} + \mu^{*2}W = -m\mu^*x , \qquad \psi = -\ddot{W}/(b\mu^*) \qquad (4.20)$$

The general integral of the homogeneous equation $\ddddot{W}_0 + \mu^{*2} W_0 = 0$ can be written as a linear combination of the four Kryloff functions defined in Table 4. Adding a particular solution W_* of the nonhomogeneous equation (4.20) we have a general solution of this equation for the profile (4.11)

$$W = W_* + \sum_1^4 B_i K_i(\eta\sqrt{\mu^*/2}) , \qquad W_* = -m\mu^* \sum_n A_n h \frac{1}{n^4 + \mu^{*2}} \cos n\eta .$$
$$(4.21)$$

The constants B_i must be determined by the conditions on the longitudinal edges. The conditions on the free edges (4.2) with the expressions of ϑ and ψ in terms of W from (4.19), (4.20) and κ_1 from (4.13) yield

$$\begin{bmatrix} B_1 & B_2 \\ B_3 & B_4 \end{bmatrix} = \begin{bmatrix} -K_4\left(\dfrac{\beta}{2}\right) & 0 \\ \\ K_2\left(\dfrac{\beta}{2}\right) & 0 \end{bmatrix} \frac{4}{\mu^*} \frac{mhc_v - 2\ddot{W}_B}{\sinh \beta + \sin \beta} ,$$
$$(4.22)$$
$$\beta = \sqrt[4]{3(1 - \nu^2)}B/\sqrt{R^*h} , \quad c_v = \nu[3(1 - \nu^2)]^{-1/2} .$$

Here $\ddot{W}_B$ denotes the value of the derivative $\ddot{W}_*$ on the edges. The stress resultants may be expressed in terms of W by using (2.24), (4.13),

(4.19) and (4.20), with the result

$$\frac{T_1}{Ehh^\circ} = \dot{\psi} = m\,\frac{x}{b} + \mu^*\,\frac{W}{b}\,, \qquad \frac{M_1}{D} = \kappa_1 - \nu\,\frac{\ddot{W}}{b^2}\,,$$

$$\kappa_1 = \frac{1}{R^*} - \frac{1}{R} = \frac{mh^\circ}{b}\,, \qquad \frac{M_2}{D} = -\frac{\ddot{W}}{b^2} + \nu\kappa_1\,. \tag{4.23}$$

The relation between the bending moment in the cross-section and the change of curvature of the beam is obtained by inserting expressions of T_1 and M_1 in terms of W into (2.32). With $R^* - R_m^* = x^* = x + W$ and (4.23) and after some transformations we obtain

$$M^\circ = \frac{m}{1 + 6\,\Sigma\,A_n^2}\left(F + \frac{1 - \nu^2 a}{1 - \nu^2}\right), \tag{4.24}$$

where it is denoted

$$F = 6\sum_n A_n^2 n^4\,\frac{n^4 + \mu^*\mu}{(n^4 + \mu^{*2})^2} - 12Z\chi(Z - \tfrac{1}{2}c_\nu)$$

$$\times\left[1 - \frac{m}{\mu^*}\left(\frac{Z_1}{Z} + \frac{\chi_1}{\chi}\right)\right] + 6c_\nu Za\,, \qquad a = \chi - \frac{m}{\mu^*}\,\chi_1,$$

$$Z = \sum_n A_n\,\frac{(-1)^n\mu^*n^2}{n^4 + \mu^{*2}}\,, \qquad Z_1 = \sum_n A_n(-1)^n\mu^*n^2\,\frac{\mu^{*2} - n^4}{(n^4 + \mu^{*2})^2}\,,$$

$$\chi = \frac{2}{\beta}\,\frac{C_1 - c_1}{S_1 + s_1}\,, \qquad \chi_1 = \frac{S_2 - s_2 - 2S_1c_1 + 2C_1s_1 - 4\beta S_1 s_1}{4\beta(S_1 + s_1)^2}\,;$$

$$C_n = \cosh n\beta\,, \quad c_n = \cos n\beta\,, \quad S_n = \sinh n\beta\,, \quad s_n = \sin n\beta$$

$$(\beta = \pi\sqrt{2\mu^*})\,. \tag{4.25}$$

The results of the solution with the Kryloff functions are identical with those of present chapter, §2.1, provided the number of terms retained in the Fourier series is sufficient.

2.3. Nonlinear flexure

The foregoing formulas describe the deformation of any shallow-profile represented by the series (4.11). Let us investigate at a closer range the nonlinear flexure and collapse for *cosinusoidal* profiles—for the cases, when the expression (4.11) contains only *one* term. Two profiles of this sort, those with $x = hA_n \cos n\eta$ for $n = 3$ and 1, are shown in Fig. 58(a) and (b).

For beams of this type the nonlinear relations between the bending moment and the change of curvature m are represented by graphs of Fig. 59. The straight line 1 indicates the linear relation $M° = m$ of the elementary beam formula. It is correct for beams with no initial curvature ($\mu = 0$), provided the displacements are sufficiently small.

With the growth of the elastic curvature (μ^*) the slope of the curve $M°(m)$ changes. For each beam there is a maximum in the relation $M°(m)$. It corresponds to the limit-point instability. A further increase of the curvature (μ^*) is accompanied by a decrease of the moment; the cross-section is flattened and converges ultimately to a straight-line profile $x^* = x + W = 0$. The curve $M° = M°(m)$ approaches the asymptote 2 indicating the value of the moment in a bent *cylindrical* strip.

The quantity $M°/n^2$ represented by the graphs indicates the bending moment per one of n full waves of the section. The value of $M°/n^2$ for a given curvature change m/n^2 is the larger the more waves there are in

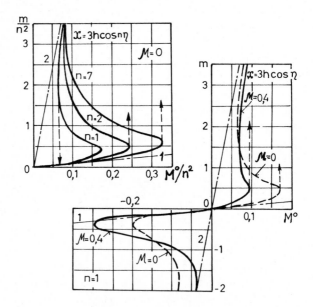

FIG. 59. Nonlinear bending (of cosinusoidal profiles) is very different for $M° > 0$.

the profile. However, for $n \geqslant 7$ the variation is insignificant. The moment has the same value as that for an infinitely *broad* corrugated plate—one with profile defined by (4.11) with $n^2 \gg 1$. The corresponding moment-curvature relation follows from (4.24) for n^2, $\beta \gg 1$, $\mu = 0$ and $\mu^* = m$,

$$M^\circ = \frac{m}{1 + 6A_n^2} \left[\frac{1}{1 - \nu^2} + \frac{6A_n^2}{(1 + m^2/n^4)^2} \right]. \qquad (4.26)$$

A similar formula was originally obtained by ASHWELL [42].

The *influence* of the free longitudinal *edges*, which causes the difference in the bending of the beams with different n-values, depends on the *direction of* the *flexure*. This nonlinear effect can be observed in the relation $M^\circ = M^\circ(m)$ of the initially straight ($\mu = 0$) beam (Fig. 59). For the profile $x = 3h \cos \eta$ the critical maximum bending moment is ca. 15% less for the flexure with $m > 0$ than for the flexure in the opposite direction (the one shown in Fig. 58(a)).

The influence of the *initial curvature* is a more fundamental phenomenon.

A small initial curvature, of merely $\mu = 0.4$, reduces the maximum bending moment to half of the value of M_{max} of the straight beam. This applies for the flexure increasing the initial curvature, i.e. for $\mu^* = \mu + m \geqslant \mu$.

In the case of *unbending* the initial curvature has a quite different effect. The maximum moment is above the value $M^\circ = -\mu$, at which the beam is straightened ($m = -M^\circ$, $\mu^* = 0$) and the cross-section assumes its initial, undeformed, shape.

The examples of Fig. 59 concern profiles with the same amplitude—three times the wall-thickness. But the calculations with formulas (4.16) or (4.24) display only a rather weak dependence of M° on the height of the profile $2hA_n$ provided $6A_n^2 \gg 1$. For the case of a broad plate this becomes evident in view of the simple formula (4.26).

In the other limiting case (the one-wave profile of Fig. 58(b)), the maximum (critical) value of M varies only by 6% for the amplitudes of the profile from $hA_n = h2$ to infinity. Thus, the graphs of Fig. 59 render an estimate of the bending moment-curvature relation for *any heights* of the profiles, provided $A_n \geqslant 2.5$ and the condition (4.10) holds.

The shape of the profile considered in this section was cosinusoidal. But the basic results are valid for *any* profiles defined by the series

(4.11). To illustrate this, let us apply the solution to the profile in the form of a shallow *arc* of a *circle*[2] (Fig. 58(c)).

The circular-arc profile may be described by the equation $x = B + C\eta^2$ with constant B and C. The Fourier series of such a function can be written in the form (4.11) as (H is denoted in Fig. 58(c))

$$x = x_0 + h \sum_n A_n \cos n\eta \,, \quad A_n = -\frac{4H}{\pi^2 h}(-1)^n \frac{1}{n^2} \quad (n = 1, 2, \ldots) .$$
$$(4.27)$$

Substituting this set of $A_1, A_2, \ldots$ into the formulas of the shallow-profile beams we find that the circular-arc profile has nearly the same displacements and relation $M°(m)$ as has the profile $x = hA_1 \cos \eta$ with an equal A_1. The values of the other Fourier coefficients appear to be in this case of secondary importance. This situation should prevail also for other profile shapes, provided the corresponding Fourier series converge at least as intensively as that in (4.27).

3. Profiles allowing a closed-form solution

The linear and nonlinear flexure is considered for those shallow profile shapes that allow an explicit closed-form solution of the governing equation (4.20).

3.1. The simplest profiles

For all shallow profiles—those satisfying the conditions (4.10) the governing equations can be reduced to the form (4.20). For those of such profiles that satisfy the condition

$$\frac{d^4}{d\eta^4} x = 0 \,,$$
$$(4.28)$$

equation (4.20) has a simple particular solution W_*. For these profiles the expression

$$W_* = -\frac{m}{\mu^*} x$$
$$(4.29)$$

satisfies, obviously, equation (4.20): $\ddddot{W}_* + \mu^{*2} W_* = -m\mu^* x$.

---

[2] We return to this problem in the next section; it has been further investigated by BENSON [219].

Such profiles can be termed the *simplest*.

For any profile, or part of a profile, satisfying condition (4.28), a general solution of the linear or nonlinear flexure problem may be sought in terms of W_* and K_i of Table 4 (Chapter 3, §6)

$$W = -\frac{m}{\mu^*} x(\eta) + \sum_{i=1}^{4} B_i K_i(\eta\sqrt{\mu^*/2}) . \tag{4.30}$$

The constants B_i are determined by four boundary conditions. With the function W known, all the stress and displacement characteristics may be easily evaluated using $\vartheta = -\dot{W}/b$, $\psi = \ddot{\vartheta}/\mu^*$ and the formulas of Chapter 2, §2. In particular, formula (2.32) for the bending moment becomes

$$M_z = \int_s (x + W)T_1 \, ds + \int_s M_1 \cos \alpha^* \, ds , \tag{4.31}$$

where the integration extends over the entire profile of the cross-section and ds is the differential length of the profile.

Typical examples of the *simplest* profiles are given by the two cross-sections shown in Fig. 60 as well as the section of Fig. 58(c).

The solution of the simplest form (4.30) can be extended to some nonshallow profiles, e.g., to those with $\alpha = $ const. For these cases it

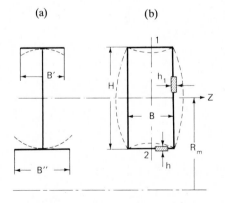

FIG. 60. Two beam profiles.

is not difficult to obtain an equation similar to (4.20)

$$\overset{....}{W} + W\mu^{*2} \cos^2 \alpha = -m\mu^* x .$$

As $\overset{...}{x} \equiv 0$, the profiles of the type Fig. 58(d) belong to the "simplest". There is, however, an important restriction: the displacements (rotations) must be small enough to make $\cos \alpha^* \approx \cos \alpha$.

A linear analysis of beams with the simplest profiles is easily carried out (cf., for instance, [147]). In the following we discuss mainly the nonlinear flexure of these profiles.

3.2. Box-beam

The profile of the beam (Fig. 60) is assumed to be symmetric with respect to the two axes and $R - R_m$ small compared to the radius of curvature R_m. Consider pure bending in the plane of curvature.

The deformation of the cross-sections is determined by the Kármán effect in the parts of the profile of the breadth B. Consider one of these. Its radial deflection is described by the general solution (4.30) with

$$x = \pm H/2 , \qquad \mu^* = \sqrt{12(1 - \nu^2)}B^2/(R_m^* h) . \qquad (4.32)$$

The dimensional parameter b is set equal to the breadth B of the profile, the coordinate η is varying from $-1/2$ to $1/2$ for $-B/2 < z < B/2$.

At the corner points of the profile there is no radial displacement ($W = 0$). Here the bending moment M_2, as well as the angle of rotation $\vartheta = -\dot{W}/B$, are equal to the moment and the rotation of the edge of the adjoining radial plate. Equating $M_2 = -D\ddot{W}/B$ (according to (4.23), without the small term $\nu\kappa_1$) to the moment $\dot{W}D_1 2/BH$, required to rotate the edge of the radial plate by the angle $\dot{W}/B$, we have the boundary conditions at the corner points of the profile (Fig. 60(b)),

$$z = \eta B = \pm B/2: \qquad W = 0 , \quad \ddot{W}D/B^2 \pm \dot{W}D_1 2/(BH) = 0 ,$$

$$D_1 = Eh_1^3/[12(1 - \nu^2)] . \qquad (4.33)$$

These conditions determine the four constants B_i of expression (4.30). The equations for the B_i are easily written out using the formulas of Table 4 (Chapter 3, §6)

$$\begin{bmatrix} k_1 & k_3 \\ -4k_3 - 4k_4 a & k_1 + k_2 a \end{bmatrix} \begin{bmatrix} B_1 & B_2 \\ B_3 & B_4 \end{bmatrix} = \begin{bmatrix} \dfrac{mH}{2\mu^*} & 0 \\ 0 & 0 \end{bmatrix} ,$$

$$a = \frac{h^3 2B}{h_1^3 H \beta} , \tag{4.34}$$

$$k_n = K_n(\beta/2) , \quad \beta = \sqrt[4]{3(1 - \nu^2)} B/\sqrt{R_m^* h} .$$

With B_i from (4.34) and W_* from (4.29) the expression (4.30) for the solution is complete. The stress resultants are easily evaluated by means of (4.23), where, for the box section, the κ_1 terms may be neglected.

The bending moment corresponding to the given change of curvature is rendered by formula (4.31) without the small M_1 term,

$$M_z = 2 \int\limits_{-1/2}^{1/2} \left[B_1 K_1(\beta\eta) + B_3 K_3(\beta\eta) + \frac{R_m^*}{R_m} x \right]$$

$$\times \frac{Eh}{R_m^*} [B_1 K_1(\beta\eta) + B_3 K_3(\beta\eta)] B \, d\eta + 2 \frac{Eh_1 H^3}{12} \left(\frac{1}{R_m^*} - \frac{1}{R_m} \right) .$$

$$\tag{4.35}$$

The bending moment in the radial walls of the beam is determined by their moment of inertia $2h_1 H^3/12$.

After integration (4.35) gives

$$M_z = \left[\left(B_1^2 + \frac{B_3^2}{4} \right) \frac{\sinh\beta + \sin\beta}{2\beta} + \left(B_1^2 - \frac{B_3^2}{4} \right) \left(\frac{K_2(\beta)}{2\beta} + \frac{1}{2} \right) \right.$$

$$\left. + B_1 B_3 \frac{K_4(\beta)}{\beta} \right] \frac{Eh}{R_m^*} B$$

$$+ 2 \frac{EhH}{\beta R_m} B \left[B_1 K_2 \left(\frac{\beta}{2} \right) + B_3 K_4 \left(\frac{\beta}{2} \right) \right] + 2 \frac{Eh_1 H^3}{12} \left(\frac{1}{R_m^*} - \frac{1}{R_m} \right) .$$

$$\tag{4.36}$$

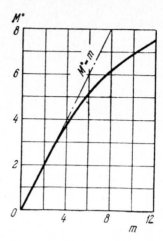

FIG. 61. Nonlinear flexure of a box-section beam.

For *initially straight* beams, when $1/R_m = 0$, the second term of formula (4.36) vanishes.

The nonlinear bending moment-curvature relation of an initially straight beam with $B = 5H$ and $h_1 = h$ is presented in Fig. 61. This graph gives also an estimate of the relation $M°(m)$ for beams with other dimensions, provided the main part of the moment of inertia of the cross-section (I) is constituted by the first term of its expression

$$I = \tfrac{1}{2} BH^2 h + \tfrac{1}{6} h_1 H^3 . \tag{4.37}$$

In the linear approximation the first term of formula (4.36) vanishes. The formula then becomes

$$M_z = \left(\frac{1}{R_m^*} - \frac{1}{R_n} \right) E \left\{ \frac{BH^2 h}{2\beta} \; \frac{\sinh \beta + \sin \beta + a(\cosh \beta - \cos \beta)}{\cosh \beta + \cos \beta + (a/2)(\sinh \beta + \sin \beta)} \right.$$

$$\left. + \frac{h_1 H^3}{6} \right\}, \tag{4.38}$$

with a defined in (4.34).

For beams with *small* curvature, i.e. with $\beta \ll 1$, the expression (4.38) approaches

$$M_z = \left(\frac{1}{R_m^*} - \frac{1}{R_m} \right) EI .$$

3.3. Shallow arc profile

The flexure of a beam with the cross-section in the shape of a shallow arc of a circle[3] was discussed briefly in present chapter, §2.3 as a case of the Fourier-series solution. But this profile can be treated as a "simplest" one. Indeed, the equation of a shallow circular arc shown in Fig. 58(c) may be presented in the form

$$x = -\frac{z^2}{2R_2} = -\eta^2 4H, \qquad \eta = \frac{z}{B}, \qquad -\tfrac{1}{2} \leq \eta \leq \tfrac{1}{2} \quad (b = B). \quad (4.39)$$

This function $x(\eta)$ satisfies condition (4.28). Hence, the solution function W is obtained by substituting $x(\eta)$ into expression (4.30) and into equations (4.22), which render the constants.

The nonlinear relation between the bending moment and the change of curvature is obtained by inserting expressions for T_1 and M_1 in terms of W into formula (4.31). Using (4.30) and (4.23) we have

$$\frac{M_z}{BD} = \left(\frac{1}{R^*} - \frac{1}{R}\right)\left\{1 - \frac{\nu R^*}{R_2} - \nu\left(\nu - \frac{R^*}{R_2}\right)x + \left(1 - \frac{R^*}{R}\right)\left(\frac{R^*}{R_2} - \nu\right)^2 x_1\right.$$

$$\left. + \frac{R^*}{R}\frac{R^*}{R_2}\left(\frac{R^*}{R_2} - \nu\right)(1 - x)\right\}. \quad (4.40)$$

The functions $\chi(\beta)$ and $\chi_1(\beta)$ are determined by (4.25) for β as defined in (4.22).

For the initially *straight* beam, i.e. $1/R = 0$, the relation (4.40) reduces to the WUEST [51] formula

$$\frac{M_z}{BD} = \frac{1}{R^*}\left\{1 - \nu\frac{R^*}{R_2} + \nu\left(\frac{R^*}{R_2} - \nu\right)\chi + \left(\frac{R^*}{R_2} - \nu\right)^2 \chi_1\right\}. \quad (4.41)$$

The behaviour of these beams, as implied by (4.40) and (4.41), does not differ fundamentally from the nonlinear flexure of the cosinusoidal profiles, illustrated in Fig. 59. The extremum values of the bending moment parameter $M°$ of the shallow arc profile are comparatively close to the values of $M°_{max}$ and $M°_{min}$ of the $\cos \eta$ profile. In the case of bending shown in Fig. 58(c), the critical moment is considerably lower

---

[3] The best-known example thereof is probably the household metal tape measure.

than the one for bending of the (initially straight) beam in the opposite direction. This effect is more pronounced than for the cosinusoidal profile (the broken-line graph in Fig. 59).

In the postcritical range the cross-section becomes straight. This state is characterized by large values of the curvature parameter β. The corresponding values of $\chi(\beta)$ and $\chi_1(\beta)$ are small and the respective terms in (4.41) or (4.40) may be neglected. Equation (4.41) becomes

$$M_z = DB\left(\frac{1}{R^*} - \frac{\nu}{R_2}\right). \tag{4.42}$$

As an immediate consequence of complete straightening of the cross-section the relation (4.42) follows directly from (4.31) when we insert the values of the curvature parameters and the moment M_1 corresponding to the straightening

$$1/R_2^* = 0, \quad x^* = x + W = 0; \quad \kappa_2 = -1/R_2, \quad \kappa_1 = 1/R^*,$$

$$M_z = BM_1 = BD(\kappa_1 + \nu\kappa_2).$$

Thus, even the nonlinear flexure of the "simplest" profiles can be described by comparatively simple explicit formulas. (Of course, this is made possible by the Saint-Venant problem-setting.)

The minimum beam-length needed to render the Saint-Venant solution applicable in the midlength part, away from unavoidable end constraints, can be assessed with the help of the semi-momentless equations (Chapter 2, §§4, 5). A rough estimate [176] is that the length of the beam must satisfy the condition

$$\left(\frac{L}{B}\right)^2 \gg 15\,\frac{H}{h}\,.$$

4. Bimetallic strip

The possibility of extending the analysis of the open-section beams to thermoelastic problems and nonhomogeneous materials is discussed briefly. The linear and nonlinear flexure caused by thermal expansion is considered more closely for a bimetallic strip.

4.1. Thermoelastic deformation

The deformation of a shell with nonhomogeneous elastic properties, due to a variable thermal expansion, can be determined as the deformation of a *reduced homogeneous* shell with elasticity relations (1.88), loaded with the additional *thermal loading* defined in Chapter 1, §7.

From the standpoint of practical applications, the multilayer and, in particular, bimetallic double-layer shells, deformed by thermal expansion varying only through the thickness of the shell wall, are of special interest.

Besides the restrictions imposed by edge constraints, this thermal deformation is determined by two simple factors: a constant-over-the-shell expansion defined in (1.111) and a deformation corresponding to the temperature-load moment M^t—according to (1.108)—applied all along the edge of the shell (Fig. 62).

Thus, the extension of most of the results obtained for homogeneous shells to the wide class of nonhomogeneous elastic materials and thermal expansion is straightforward. First, it requires the introduction of the reduced elasticity characteristics D, E', v', E'' and v'', which are defined by (1.88) and written out for the double-layer shell in (3.247). Further, the boundary conditions must take into account the moment M^t distributed along the edges. For the double-layer shell the value of M^t is determined by (1.108) with ζ_- and ζ_+ from (3.246).

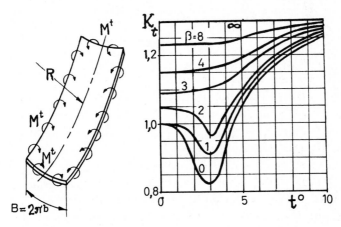

FIG. 62. Temperature loading and factor K_t determining nonlinear flexure.

The thermal deformation of the open-profile beams is generated by the moment M^t applied along the longitudinal edges and at the free end of the beam. The boundary conditions (4.2) on the edges are replaced by

$$M_2 = D\left(\frac{\dot{\vartheta}}{b} + \nu''\kappa_1\right) = M^t, \qquad \psi = 0.\qquad (4.43)$$

The relations (2.32) or (4.31) (representing the factual Saint-Venant boundary conditions on the beam ends) are supplemented with the term reflecting the contribution of the "thermal-load moment" M^t on the end

$$M_z + \int_s M^t \cos\alpha^* \, ds = \int_s (x + W)T_1^e \, ds + \int_s M_1^e \cos\alpha^* \, ds.\qquad (4.44)$$

Recall that T_1^e and M_1^e are the values of the stress resultants determined by the actual strain as if it were purely elastic—without any thermal expansions—caused by the loading (actual plus "thermal") alone (Chapter 1, §7.4). Thus, (2.24) and (4.23) also apply for the T_1^e and M_1^e.

The further solution differs only marginally from that discussed for purely static problems of homogeneous beams. It remains to determine the constants (the C in (4.3), the B_i in (4.21) or in (4.30)) with the aid of the boundary conditions (4.43).

The process just sketched leads to the relation following from (4.44)

$$M^\circ = M^\circ(m, M^t, A_n E'/E'', \beta, \nu'').\qquad (4.45)$$

For homogeneous material ($E' = E''$, $\nu'' = \nu$) and zero thermal influence ($M^t = 0$), the relation (4.45) coincides with (4.24). The value of F is determined by the formulas (4.25) with one change: the quantities A_n must be replaced by $A_n(E'/E'')^{1/2}$. For E', E'' and ν'' see Chapter 3, §12. The explicit relation (4.45) and a discussion of the deformation of a bimetallic beam can be found in the article [65].

The case of a bimetallic strip, i.e. of a straight-line profile $A_n = 0$, is of particular interest in applications. It provides the highest sensitivity—the largest change of curvature of the bimetallic instrument. The case is discussed in the next section.

4.2. Nonlinear thermal flexure

The nonlinear relation (4.45) between thermal expansion and change of curvature can be written for a bimetallic strip in a form

$$\frac{1}{R^*} - \frac{1}{R} = K_t(\beta, t^\circ) \, \frac{3}{2} \frac{\beta_1 - \beta_2}{hk_{12}} \, t, \tag{4.46}$$

where β_1 and β_2 denote the coefficient of the linear thermal expansion of the two layers of the bimetallic sheet. The nondimensional parameters β and t° represent the initial curvature of the bimetallic band and the effect of the difference in the thermal expansion of the two layers, when heated to an equal temperature change t,

$$[\beta \quad t^\circ] = \sqrt[4]{3(1 - \nu^2)} \, \frac{B}{h} \left(\frac{E'}{E''}\right)^{1/2} \left[\sqrt{\frac{h}{R}} \quad \sqrt{\frac{3}{2} \frac{\beta_1 - \beta_2}{k_{12}}} \, t\right]. \tag{4.47}$$

The factor K_t represents in (4.46) the nonlinear dependence of the change of curvature on t. It is set forth by graphs in Fig. 62, which reflect the evaluation of m according to the relation (4.45). In linear approximation K_t varies between 1 and $1 + \nu$.

The quantity k_{12} indicates (as the E'/E'' does) the influence of the difference of the moduli E_1 and E_2 and of the thicknesses h_1 and h_2 of the two layers

$$k_{12} = 1 + \frac{(1 - en^2)^2}{4en(1 + n)^2} \quad \left(e = \frac{E_2}{E_1}, \quad n = \frac{h_2}{h_1}\right). \tag{4.48}$$

The relation E'/E'' is determined, for equal values of the Poisson coefficients of the layers $\nu_1 = \nu_2$, by the formula following from (3.247)

$$\frac{E'}{E''} = \frac{(1 + n)^2(1 + en)}{4(1 + en^3) - 3(1 - en^2)^2(1 + en)^{-1}}. \tag{4.49}$$

The graphs of Fig. 62 indicate a considerable dependence of the ratio of change of curvature to thermal expansion on the initial curvature parameter β.

For small values of t° (in practical terms $t^\circ < 1$), the coefficient K_t depends only on the parameter β and the formula (4.46) gives the change of curvature proportional to the thermal expansion. When

$t^\circ < 1$ the linear approximation applies also for stresses. On a linear model, the factor K_t varies from the minimum $K_t = 1$ to $K_t = 1 + \nu \approx 1.3$. For initially straight or narrow bimetallic band, i.e. for $\beta < 1$ and $t^\circ < 1$ we have $K_t = 1$ and (4.46) reduces to the formula due to TIMOSHENKO [8].

This case corresponds to small deflections of a bimetallic plate, for which we have (from (1.110)) at any point of the plate $M_1^e = M_2^e = M'$, and with the elasticity relations (1.88)

$$\frac{1}{R_1^*} = \frac{1}{R_2^*} = \kappa_1 = \kappa_2 = \frac{M'}{(1 + \nu'')D} \, . \tag{4.50}$$

For a bimetallic plate this gives in fact the value of $1/R^*$ equal to that of (4.46) for $1/R = 0$ and $K_t = 1$.

The longitudinal curvature ($1/R$, Fig. 62) does not constrain the lateral deformation when the breadth B is *small* compared to $(Rh)^{1/2}$, i.e., when $\beta < 1$. This leads to relations (4.50) and to the formula (4.46) with $K_t = 1$. The other limiting case is constituted by *broad* bimetallic strips with $B \gg (R^*h)^{1/2}$. Away from the edge zone, the lateral deformation is eliminated ($\kappa_2 = 0$) and we obtain, similarly to (4.50), $\kappa_1 = M'/D$, which is equivalent to (4.46) with $K_t = 1 + \nu''$.

The nonlinear characteristics of the thermal flexure (Fig. 62) persist to $K_t < 1$. The deformation of the cross-section can substantially increase the flexural stiffness of the bimetallic strip.

The foregoing applies, of course, only for sufficiently long strips or bands. When the length is of the same order as the breadth, the Saint-Venant setting of the problem is inadequate. As is often the case, the end fixtures restrict the deformation of the cross-sections of a substantial part of the bimetallic strip. In the limiting case of a *short broad* strip this leads to a relation similar to (4.46) with $K_t = 1 + \nu''$.

An analysis of the middle-length bimetallic strips may be found in the articles [62, 65].

Finally, it should be noted that deviations of the cross-section of a bimetallic strip from the straight-line profile may cause substantial variations of the stress state and deformation. This is evident on inspection of equation (4.44). The deviation of profile with an amplitude $\sim h$ drastically changes the flexural stiffness and may involve a snap-through by large displacements. These effects can be useful in certain instrument applications.

CHAPTER 5

FLEXIBLE SHELLS OF REVOLUTION

CHAPTER 5

FLEXIBLE SHELLS OF REVOLUTION

1. Basic problems

The most flexible type of shells of revolution is a bellows—tube with circumferential corrugations (Fig. 63).

With a sufficient number of corrugations the bellows allows practically unlimited elastic displacements and rotations. It can sustain large external or internal pressure.

Besides the *seamless* bellows shown in Fig. 63, there are widely used bellows composed of ring shells, welded or brazed together, as in Fig. 75. The shape of the seamless bellows is mostly set forth by detail drawings as indicated schematically in Fig. 64(b). The profile is composed of straight lines and circle arcs. It is often overlooked that the *real* shape of a bellows can be substantially different. An example of such a profile is presented in Fig. 64(a).

The wall-thickness of the seamless bellows varies along its profile. This variation is determined (as the real profile also is) by manufacturing technology. An example is provided by the ground edge in Fig. 65. The following elementary analysis of the plastic deformation accompanying the forming of a bellows out of a tube is intended to give a rough assessment of the wall-thickness variation.

For the hydraulic shaping of a seamless bellows a tube of radius R_0 and wall-thickness h_0 is extended by an inner pressure and a simultaneous axial compression. To estimate the change of the wall-thickness during this process we assume two features of plastic deformation. First, we suppose the *volume* of the material to remain *constant*. This implies for the ring element with initial dimensions ds_0, R_0 and h_0,

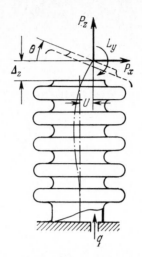

Fig. 63. Bellows.

which becomes in the finished bellows a ring with dimensions ds, R and h, respectively (both elements are shown in Fig. 64), the relation $hR \, ds = h_0 R_0 \, ds_0$.

The next assumption is less evidently true. We suppose the ring element which has its radius gradually extended from R_0 to R to reduce its cross-section dimensions in *equal* measure in each of *two directions*. This implies $h/h_0 = ds/ds_0$. The two assumptions lead to the following estimate of the wall-thickness (put forward in [99]):

$$h = h_0 \sqrt{R_0/R} \, . \tag{5.1}$$

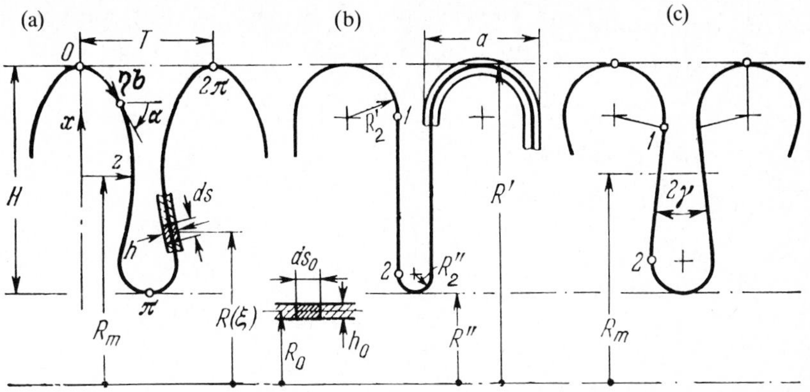

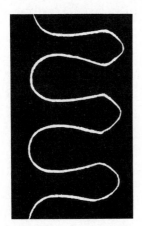

Fig. 64. Real profile and its two "nominal" versions.

The relation (5.1) is usually confirmed by measurements of real bellows (cf. *Standards of EJMA*, 5th Edition (1985) Sect. C-5.2, and Fig. 4 in [165]).

However, it must be noted that (5.1) cannot be applied to bellows shaped by rolling. This is illustrated by the profile shown in Fig. 65. The left-hand side of the picture corresponds to (5.1). The right-hand side shows the part of the profile shaped by rolling. In this case, the *smaller* the radius $R(\eta)$, the *thinner* the wall.

The *applications* of bellows, though manifold, belong mostly to one of the two main classes: (1) pressure-measuring instruments, which be-

Fig. 65. A ground edge of bellows.

come extended by an increase of inner pressure; (2) elastic tubes allowing large displacement and rotation of one end with respect to the other.

The analysis of elastic bellows used as *measuring instruments* must include the relation between the applied pressure q and the corresponding displacement Δ_z (Fig. 63).

To measure the pressure q the reactions P_q at the *fastened ends* of bellows can be used. Besides P_q, the critical inner pressure, causing the sidewards bulging of bellows, shown in Fig. 66, must be evaluated.

For bellows used as highly *elastic tubes* the moments and forces (Fig. 63) caused by displacement and rotation of an end of the bellows must be evaluated. Of course, in both cases the design stresses have to be determined.

One of the first theoretical investigations of bellows is due to DONNELL[1]. An analysis used in industry for decades has been developed by FEODOS'EV [32] on the basis of a Ritz solution for the profile of Fig. 64(c) with $R_2' = R_2''$ and $h_0 = h_0 R_0 / R(\eta)$. Later work was concerned mostly with the "nominal" profile of the type of Fig. 64(b), composed of plates and toroidal parts. HAMADA and TAKEZONO [96] investigated this U-shaped bellows with the aid of a finite-difference solution of the linear Reissner–Meissner equations and provided a parametric study represented by graphs of the stress and strain for $R_2' = R_2''$ and $\gamma = 0$. (Similar results were published by ANDREEVA et al. [165].) A numerical solution of the nonlinear Reissner–Meissner

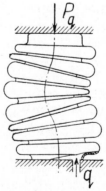

FIG. 66. Instability by inner pressure.

[1] *Trans. ASME, J. Appl. Mech.* **2** (1932).

equations was obtained for the type of bellows just indicated by HAMADA and SEGUCHI [115].

The Fourier-series solution of the linear and nonlinear Reissner–Meissner equations [99] allowed the analysis for the real profile shape and wall-thickness variation. The corresponding parameter study has been presented by SAVKIN [128]. Graphs allowing a simplified evaluation of the stresses and displacements (for any R_2'/R_2'' and γ) can be found in [130]. A similar analysis is presented in §2.

Apparently, the first investigation of *welded bellows* is due to SINTO, SEGUCHI and JOKODA [134].

In §3 we consider an arbitrary profile and large displacements including the contact of the outer surfaces of the bellows.

The first investigation of the *nonaxisymmetric* deformation of bellows has been accomplished by HARINGX [46] by introducing a scheme presenting a bellows as a system of annular plates (Fig. 70).

More exact (thin-shell theory) solutions of the pure bending problem have been obtained by HAMADA et al. [138] by using numerical integration and by IOKHELSON [139] and VASIL'EV [142]—in the Fourier-series form. The parametric study of the pure bending of bellows [144] provides a system of graphs for the stiffness and stress factors.

In contrast to the *periodic* problem of the pure bending of bellows, their cross-bending is a *boundary* problem. Its solution, set forth in §4, was obtained by the late V.V. VASIL'EV in collaboration with the present author.

The analysis of the *expansion joints* has essentially the same aims as that of bellows (excepting the problem of sensitivity).

The expansion joints are diverse in shape and dimensions. The profile shown in Fig. 67(a) has been comprehensively investigated (in the linear approximation) in the monographs of CHERNINA [114] and CHERNYKH [86] as a boundary problem of the circular torus and plate parts of the shell. The series solution of §3 encompasses also the profiles differing from the "torus-plate" shape.

The circular-torus shells and, in particular, the expansion joints of this form (Fig. 67(b)) have attracted much attention in the literature. The linear analysis of this problem and references to the earlier work can be found in the book of SEIDE [173].

Progress in the investigation of these shells with the aid of Fourier series is due to TUMARKIN (1952) and DAHL (1953). The analysis of the

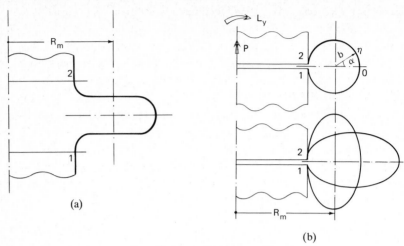

FIG. 67. Expansion-joint profiles.

nonsymmetric deformation apparently was first carried out by TURNER [72], further results are due here to CHERNYKH [86] and CHERNINA [114]. The peculiarities of the stress state in the zone of the toroidal shell, where $\alpha \approx \pi/2$ and the Gauss curvature is nearly zero, were clarified by JORDAN [154].

Outside the scope of this book are the snap-through shells, homogeneous and bimetallic. The conical and spherical shells are treated in [211, 228] and in Liu Ren-Huai, *Internat. J. Non-Linear Mech.* **18** (1983) 409–429, any polynomial meridian shape in E.L. Axelrad, *Raschet Prostranstv. Konstr.* **6** (1961) 275–298 (in Russian). On relevant general methods cf. [221, 222].

2. Bellows under axial tension and normal pressure

Linear and geometrically nonlinear axisymmetric deformation is considered for bellows of various shapes and a given variation of the wall-thickness. For simplified evaluation, graphs of stress and flexibility factors are presented.

2.1. Solution of the axisymmetric problem

Introduce the coordinates according to Figs. 64(a) and 5. The parameter $b = \text{const.}$ is fixed to make the variation of the coordinate η inside one convolution equal to 2π. Consider the deformation of one of the convolutions sufficiently distant from the edges of the bellows to make the deformation identical to that of its two immediate neighbours. (The influence of edges will be discussed in present chapter, §3.) Thus, the deformation of the corrugations varies *periodically* with η.

Choosing the origin of the coordinate η in the plane of symmetry of a corrugation we can represent the solution by the sine series (2.170). The coefficients are determined by equations (2.171). It is known from experience that the nonlinearity of deformation of most bellows remains comparatively limited within the elastic range. This suggests the perturbation-method solution represented by expansions in the load parameters P° and q°,

$$\vartheta = P^\circ \vartheta_P + q^\circ \vartheta_q + P^{\circ 2} \vartheta_{PP} + \cdots, \qquad \psi = P^\circ \psi_P + q^\circ \psi_q + P^{\circ 2} \psi_{PP} + \cdots. \tag{5.2}$$

Substitute the corresponding expansions of matrices of Fourier coefficients into the system (2.171). Retaining only linear terms, we obtain for the two cases of loading the equations

$$B \begin{bmatrix} \psi_P \\ \vartheta_P \end{bmatrix} = \begin{bmatrix} 0 \\ -s \end{bmatrix}, \qquad B \begin{bmatrix} \psi_q \\ \vartheta_q \end{bmatrix} = - \begin{bmatrix} 0 \\ s_+ \Lambda^{-1} r_- s - c_- \Lambda_+^{-1} r_+ c \end{bmatrix},$$

$$B = \begin{bmatrix} \Lambda_+ r_+ t_+^{-1} \Lambda & -\mu c_- \\ \mu c_- & \Lambda_+ r_+ t_+^3 \Lambda \end{bmatrix}, \tag{5.3}$$

$$[\psi_P \; \vartheta_P \; \psi_q \; \vartheta_q] = \begin{bmatrix} b_1^P & a_1^P & b_1^q & a_1^q \\ b_2^P & a_2^P & b_2^q & a_2^q \\ \cdots & \cdots & \cdots & \cdots \end{bmatrix}.$$

In the following the *nonlinear* problem will be considered in detail only for the axial extension. The nonlinearity is more pronounced in this case than for a normal-pressure loading.

The systems of equations for the second and higher approximations are obtained by substituting the expansions (5.2) into (2.171), where also the c'_-, c_-^* and s^* are represented by their expansions (2.172) and (2.174).

Taking into account in (2.171) the relations (5.3), dividing both sides of each equation by P^{o2} and after that assuming $P^o = 0$ we obtain equations for ψ_{PP} and ϑ_{PP}. These equations are

$$B\begin{bmatrix} \psi_{PP} \\ \vartheta_{PP} \end{bmatrix} = \mu \begin{bmatrix} -\frac{1}{2}s_+\bar{\vartheta}_P\,\vartheta_P \\ s_+\bar{\vartheta}_P\,\psi_P \end{bmatrix} - \begin{bmatrix} 0 \\ c_-\vartheta_P \end{bmatrix}, \tag{5.4}$$

where $\bar{\vartheta}_P$ is composed similarly to $\bar{a}_s$ in (2.148). After the determination of ψ and ϑ the stress and strain resultants as well as the stresses may be evaluated by means of explicit formulas. Formulas (2.24) and (2.39) render T_1, T_2 and

$$\kappa_1 R = -s'\vartheta \,, \quad \kappa_2 b = \dot{\vartheta} \,,$$

$$\begin{bmatrix} \sigma_1 \\ \sigma_2 \end{bmatrix} = \frac{Eh/2}{1-\nu^2} \begin{bmatrix} 1 & \nu \\ \nu & 1 \end{bmatrix} \begin{bmatrix} \kappa_1 \\ \kappa_2 k_R \end{bmatrix} + \frac{1}{h}\begin{bmatrix} T_1 \\ T_2 \end{bmatrix}; \quad k_R = \frac{1\pm h/6R_2}{1\pm h/2R_2}\,. \tag{5.5}$$

Recall that the matrices s and c of the Fourier coefficients c_j and s_j of $\cos \alpha$ and $\sin \alpha$ represent the shape of the meridian. The coefficients are computed with the use of the Fourier formulas as indicated in (3.59) and

$$c = \tfrac{1}{2}c_0 + \sum_j c_j \cos j\eta \,, \quad s = \sum_j s_j \sin j\eta \quad (j=1,2,\ldots)\,. \tag{5.6}$$

Since the profile is represented by its coordinates $x(\eta)$ and $z(\eta)$, it is convenient to calculate the c_j and s_j by taking advantage of the relations $s = -\dot{x}/b$ and $c = \dot{z}/b$. The Fourier formulas then yield (cf. Fig. 64)

$$\frac{c_0}{2} = \frac{T}{2\pi b}\,; \quad c_j = \frac{2j}{\pi b}\int_0^\pi \left(z - \frac{c_0}{2}\,b\eta\right)\sin j\eta \, d\eta \,,$$

$$\tag{5.7}$$

$$s_j = \frac{2j}{\pi b}\int_0^\pi x \cos j\eta \, d\eta \,.$$

For the "nominal" profiles of Fig. 64(b), (c), formulas (5.7) lead to expressions for c_j and s_j similar to those presented in Chapter 3, §3.2 for the flat-oval profile of a tube.

Of course, the effectiveness of the series solution depends on the convergence of the series for c, s and ψ, ϑ. A perception of the convergence gives Table 10. For the profiles of Fig. 64(b), (c) it is usually sufficient to retain in the series (5.6) the terms with $j \leqslant H/(4|R_2|_{\min}) + 2$. For the ψ and ϑ series (2.170) it is necessary to retain at least $H/(4|R_2|_{\min}) + 2$ or $\mu^{1/2} + 2$ terms, whichever of the two quantities is larger.

For most real bellows it is sufficient to retain in the series of c, s, ψ and ϑ not more than eight terms. This assures a simplicity of the series solution in comparison with the purely numerical methods.

We note, finally, that the "real" profile (that of Fig. 64(a)) is represented by Fourier series, which converge at least as effectively as the series for the corresponding "nominal" profiles (b) and (c).

2.2. Influence of the profile shape and wall-thickness variation

The results of the analysis of five versions of a bellows are presented in Table 10. The five versions correspond to one and the same detail drawing, fixing (as is often the case) only the dimensions R', R'', T, a, R_0 and h_0 and depicting the profile of Fig. 64(b). For all five examples

$$[R' \ R'' \ T \ a \ R_0 \ h_0] = [50 \ 37.75 \ 7.3 \ 5.2 \ 38 \ 0.16] \, \text{mm} .$$

Figure 68 presents for the cases a_1, b_1, b_2 and c_1 the maximum over the wall-thickness meridional stress. The notations a, b and c of Table 10 and Fig. 68 correspond to the shapes labelled with these letters in Fig. 64. The numbers 1, 2 and 3 in the indices ($a_1, b_1, \ldots$) reflect the three distributions of the wall-thickness represented by the Fourier coefficients t_j in the lines 7, 8 and 9 of the table. The distributions are (1) the $t = h(\eta)/h_m$, according to (5.1); (2) the wall-thickness, corresponding to Fig. 65, produced by a combination of hydraulic shaping and rolling; and (3) uniform wall-thickness.

The results of calculation of the examples show a weaker dependence of stress on the wall-thickness variation than on the profile shape. The most substantial difference between the five cases can be observed

TABLE 10.

#	j		0	1	2	3	4	5	6	7	8
1	$s_j \cdot 10^4$	a	—	12 340	−453	3 310	−697	1 250	−227	599	−184
2		b	—	12 180	−710	2 880	−954	963	−719	405	−348
3		c	—	12 055	−441	2 543	−526	525	−282	11	−19
4	$c_j \cdot 10^4$	a	2 600	1 810	3 840	−916	2 240	−875	1 010	−220	321
5		b	2 540	1 910	4 030	385	1 970	−940	702	−995	476
6		c	2 617	990	4 750	−61	2 144	−773	315	−439	−149
7	$t_j \cdot 10^4$	1	10 084	−917	68,5	−55	−5	4	1	−3	−9,2
8		2	9 670	−621	−387	893	10	35	−52	25	−10
9		3	10 000	0	0	0	0	0	0	0	0
10	$ja_j^p \cdot 10^4$	$a1$	—	7 200	−2 780	−1 002	−720	−161	−82,6	−56	9
11		$b1$	—	8 710	−2 210	−1 480	−1 530	−760	−353	−70	128
12		$b2$	—	9 310	−3 030	−743	−2 050	−575	−567	−7	43
13		$b3$	—	9 130	−3 440	−948	−1 410	−560	257	60	131
14		$c1$	—	9 043	−977	−1 470	−925	−1 040	−360	−300	146
15	$ja_j^q \cdot 10^4$	$a1$	—	−1 290	2 940	−366	488	−203	158	−43	16
16		$b1$	—	−1 110	3 180	−605	115	−383	2,15	−90	34
17		$b2$	—	−1 450	3 800	−977	494	−610	122	−181	95
18		$b3$	—	−1 510	3 630	−1 040	354	−424	69	−98	42
19		$c1$	—	−619	2 894	−343	−199	−187	−194	28	−55
20	$ja_j^{pp} \cdot 10^4$	⎫	—	−6 361	−979	2 640	710	785	407	220	116
21	$ja_j^{ppp} \cdot 10^4$	$b1$	—	−15 076	12 950	13 700	−1 865	515	−1 160	−148	−692
22	$ja_j^{qq} \cdot 10^4$	⎭	—	207	−396	164	74	105	31	6	9
23	$a_j^p \cdot 10^4$	⎫	—	12 410	107	446	−21,3	95,3	−12,3	33,4	−2,8
24	$a_j^{pp} \cdot 10^4$	Fig. 69	—	−365	−474	20,5	−48,6	−4,2	−11,5	−1,0	−3,2
25	$a_j^{ppp} \cdot 10^4$	⎭	—	−61 200	1 500	13 140	81,9	1 300	38	413	−11

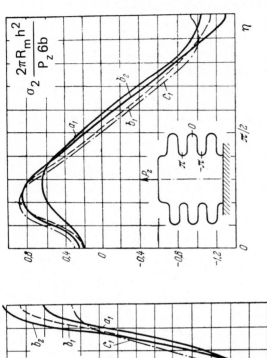

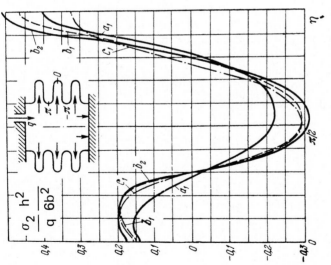

FIG. 68. The shape of profile and the wall-thickness variation influence the stresses.

in the stresses at the inner radius ($R = R''$, $\eta = \pi$) of the corrugation. This is the case for both the axial extension force P_z and the normal pressure. The design stress is in case b_2 approximately 25% higher than in case a_1.

Consider now the nonlinear case of *large displacements* of bellows under an axial force P_z.

The solution described in the preceding section (§2.1) yields the Fourier coefficients (Table 10, lines 20–22) of the second and third-power terms of the expansion (5.2). The nonlinearity of the comparatively thick-walled bellows of Fig. 64 is weak, at least within the limits of elasticity of available metals.

Another example (the bellows with deep, shallow and thin corrugations) is shown in Fig. 69. It is capable of large elastic displacements inside one corrugation and, thus, displays a more considerable geometric nonlinearity.

The corresponding Fourier coefficients are presented in the lines 23–25 of Table 10. The nonlinear relation between the axial force and extension of the bellows is illustrated by the graphs in Fig. 69. The

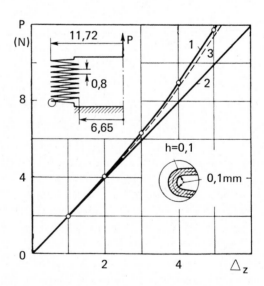

FIG. 69. Experiment—1, linear approximation—2 and the nonlinear solution—3. Dimensions in mm.

extension Δ_z is calculated through (2.32) and (2.25) with ϑ from (5.2) and (5.3). The nondimensional longitudinal load P° is expressed in terms of P_z by (2.22). For the D_m and $R_m = (R' + R'')/2$, according to Fig. 69, $E = 130\,\text{kN}/\text{mm}^2$ and $\nu = 0.3$ we find $P^\circ = 0.0047P$ (N). In the elastic range for this bellows the values are $P < 12\text{N}$ and $P^\circ < 0.056$. The nonlinearity of the relation $P(\Delta_z)$ is $\sim 20\%$. A satisfactory convergence of the expansion (5.2) in powers of P° is observed even for the upper-limit value of P°. Though the $P^{\circ 2}$ and $P^{\circ 3}$ terms are of equal order of magnitude, the $P^{\circ 4}$ term is of negligible value. We note in Table 10 also the satisfactory convergence of the Fourier series of the functions ϑ_{PP} and ϑ_{PPP}. It is at least as good as the convergence of the series for $\vartheta_P(\eta)$.

The solution just described can be realized with a desk computer. The alternative successive-approximations solution on the lines of (3.36), (3.37) has its specific advantages.

From the standpoint of practical applications an overall graphic view of the relations between the strength and the stiffness of bellows and its geometry is required. It is presented in present chapter, §2.4 based on computations made with the aid of the trigonometric-series solution. Let us first consider a class of bellows, allowing explicit formulas for stresses and displacements.

2.3. Elementary formulas

The high deformability of bellows is achieved by the use of profiles of the type shown in Fig. 69. These profiles are the limiting case of the one in Fig. 64(b) when $H = \frac{1}{2}(R' - R'')$ is large compared to the pitch T. Such a bellows is in fact composed of small toroidal parts (of radius R_2' or R_2'') connecting the edges of comparatively broad annular plates. For larger values of $H/|R_2'|$ and $H/|R_2''|$, the series (5.6) representing the shell form and those representing the stress and displacements functions ψ and ϑ converge less effectively. On the other hand, these profiles with $H/|R_2| \gg 1$ can be approximated by a scheme set forth in Fig. 70.

The deformability of the small toroidal parts can be neglected compared to that of the large plane-plate parts of the bellows. Consider this limiting case of bellows in the linear approximation, for uniform wall-thickness. For a plate, i.e. $\alpha^* = \alpha = \pi/2$, the system (2.18)

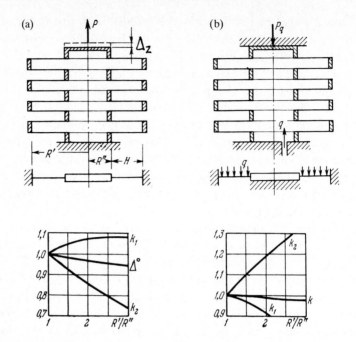

FIG. 70. Simplified model of bellows. Stress, displacement and reactive force factors.

decouples into two separate equations. The second of the equations includes only one unknown, $\vartheta(\eta)$. Setting in this case the dimensional parameter b equal to R', the equation presents itself in the form

$$(r\dot\vartheta)^{\cdot} - \frac{\vartheta}{r} = P^{\circ} + \tfrac{1}{2}q^{\circ}r , \quad r = \frac{R}{R'} , \quad ()^{\cdot} = \frac{\mathrm{d}}{\mathrm{d}r}() ,$$

$$P^{\circ} = \frac{PR'}{2\pi D} , \quad q^{\circ} = \frac{qR'^{3}}{D} . \tag{5.8}$$

The general solution of this equation is

$$\vartheta = C_1 r + C_2 \frac{1}{r} - \tfrac{1}{2}P^{\circ}r \ln r + \tfrac{1}{16}q^{\circ}r^{3} . \tag{5.9}$$

For the case of axial extension by a given force P, shown in Fig. 70(a), it remains to determine the constants with the aid of conditions of no rotation of the edges of the plate:

$$r = \frac{R''}{R'} \, , 1: \quad \vartheta = 0 \, .$$

For the bellows loaded by inner pressure and *restricted* in *its extension*, besides C_1 and C_2, the reaction P_q resisting the extension must be determined. This is done by satisfying a condition of zero extension of the bellows, which means a zero displacement of inner edge of each plate with respect to its outer edge (Fig. 70(b)). In general, this displacement is determined with the aid of (2.25) with $s' = s = 1$, $c^* = c = 0$ and $b = R'$, $\eta = r$,

$$\frac{\Delta_z}{2n} = \int_{R''/R'}^{1} \vartheta R' \, dr \, , \tag{5.10}$$

where Δ_z is the extension of the bellows composed of n corrugation waves and, thus, of $2n$ plates.

The maximum stress occurs on the inner edge of the "plate". It is evaluated by inserting the expression (5.9) into formulas (5.5) for $r = R''/R'$. Here it is necessary to take into account the real form of the toroidal part of bellows, calculating the stress with the factor k_R.

The resulting formulas for the design stress can be easily written out. But they are somewhat involved. Detouring around this complication we construct auxiliary formulas for the stresses in an elementary way for the case when the annular plates are narrow, i.e. when the relation R''/R' is near unity. Such a plate bends as a beam of length $H = R' - R''$ with both ends built-in and a rectangular cross-section with moment of inertia $2\pi R \, h^3/12$. For this beam the formulas of maximum stress at $R = R'$, R'' and of deflection are

$$\sigma_2(R') = k_1 0.478 \, \frac{PH}{h^2 R'} \, , \qquad \sigma_2(R'') = k_2 \, 0.478 \, \frac{PH}{h^2 R''} \, ,$$

$$\frac{\Delta_z}{2n} = \Delta^\circ \, \frac{0.29}{R' + R''} \, \frac{PH^3}{Eh^3} \, . \tag{5.11}$$

The coefficients k_1, k_2 and $\Delta°$ take into account the actual annular-plate ratio R'/R''. The values of k_1, k_2 and $\Delta°$ calculated with the aid of the solution (5.9) and formula (5.10) are presented in Fig. 70.

In a similar way, one obtains for the scheme of Fig. 70, b the formulas

$$\sigma_2(R') = k_1 \frac{qH^2}{2h^2} , \qquad \sigma_2(R'') = k_2 \frac{qH^2}{2h^2} ,$$

$$P_q = kq\pi \left(\frac{R' + R''}{2} \right)^2 . \tag{5.12}$$

For most real bellows, the ratio $R'/R'' \leqslant 1.5$ and for them the factors k_1, k_2 and k are not much different from unity (Fig. 70).

Formulas (5.11) and (5.12) give the maximum absolute value of the stress on the inner and outer surface of the bellows. It occurs in that part of the bellows where the curvature $1/R_2$ of the profile is large. This should be taken into account by multiplying the value of σ_2 from (5.11) and (5.12) with the factor k_R of (5.5). The signs in the k_R formula are negative on the concave surface.

We can now return to the bellows of a more general form to determine for them stress and flexibility by means of a set of graphs.

2.4. Evaluation of stresses and displacements

As was mentioned, there are three basic types of loading of bellows: (1) extension by an axial force, (2) loading by a normal pressure when both ends of the bellows are fixed and (3) loading with normal pressure when the ends of the bellows are free.

For the *first case* we compose formulas for the design stress and the extension of bellows Δ_z as a generalization of the formulas (5.11),

$$\sigma_2 = \pm F_2 \frac{PH}{h_m^2 R_m} k_R , \qquad \sigma_1 = F_1 \frac{PH}{h_m^2 R_m} + \nu\sigma_2 ,$$

$$\frac{\Delta_z}{n} = F_w P \frac{H^3}{R_m E h_m^3} \qquad \left(R_m = \frac{R' + R''}{2} \right) . \tag{5.13}$$

The plus sign in the formula for the wall-bending stress σ_2 corresponds to the inner surface of the bellows for $R = R'$ and to the outer surface for $R = R''$ (Fig. 64).

The factors F_i represent the shape of the bellows. To define the shape using only a reasonable number of parameters we must restrict the analysis to a certain type of profile and of wall-thickness variation. Consider the profiles of the type of Fig. 64(c). They assure enough variability and are defined by only six dimensions or *five* dimensionless parameters. The five will be

$$\frac{H}{T}, \quad \mu_T, \quad \frac{R_2'}{R_2''}, \quad \frac{b}{R_m}, \quad \gamma \quad \left(\mu_T = \frac{T^2}{R_m h_m} \right). \tag{5.14}$$

Evaluation with (5.13) will become simple if we determine the factors F_i in terms of the parameters (5.14) in a simple enough way. This will be done by representing the functions F_i (computed on the basis of the series solution) with graphs. It is made decisively easier by a circumstance known on account of the computed values of F_i. Namely, the last three of the five parameters (5.14) are of *secondary importance*. The influence of the value of R_2'/R_2'', b/R_m and γ on the factors F_i is comparatively weak. This allows us to represent the F_i by their Taylor series with respect to the three parameters and to retain in these series only the linear terms

$$F_i = S_i + \frac{b}{R_m} A_i + \left| \frac{R_2'}{R_2''} - 1 \right| B_i + \gamma C_i \quad (i = 1, 2, W). \tag{5.15}$$

The factors S_i, A_i, B_i and C_i depend on only *two* geometry parameters: H/T and μ_T. The values of these quantities (calculated with the aid of the series solution) are presented by graphs of Figs. 71 and 72.

As the circumferential stress σ_1 is smaller than σ_2, it may be determined with less accuracy. The A_1, B_1 and C_1 terms in (5.15) may be disregarded. Correspondingly, the graphs for the A_1, B_1 and C_1 terms are not presented.

The maximum stresses determined by (5.13) and (5.15) occur mostly at the outer- and inner-radius points $\eta = 0$, π (Fig. 64). The solid lines of the S_2 graph pertain to both these points. When the maximum stress occurs *outside* the points $\eta = 0$ or $\eta = \pi$, the S_2 is determined by the dashed lines of Fig. 71. The factors A_2, B_2 and C_2 are set forth by the solid line graphs for the points with $R > R_m$, and by the dashed lines for $R < R_m = (R' + R'')/2$.

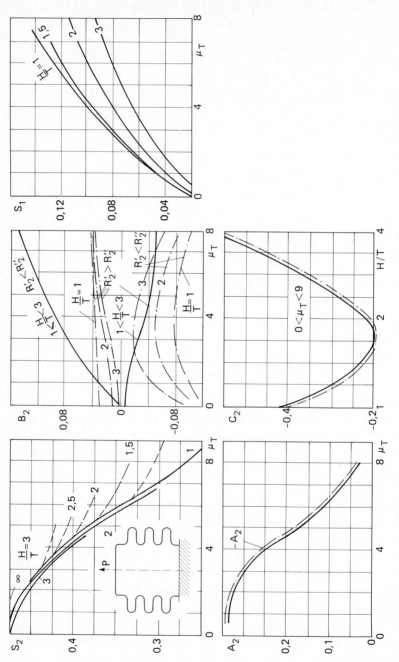

FIG. 71. Stress factors. Dashed lines for A_2, B_2 and C_2 determine stresses at $R < (R' + R'')/2$.

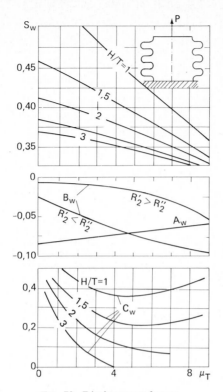

FIG. 72. Displacement factors.

Turning to the case of bellows with *fixed ends* loaded by an inner *pressure* (Fig. 68), we determine the maximum stresses σ_2, which occur near the outer- and inner-radius points, by formulas similar to (5.12),

$$\sigma_2 = \pm F_2 q \, \frac{H^2}{h_m^2} \, k_R \,, \qquad \sigma_1 = F_1 q \, \frac{H^2}{h_m^2} + \nu \sigma_2 \quad (F_1 \approx S_1) . \quad (5.16)$$

The factors F_2 and F_1 are determined by (5.15). However, the values of S_i, A_i, B_i and C_i *differ* from those of the preceding case. The parameters are represented by graphs of Fig. 73: solid lines for the area $R > R_m$, dashed lines for $R < R_m$. As in the first case, the lower-level stress component σ_1 is calculated with $F_1 \approx S_1$, neglecting the influence of the secondary geometry parameters in (5.15).

For the third case (bellows with a *free end*, loaded by a uniform normal *pressure* q), the stresses and displacements can be found as a

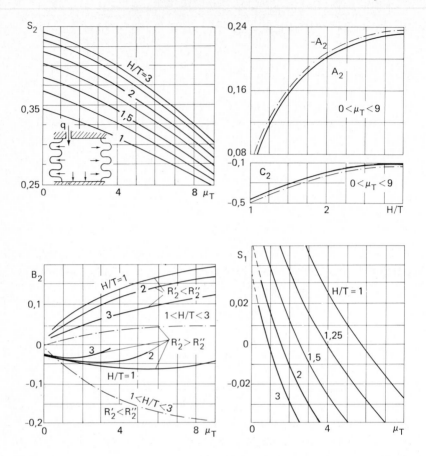

FIG. 73. Stress factors.

superposition of the first two cases. The stresses caused by the inner pressure, evaluated according to (5.16) are added to the stresses caused by the axial force, rendered by (5.13) for axial force $P = P_q$ from (5.12). The force $P = P_q$ also determines, through (5.13), the extension Δ_z of the bellows, caused by pressure.

The error in the evaluations is due mainly to the linearization in (5.15) of the relations between F_i and the three secondary geometry parameters. This error does not exceed the maximum of $b/2R_m$, $|R'/R'' - 1|/15$ and $|\gamma|/5$.

3. Bellows with edge constraints

The stressed state of bellows was discussed in present chapter, §2 without any allowance for the influence of the edges. In actual fact, the foregoing was concerned with the convolutions of the middle part of the bellows, away from its ends.

Consider now bellows with a relatively small number of convolutions taking into account the conditions on edges. The stressed state is supposed to be symmetric with respect to the middle cross-plane of the bellows (or of an expansion joint, as in Fig. 67). Only the *linear* approximation is considered.

3.1. Solution of the boundary problem

Under the conditions of symmetry of the shell shape and of the boundary conditions the solution of the axisymmetric problem has the form (2.175) with two of the four constants equal to zero,

$$\psi = \bar\psi + C_1\psi_1 + C_3\psi_3 , \qquad \vartheta = \bar\vartheta + C_1\vartheta_1 + C_3\vartheta_3 . \qquad (5.17)$$

The functions ϑ_1, ψ_1, ϑ_3 and ψ_3 are determined by the boundary conditions

$$\eta = \pm\pi : \qquad \vartheta_1 = \pm 1 , \quad \psi_1 = 0 ; \qquad \vartheta_3 = 0 , \quad \psi_3 = \pm 1 . \qquad (5.18)$$

That is, ϑ_1, ψ_1 is the particular solution describing the deformation of the shell caused by a symmetric rotation $\vartheta = \pm 1$ of both edges; ϑ_3, ψ_3 is the solution describing the deformation caused by radial forces (shown in Fig. 74), distributed on both edges. The functions ϑ_i and ψ_i can be found in trigonometric-series form, as set forth in Chapter 2, §8.2.

The particular solution $(\bar\psi, \bar\vartheta)$ describes the deformation of the shell by a distributed load and the longitudinal force P_z. It can be a solution corresponding to any (but in context of (5.17) a symmetric) set of boundary conditions. The edge conditions are satisfied by an appropriate choice of the C_1 and C_2 values. The pair $\bar\vartheta$ and $\bar\psi$ can be set equal to the periodic solution discussed in present chapter, §2. However, it should be recalled that the coordinate η was defined to extend over 2π inside *one* corrugation. On the other hand, the trigonometric-series

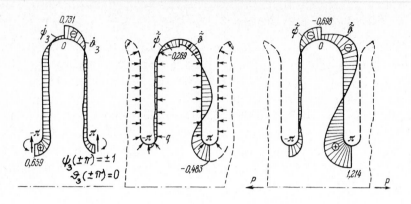

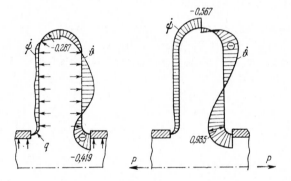

FIG. 74.

presentation of ϑ_i and ψ_i set forth in Chapter 2, §8.2 implies $\eta = \pm\pi$ on the edges of the *entire* shell.

Consider the case of bellows, welded on both ends to some shells of revolution, as in Fig. 67. The continuity conditions (Chapter 1, §7.3) require equality, at the welded edges of the two shells, of four quantities: the circumferential extension, the angle of rotation, bending moment and the radial forces. With the quantities pertaining to the adjoining shell denoted by a prime, the four boundary conditions on an edge are

$$\eta = \pm\pi: \qquad \varepsilon_1 = \varepsilon_1', \quad \vartheta = \vartheta', \quad M_2 = M_2', \quad T_r = T_r'. \quad (5.19)$$

For each of the shells the values of ε_1, M_2 and T_r must be expressed in terms of the functions ϑ and ψ. With the aid of (2.23), (2.24) and (1.84) we have

$$\varepsilon_1 = \left(\dot\psi + \nu \frac{\lambda}{r} s\psi - \nu \frac{h^\circ}{r} F_1^\circ \right) \frac{h^\circ}{t} ,$$

$$\frac{M_2}{D} b = \dot\vartheta - \nu \frac{\lambda}{r} s\vartheta , \qquad\qquad (5.20)$$

$$T_r = \frac{\lambda}{r} Eh_m h^\circ \psi + \frac{\lambda}{r} \int (qc^* - q_2 s^*) r \, \mathrm{d}\eta .$$

The ν terms can in most cases be neglected here.

Substituting into the boundary conditions (5.19) the expressions (5.20) and (5.17) for both the bellows and the adjoining shells, provides four equations for the constants C_1, C_3 and C_1', C_3'.

The severest constraint of the shell deformation is obviously imposed by built-in edges. In this case there are on an edge only two geometrical conditions $\varepsilon_1 = 0$ and $\vartheta = 0$ (cf. Chapter 1, §7). This yields two equations for the constants C_1 and C_3.

$$\eta = \pi : \quad C_1 = -\tilde\vartheta , \quad C_1 L(\psi_1) + C_3 L(\psi_3) = -L(\tilde\psi) + F_1^\circ \nu \frac{h^\circ}{r} ,$$

$$L(\ ) = (\ )^{\displaystyle\cdot} + (\ )\nu \frac{\lambda}{r} s . \qquad\qquad (5.21)$$

The case of $\tilde\vartheta = 0$ along the edges $\eta = \pm\pi$ is of particular interest. (Such a profile is shown in Fig. 67.) With $C_1 = 0$ and C_3, determined by (5.21), the resolving functions (5.17) become

$$\psi = \tilde\psi + C_3 \psi_3 , \qquad \vartheta = \tilde\vartheta + C_3 \vartheta_3 ;$$

$$C_3 = \left\{ \left[-L(\tilde\psi) + F_1^\circ \nu \frac{h^\circ}{r} \right] \Big/ L(\psi_3) \right\}_{\eta=\pi} . \qquad\qquad (5.22)$$

3.2. Deformation near an edge

Let us compare the deformation of an "endless" bellows, determined by a periodic solution of present chapter, §2, with that of a shell in the shape of one convolution wave of that bellows and both edges built-in undeformably.

Consider the bellows with the profile of Fig. 64(b), defined in the lines 2, 5 and 7 of Table 10.

The particular solution ψ_3, ϑ_3 is presented in Table 11. The ψ_3, ϑ_3 as well as the periodic solution $\dot\psi$, $\dot\vartheta$ for the cases of inner pressure and axial extension (those described by the lines and curves marked $b1$ in Table 10 and Fig. 68) are illustrated by the graphs of Fig. 74. Table 11 and Fig. 74 show satisfactory convergence of the series for ψ_3 and ϑ_3. The graphs of ψ_3 and ϑ_3 in Fig. 74 correspond to only 8 terms in the sine series of ψ, ϑ. The oscillation of the graphs becomes indiscernible when 10 terms are retained in the series. (The same effect is achieved by introducing into the series of ψ_3, ϑ_3 the σ-factors of LANZCOS [112].) The values of $|\dot\psi_3|_{max}$ and $|\dot\vartheta_3|_{max}$ are determined by the 8 terms with a satisfactory accuracy.

In this case the influence of the edge forces is not localized but causes a stressed state extending on the entire shell. (It is a consequence of the small curvature of the profile ($1/R_2$) in the edge zone.) The resultant deformation of the shell (the curves $\dot\psi$ and $\dot\vartheta$ in Fig. 74) deviates noticeably from the deformation of the *long* bellows away from its edges, represented by the curves $\bar\psi$ and $\bar\vartheta$.

The expansion joint of the type shown in Fig. 67(a) does not differ substantially from the bellows just discussed (Fig. 74); the difference lies mainly in the quantitative characteristics of the profile. In particular, this shell is larger, compared to the wall-thickness.

The torus part of the shell is relatively thin-walled. The quantity $R_2^2(\pi)/hR(\pi)$ characterizing this part of the shell as the parameter μ does for a closed torus-shell, is, as a rule, considerably above 1.

In such a shell edge forces cause only a more or less local deformation. Away from the edges the stressed stated is described with sufficient accuracy by the *periodic* solution $\dot\psi$, $\dot\vartheta$.

Besides, in the zone near a built-in edge the level of stress is mostly *under* that of a periodic solution.

TABLE 11. ψ_3, ϑ_3.

j	1	2	3	4	5	6	7	8
$a_j \cdot 10^4$	−1931	−598	−416	−139	−111	−7	−20	16
$b_j \cdot 10^4$	−2073	−371	−396	−62	−115	21	−29	28

All this indicates the possibility of evaluating the stresses in bellows and expansion joints, as well as their stiffness with respect to axial extension, with the aid of the graphs and formulas of present chapter, §2.

3.3. Welded bellows

The bellows with a profile shown in Fig. 75 has the important merit of being well secured against overloading by inner or outer pressure or axial compression. In case of overloading the side surfaces of the corrugations lean on each other. This is shown for the case of large inner pressure in Fig. 75. The bellows welded of single annular shallow shells have also considerable technological qualities making them in many cases preferable to the seamless bellows, discussed in the foregoing.

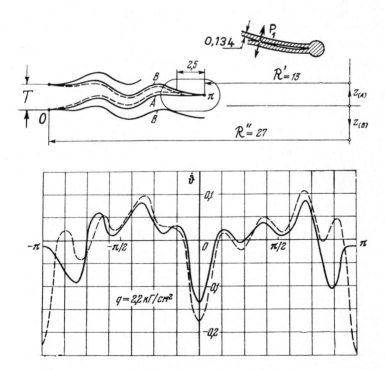

FIG. 75. Profile of bellows (mm) and its deformation with the effect of the contact and without it (broken line).

The analysis of stress and strain of the welded bellows has the following three substantial differences to that of the expansion joint.

First, the deformation of a wave of corrugation is mostly *not symmetric* with respect to a plane $z = $ const.; the corrugation has no symmetry with respect to such a plane (as in Fig. 75).

Second, there are rib-like *rings*, caused by the welding seams or introduced with an aim of increasing the strength of the bellows.

Third, it is necessary to take into account the *contact* forces between the side surfaces of the corrugations. This contact is made possible by the profile form and plays, as was mentioned, an important role in assuring the strength of the bellows.

Consider the main stages of the evaluation of the stress and strain taking into account these three features of the welded bellows. The contact of the side surfaces leads to a considerable geometric non-linearity of the deformation. But the angles of rotation (ϑ) will be assumed small enough to make $c^* \approx c' \approx c$ and $s^* \approx s$.

The solution of the axisymmetric problem (the functions ϑ and ψ) is sought in the form of $\sin j\xi$-series. But the basic stages of the solution, described in the following, apply to other methods of integration.

The function $\vartheta(\eta)$ (the expression for $\psi(\eta)$ is analogous) will be determined in the form

$$\vartheta = P^\circ \vartheta_P + q^\circ \vartheta_q + \sum_{n=1,2,..} P_n \vartheta_n^* + \sum_{i=1}^{4} C_i \vartheta_i \, ,$$

$$[\vartheta_P \ \ \vartheta_q \ \ \vartheta_n^*] = \sum_j [a_j^P \ a_j^q \ a_j^n] \sin j\xi \, .$$

(5.23)

The function ϑ_P and ϑ_q and the corresponding ψ_P and ψ_q represent the deformation caused by the axial force P and normal pressure q. The series for ψ_P, ψ_q, ϑ_P and ϑ_q are determined by (5.3).

The term $P_n \vartheta_n^*$ represents the deformation caused by the force P_n occurring at the *line of contact* $\eta = \eta_n$ of the inner or outer surfaces of the bellows. (The force on the first line of contact is indicated in the scheme of Fig. 75.) The functions ϑ_n^* and ψ_n^* are determined by equations following from (2.30),

$$\left(\frac{r}{t}\dot\psi\right)^{\cdot} + \mu c\vartheta = 0 \, , \qquad (rt^3\dot\vartheta)^{\cdot} - \mu c\psi = f_n(\eta)s \, .$$

Here $f_n(\eta)$ denotes the axial force acting in the section $\eta = $ const. as a result of an external axial force being applied at the line $\eta = \eta_n$. The function $f_n(\eta)$ has a constant value for $0 < \eta < \eta_n$, a jump at $\eta = \eta_n$, increasing the value of $f_n(\eta)$ by 1.0, and a (new) constant value for $\eta_n < \eta < \pi$. This function can be represented by the Fourier series

$$f_n(\eta) = \sum_{j=1,2,\ldots} \frac{\sin j\eta_n}{j} \cos j\eta \,. \tag{5.24}$$

It is expedient to introduce in the f_n-series the σ-factors of LANCZOS [112] (similar to the series of η_σ in (2.160)). This ameliorates the convergence. The functions $f_n(\eta)$ so obtained represent forces distributed over a strip of *finite* breadth, which corresponds to the actual contact situation.

The four particular solutions ϑ_i ψ_i and the method for their determination are discussed in Chapter 2, §8.2.

In the simplest case of a corrugation-wave symmetric with respect to the plane $\eta = 0$ and deformed identically to the two adjacent waves, the solution (5.23) retains only the first two terms and is valid for the entire profile. It is the periodic solution of present chapter, §2.

For the general case, shown in Fig. 75, the *full* expression (5.23) together with the corresponding one for $\psi(\eta)$ are required for *each* of the two parts A and B of one corrugation.

Consider the evaluation of the constants, determining the solution (5.23): the four constants $C_1, \ldots, C_4$ for each of the parts A and B and the constants $P°, P_1, P_2, \ldots$ valid for both parts.

We begin with the condition expressing the relation between the axial force, normal pressure q and the extension of one corrugation wave in the axial direction $\Delta(\pi)$. According to (2.25) without the ε_2 term, this relation has the form

$$[U_Z(\pi)]_A + [U_Z(\pi)]_B = \Delta(\pi) \,, \qquad U_Z(\eta) = -\int_0^\eta \vartheta \sin \alpha b \, d\eta \,. \tag{5.25}$$

The two parts of a convolution denoted A and B in Fig. 75 are related to their respective axes z_A and z_B. The angles α and $\vartheta = \alpha^* - \alpha$ are measured with reference to the appropriate of the two axes.

At the lines $\eta = 0, \pi$ the angles of rotation ϑ and the radial displacements of the adjoining edges of the parts A and B must be equal to each other and to the angular and radial displacements of the respective ring. The difference between the bending moments M_2 and radial stress resultants T_r on the edges of the parts A and B comprise the moment and the distributed radial force N_0 or N_π applied to the ring element (and deforming it).

The eight conditions at the lines $\eta = 0$, π, where the parts of the shell A and B are connected to each other and to the stiffening rings, are

$$C_{1B} = -C_{1A}, \qquad C_{3B} = -C_{3A};$$

$$\eta = 0: \qquad D\,\frac{1}{b}\,(\dot{\vartheta}_A - \dot{\vartheta}_B) + m_0 C_{3A} = 0;$$

$$\eta = \pi: \qquad D\,\frac{1}{b}\,(\dot{\vartheta}_A - \dot{\vartheta}_B) + m_\pi C_{1A} = 0; \qquad (5.26)$$

$$\eta = 0: \qquad \dot{\psi}_A = \dot{\psi}_B = N_0 \delta_0, \quad N_0 = C_{2A} + C_{2B};$$

$$\eta = \pi: \qquad \dot{\psi}_A = \dot{\psi}_B = N_\pi \delta_\pi, \quad N_\pi = C_{4A} + C_{4B}.$$

The indices A and B refer to the respective parts of the shell. The quantity δ_0, is defined as the radial displacement of the elastic ring adjoining the line $\eta = 0$, caused by a radial force of intensity $Ehh°$ distributed over the ring; m_0 denotes the moment per unit length of the ring, which would cause a rotation of each of its meridional sections, equal to an angle of 1 radian. The quantities δ_π and m_π are defined similarly for the ring adjoining the line $\eta = \pi$. The values of the characteristics of the rings can be determined in a straightforward manner.

Equations (5.26) are sufficient to determine all the constants of the general solution (5.23) as long as there are no contacts between the side surfaces of the parts A and B and thus the contact-force parameters P_n are equal to zero.

We have to check whether the contact has taken place and, if the contact exists, to find the coordinate η_1 of the first contact line. To find η_1 we use the condition that the contact appears on that line $\eta = \eta_1$, where the displacement of the approach of the two surfaces, measured by $[u_z(\eta)]_A + [u_z(\eta)]_B$, divided by the distance between the same points in the undeformed shell, reaches a maximum.

For the case when the contact comes about between the outer surfaces (this situation, shown in Fig. 75, is caused by inner pressure deforming a bellows with both ends fixed, i.e. $\Delta(\pi) = 0$), the condition at the point of eventual contact is

$$\frac{[u_z(\eta)]_A + [u_z(\eta)]_B}{T - z_A(\eta) - z_B(\eta) - h(\eta)} = F(\eta) = \max . \tag{5.27}$$

Here the displacement $u_z(\eta)$ is expressed in terms of $\vartheta(\eta)$ according to the (5.25) and (5.23). The actual contact occurs when the quantity (5.27) reaches unity (at a point $\eta = \eta_1$).

In the algorithm used, the values of the ratio (5.27) were calculated and compared at 28 points of the profile $0 \leq \eta \leq \pi$.

The value of η_1 determines the function $f_1(\eta)$. The deformation caused by the contact force is characterized by functions ψ_1^* and ϑ_1^* defined by equations of the type

$$P\begin{bmatrix} \psi_1^* \\ \vartheta_1^* \end{bmatrix} = \begin{bmatrix} 0 \\ f_1^- s \end{bmatrix} . \tag{5.28}$$

Inserting $\vartheta(\eta)$ from (5.23) with the $\vartheta_1^*(\eta)$ from (5.28), and the analogous expression for $\psi(\eta)$ into equations (5.25) and (5.26) and into the equation of contact $F(\eta_1) = 1$, corresponding to (5.27), yields a system determining all the constants, including the contact force P_1.

To find the coordinate of the second contact we construct a condition similar to that of (5.27). The difference is that the displacements u_z and the coordinates z_A and z_B are measured not from the initial shape of the shell but from the shape corresponding to the *first touch* of the surface at $\eta = \eta_1$.

The process just described can be extended to three and more lines of contact. However, in the elastic range such situations are not often to be expected.

The results of the evaluation of an example[2] are presented in Fig. 75. It is a bellows with fixed ends, loaded by the inner pressure q. The first contact occurs near the inner welded seam $\eta = \pi$ when $q = 10.2\,\text{N/cm}^2$. By $q = 44\,\text{N/cm}^2$ the contact of the outer surfaces de-

[2] All calculations were carried out by M.K. Alipbaev.

creases the bending stress at $\eta = 0$ from $\sigma_{M2} = -1150$ to $-760\,\text{N}/\text{mm}^2$. At the point $\eta = \pi$ the contact leads to a change of sign of the stresses σ_{M2}. For the bellows in the example the second contact line is not possible within the elastic range.

In the cases similar to this example, when the pitch T is small and the extension of the bellows is prevented, a small-displacement approximation ($\alpha^* = \alpha$) is sufficient.

4. Lateral bending and stability of bellows

The deformation of bellows by any lateral forces and tilting moments can be considered in linear approximation as a superposition of two basic states. These are the pure flexure (Fig. 76) and the parallel displacements of the edges in the planes orthogonal to the axis of symmetry (Fig. 77). The basic cases are considered in §§4.1, 4.2. The boundary conditions are fulfilled in the sense of Saint-Venant. Next (§4.3), the flexure of bellows is evaluated on the basis of the simple annular-plates scheme of Fig. 70. This provides explicit formulas for stresses and displacements.

4.1. Pure bending

The loading of a bellows by tilting moments L at the ends produces in any cross-section the same bending moment $L_y(\eta) = L$. All corrugations sufficiently distant from the shell edges have identical stress and strain. Deformation of such parts of bellows is described by a *periodic* solution of the Schwerin–Chernina equations of Chapter 2, §3. Equations (2.59) will be used in the form

$$\left(\frac{r}{t}\,\dot{\psi}\right)^{\cdot} - \psi 2\lambda^2\,\frac{1+s^2}{rt} + \mu\vartheta c = 0\,,$$

$$(rt^3\dot{\vartheta})^{\cdot} - \vartheta 2\lambda^2\,\frac{1+s^2}{r}\,t^3 - \mu\psi c = cP_x^\circ - \frac{s}{r}\,L_y^\circ\,; \qquad (5.29)$$

$$P_x^\circ = P_x\,\frac{b^2}{\pi R_m D_m}\,, \qquad L_y^\circ = L_y\,\frac{b^2}{\pi R_m^2 D_m}\,.$$

In comparison with equations (2.63), equations (5.29) retain additionally the λ^2 terms which may be substantial for plates (when those terms of (2.59) dropped in (5.29) disappear).

The periodic Fourier-series solution is found similarly to the axisymmetric case, discussed in Chapter 2, §8.1. Setting $\eta = 0$ in the plane of symmetry of a corrugation wave (Fig. 76) we have for the functions c, s, t and r the series of the form (2.166) and (2.167), and equations (5.29) with $P_x = 0$ can be satisfied by the series

$$\psi = L_y^\circ \psi_L = L_y^\circ \sum_j b_j \sin j\eta \,, \qquad \vartheta = L_y^\circ \vartheta_L = L_y^\circ \sum_j a_j \sin j\eta \,. \quad (5.30)$$

Inserting the Fourier expansions of all functions into equations (5.29) we obtain a system of algebraic equations for the Fourier coefficients of

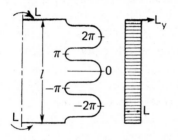

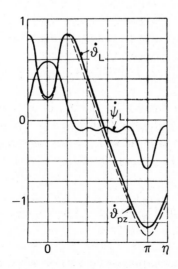

Fig. 76. Pure flexure.

the ψ_L and ϑ_L. It can be written directly as a matrix form of equations (5.29)

$$\begin{bmatrix} \Lambda_+ r_+ t_+^{-1} \Lambda + L_{(-1)} & -\mu c_- \\ \mu c_- & \Lambda r_+ t_+^3 \Lambda + L_{(3)} \end{bmatrix} \begin{bmatrix} \psi_L \\ \vartheta_L \end{bmatrix} = \begin{bmatrix} 0 \\ r_-^{-1} s \end{bmatrix}, \quad (5.31)$$

with

$$L_{(n)} = 2\lambda^2 r_-^{-1} t_-^n (E + s_+ s_-) \quad (n = -1, 3) .$$

The character of the stressed state due to the pure flexure is illustrated by an example in Fig. 76. It is the bellows described in Table 10, rows b, 1 and $b1$. Figure 76 indicates that the stressed states due to the axial extension by the force P_z and those due to flexure by moments L_y are similar for $P° = L_y°$. The difference is of the order of magnitude of $\lambda|\psi|$ or $\lambda|\vartheta|$. (This corresponds to the similarity of the respective governing equations.)

With this accuracy the stresses σ_2 and σ_1 due to bending moment L_y are equal to the stresses caused by the axial extension force $P_z = 2L_y/R_m$ ($R_m = (R' + R'')/2$). Consequently, we can use the formulas and graphs of the axisymmetric case (present chapter, §2) for evaluation of the stresses caused by the flexure. This applies also to the nonperiodic stress state in the parts of bellows near to its edges. The stresses in the vicinity of the built-in edges of a bellows, loaded by pure bending, can be evaluated as stresses under axial extension. This is achieved with an error of the order of magnitude of $\lambda/3$ when in the analysis of the axisymmetric case (present chapter, §3) the value of P_z is replaced by $2L_y/R_m$.

The angle θ of rotation of a line $\eta = $ const. with respect to some line $\eta = \eta_1$ is equal to $u_z^1(\eta)/R(\eta)$ as determined by formula (2.67), where in the case of bellows the term with ε_2 can be dropped. The formula becomes

$$\theta(\eta) = \frac{u_z^1}{R} = -\int_{\eta_1}^{\eta} \vartheta \sin \alpha \, \frac{b}{R} \, d\eta . \quad (5.32)$$

Hence, the angle of rotation of the section $\eta = \pi$ with respect to the line $\eta = -\pi$ can be expressed in terms of the Fourier coefficients of the function $\vartheta(\eta)$ by the formula

$$\theta = -\frac{\pi b}{R_m}\,\vartheta^T r_-^{-1} s\,.$$ (5.33)

For $r \approx 1$, formula (5.33) becomes very simple. For a bellows of n waves of corrugation the angle of bending is given by

$$\theta = -n\,\frac{\pi b}{R_m}\sum_k a_k s_k\,.$$ (5.34)

Formula (5.33) with $r = 1$ is fully similar to the formula (2.25) without its ε_2 term, used in the foregoing to determine the axial extension of bellows.

This indicates the possibility of a simple calculation of the angle of flexure θ. With an error of the order of magnitude of $(R/R_m) - 1$ (practically, of only $\lambda/3 = b/3R_m$) the formula for θ is obtained from (5.13) and (5.15) upon setting $\Delta_z = \theta R_m$, $P = 2L_y/R_m$, $A_w = -S_w$:

$$\theta = L_y\,\frac{2H^3(1-\lambda)}{ER_m^3 h_m^3}\,n\left(S_w + \left|\frac{R'}{R''} - 1\right| B_w + \gamma C_w\right).$$ (5.35)

The factors S_w, B_w and C_w are defined by graphs of Fig. 72.

It is convenient for engineering calculations to regard the flexure of a bellows as that of a beam. The corresponding flexural stiffness factor EI' can be determined by comparison with (5.35). A beam of length $l = nT$, bent by moments L_y at the ends has the angle of rotation of one end with respect to the other

$$\theta = \frac{L_y l}{EI'}\,.$$ (5.36)

According to (5.33) we obtain as the expression of the flexural stiffness of a bellows in terms of the solution (5.30)

$$\frac{EI'}{l} = \frac{L_y}{\theta} = \frac{D_m R_m^3/b^3}{|(\vartheta_L)^T r_-^{-1} s|}\,.$$ (5.37)

4.2. Bending by lateral forces

Under the action of lateral forces P and moments $Pl/2$ applied to the ends (Fig. 77) a bellows bends with a parallel displacement of the edges. This deformation superimposed on the pure bending covers the general case of lateral bending of bellows.

This case may be described by a periodic solution of the Schwerin–Chernina equations (5.29). Different versions of this solution, correspond to possible representations of the bending-moment term on the right-hand side of the governing equations (5.29).

First, the entire length of the bellows can be regarded as a period of

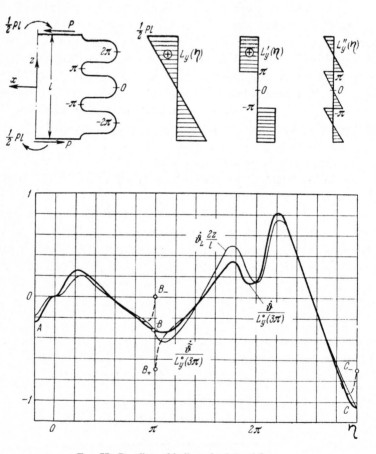

FIG. 77. Bending of bellows by lateral forces.

variation of L_y. In the corresponding series solution the entire profile of the bellows must be regarded as one cycle of variation of the shape with coordinates $\eta = \pm\pi$ at the ends.

To describe a deformation of a bellows with n convolutions (for instance, in Fig. 77, $n = 3$) n-times more terms must be retained in the series solution than are required for pure bending (when a cycle of the variation of stresses is contained in a single corrugation).

Another version involves the solution of the governing equations for each single corrugation. The corresponding periodic solution is easy to find. But the solutions do not pass continuously between the individual corrugations. The jumps in ϑ, $\dot{\vartheta}$, ψ and $\dot{\psi}$ can be eliminated with the aid of the general solution with four constants for each corrugation.

Consider the second version of the solution with reference to specific example presented in Fig. 77. The profile of this bellows is described in lines b and 1 of Table 10. The analysis starts with resolving the bending moment $L_y(\eta)$ into the two parts L_y' and L_y'', shown in Fig. 77. The moment L_y' generates inside each corrugation the periodic solution of pure bending. However, inside the adjacent corrugations of a bellows the value of L_y' has *different* values. The moment L_y'' varies in a similar way on all of the corrugations (Fig. 77); but the function $L_y''(\eta)$ has jumps on the ends of each corrugation.

The solution begins with the series representation of the moment $L_y''(\eta)$, which is equal to Pz inside the corrugation period $-\pi < \eta < \pi$ with a plane of symmetry $z = 0$ and $\eta = 0$. Expressing z in terms of $c = \cos\eta$ according to $\dot{z}/b = c$ and inserting the Fourier-series expansion of $c(\eta)$, we have

$$L_y'' = P \int_0^\eta cb \, \mathrm{d}\eta = -Pb\left(\eta_s \frac{c_0}{2} - \sum_{j=1,2,\ldots} \frac{c_j}{j} \sin j\eta\right). \qquad (5.38)$$

The function η_s is defined by formula (2.159). The symmetry conditions and the form of the right-hand side of (5.29), where $P_x^\circ = \text{const.}$, $L_y = L_y''$ indicate a cosine series for the resolving functions

$$\frac{1}{P_x^\circ}\begin{bmatrix} \vartheta \\ \psi \end{bmatrix} = \begin{bmatrix} \vartheta_{Px} \\ \psi_{Px} \end{bmatrix} = \begin{bmatrix} a_0/2 \\ b_0/2 \end{bmatrix} + \sum_{j=1,2,\ldots} \begin{bmatrix} a_j \\ b_j \end{bmatrix} \cos j\eta. \qquad (5.39)$$

Inserting these expansions, together with Fourier series of the other functions, into equations (5.29) we obtain (in the way described in Chapter 2, §8) the system for a_j and b_j,

$$\begin{bmatrix} \Lambda r_- t_-^{-1}\Lambda_+ + L_{(-1)}^+ & -\mu c_+ \\ \mu c_+ & \Lambda r_- t_-^3\Lambda_+ + L_{(3)}^+ \end{bmatrix}\begin{bmatrix} \psi_{Px} \\ \vartheta_{Px} \end{bmatrix}$$

$$= \begin{bmatrix} 0 \\ c - \lambda r_+^{-1}s_- \left(\dfrac{c_0}{2}\,\eta_s - \lambda\Lambda_+^{-1}c\right) \end{bmatrix},$$

$$(L_{(n)}^+ = 2\lambda^2 r_+^{-1}t_+^n(E + s_-s_+),\quad n = -1, 3).\qquad (5.40)$$

Inside the middle convolution, i.e. for $-\pi < \eta < \pi$, the bending moment $L_y(\eta)$ is equal to its periodic component L_y''. The periodic deformation for the case $L_y = L_y''$ has been found by solving (5.40). The corresponding curve of the *periodic* solution for the middle corrugation "wave", $-\pi < \eta < \pi$, namely,

$$\frac{\tilde\vartheta}{L_y^\circ(3\pi)} = \frac{R_m}{l/2}\,\vartheta_{Px}\qquad (5.41)$$

is plotted in Fig. 77. It is the AB_- (coincident for the most part with the thick-line graph).

In the next corrugation wave, $\pi < \eta < 3\pi$, the periodic solution $\tilde\vartheta$ is composed by superposition of two periodic parts. One of these is due to $L_y''(\eta)$, the other to the *constant* bending moment (Fig. 77)

$$L_y'(\eta) = \tfrac{2}{3}L_y(3\pi),\quad \pi < \eta < 3\pi.$$

Adding the function $\vartheta = \tfrac{2}{3}\vartheta_L$, corresponding to this moment, to that of (5.41) we have the periodic solution

$$\frac{1}{L_y^\circ(3\pi)}\,\tilde\vartheta(\eta) = \frac{R_m}{l/2}\,\vartheta_{Px}(\eta) + \tfrac{2}{3}\vartheta_L(\eta),\quad \pi < \eta < 3\pi.\qquad (5.42)$$

The graph of this function is plotted in Fig. 77 as a curve B_+C_-. The curve $AB_-B_+C_-$ represents the particular solution $\tilde\vartheta(\eta)$. It is composed of the particular solutions obtained for each corrugation wave separately and, of course, displays jumps at $\eta = \pm\pi$. It remains to superimpose inside each of the convolutions the contributions due to the moments and the radial stress resultants on the lines dividing this part of the bellows from the adjacent part. Upon evaluation of the moments and forces by fulfilling the continuity conditions, as described in present chapter, §2, we obtain a continuous solution for the entire bellows. It is represented in Fig. 77 by the thick curve $\dot{\vartheta}/L_y(3\pi)$.

The difference between the solution ϑ and the particular solution $\bar{\vartheta}$ is substantial only in the vicinity of the points $\eta = \pm\pi, 3\pi$. It can often be disregarded.

The analysis of cross-bending is thus more complicated than that of pure bending. It is attractive to extend (on the pattern well known for beams) the results of the pure-bending analysis to the cross-bending of bellows. It means equating the stresses in a cross-section, where a moment $L_y(\eta)$ acts, to the stresses caused by *pure* bending with the same (but constant all along the bellows) value of L_y. The graph of the corresponding function $\dot{\vartheta} = \dot{\vartheta}_L 2z/l$ is presented in Fig. 77 by the thin line. This example justifies somewhat the simplified evaluation. The approximate equation

$$\dot{\vartheta}(\eta) \approx L_y^\circ(\eta)\, \dot{\vartheta}_L(\eta) \tag{5.43}$$

is substantiated even for three corrugation waves. For longer bellows this simplified analysis is for practical purposes exact. The lateral displacement of the end of the bellows (U in Fig. 63) can be calculated like that for a beam with the stiffness EI' determined from (5.37). In particular, for the example of Fig. 77, the lateral displacement without a rotation of one end of the bellows with respect to the other is

$$U = \frac{Pl^3}{12EI'} . \tag{5.44}$$

The lateral displacement U of an end of a bellows is determined more exactly with the aid of (2.68) and (2.66). Neglecting the small terms representing the contribution of the extension ε_2 of the meridian, the displacement of the edge $\eta = \pi$ with respect to the edge $\eta = -\pi$ is

$$U = -u_x^1(\pi) + u_x^1(-\pi) = \int_{-\pi}^{\pi}\left[\int_{-\pi}^{\eta}(r\vartheta)\cdot\frac{d\eta}{r} + \vartheta_2^1(-\pi)\right]cb\, d\eta . \tag{5.45}$$

After substitution of the Fourier series of the functions r, ϑ and c, formula (5.45) may be written in the form

$$U = \pi b [c^T \Lambda^{-1} r_-^{-1} \Lambda_+ r_+ \, \vartheta + \vartheta \tfrac{1}{2}(-\pi)c_0] . \tag{5.46}$$

4.3. Elementary formulas

Consider bellows of the type shown in Fig. 64(b), with sufficiently deep corrugations

$$H \gg T . \tag{5.47}$$

Such bellows can be studied as a system of *annular plates* (present chapter, §2.3, Fig. 70) also in the case of flexure. This provides simple explicit formulas determining the stresses and displacements for the very class of bellows (5.47), for which the convergence of the series solution becomes less effective.

The formulas to be derived follow from the analysis of bending of the plate with a rigid centre, shown in Fig. 78. For plates ($s = 1$, $c = 0$) the system (5.29) decouples into two separate equations. The bending of plates is determined by the second of the equations, which for the constant wall-thickness ($t = 1$) and the loading shown in Fig. 78 can be written in the form

$$R \frac{d^2 \vartheta}{dR^2} + \frac{d\vartheta}{dR} - \frac{4}{R} \, \vartheta = - \frac{L}{\pi RD} . \tag{5.48}$$

For the problem of the flexure of bellows we need a solution of (5.48) satisfying the conditions $\vartheta = 0$ at both edges of the plate. (Recall the definition of the function ϑ in (2.50).) It is easy to verify that the

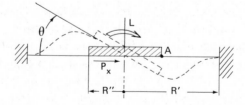

FIG. 78. Bending of a plate with a rigid centre.

required solution is presented by the expression

$$\vartheta = \frac{L}{4\pi D}\left(1 - \frac{1}{r^2}\frac{a^2 r^4 + 1}{a^2 + 1}\right), \quad r = \frac{R}{R'}, \quad a = \frac{R'}{R''}. \quad (5.49)$$

The angle θ of rotation of the rigid centre of the plate is found by substituting ϑ into (5.32). This renders

$$\theta = \frac{L}{4\pi D}\left(\ln a - \frac{a^2 - 1}{a^2 + 1}\right). \quad (5.50)$$

The maximum bending stress occurs at the inner edge of the plate. It is determined by (1.89) and (2.57) (with $T_2 = 0$, $\kappa_1 = 0$) which give

$$\sigma_{2max} = \frac{E}{1 - \nu^2}\frac{h}{2}\left[\frac{1}{R}\frac{d}{dR}(R\vartheta)\right]_{R=R''}$$

or, after substitution of ϑ from (5.49) and introduction of the "large-wall-curvature factor" k_R,

$$\sigma_{2max} = \frac{3L}{\pi h^2 R''}\frac{a^2 - 1}{a^2 + 1}k_R. \quad (5.51)$$

The factor k_R is determined by (2.39). (Of course, for a *plate* $h/R_2 = 0$ and $k_R = 1$. The factor k_R is intended for applications of (5.51) to *bellows*.)

The direct contribution of the lateral force P_x (shown in the Fig. 78) consists of a relatively insignificant plane stressed state of the plate. Its influence on the lateral displacement is small compared to the wall-bending.

In terms of the flexure of an equivalent beam the flexural stiffness of bellows can be determined by viewing the angle θ given by the formula (5.50) as an angle of bending $\theta = LT/(2EI')$ of an element of a beam with length $T/2$. This leads to a formula similar to (5.37)

$$\frac{EI'}{T/2} = 4\pi D\left(\ln a - \frac{a^2 - 1}{a^2 + 1}\right)^{-1}\left(a = \frac{R'}{R''}\right). \quad (5.52)$$

Of course, (5.50)–(5.52) can be applied only for sufficiently deep corrugations, $H > 3T$.

4.4. Stability

Under axial compression the straight-line equilibrium form of a bellows can become unstable, the bellows begins to bend (Fig. 79). When the flexural stiffness can be determined by (5.37) or (5.52) the critical axial compressive force (P_*) may be determined by the Euler formula

$$P_* = \frac{\pi^2 EI'}{(ml)^2} \, , \tag{5.53}$$

where l denotes the actual length of bellows, ml, its reduced length, equal to the length of a half a sine wave representing the axis of the bellows, deformed under the critical load P_*. The value of m depends on the conditions at the ends of the bellows (Fig. 79). For the most realistic case, when both edges of bellows are built-in (shown in Fig. 66) $m = 1/2$.

Figure 66 illustrates a particular feature of bellows—the instability under inner pressure, when the axial extension is restricted. The instability occurs when the reactive force reaches the critical value indicated approximately by (5.53). (This sort of instability is possible in principle also for a cylindrical shell (Fig. 79). The corresponding inner pressure is, however, much higher than for bellows. For realistic lengths of a cylinder shell the critical pressure is far outside the elastic range.)

The following simple formula for the evaluation of the critical pressure q_* of a bellows can be derived from (5.53)

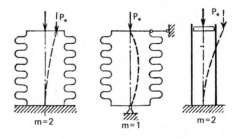

FIG. 79. Instability under axial compression.

$$q_* = q_0 + \frac{\pi EI'}{(ml)^2 R_m^2}, \quad R_m = \frac{R' + R''}{2}. \tag{5.54}$$

Here q_0 denotes the pressure, at which the end of the bellows reaches the stop. When both ends of the bellows are fixed, $q_0 = 0$.

According to the experimental data of ZVER'KOV (in the book of ANDREEVA et al. [165]) the real critical pressure deviates from that of (5.54) by up to 20%.

The axial compression may have a substantial influence on the effective resistance of a bellows to flexure. This can be evaluated in a way similar to that well known for beams. The bending stiffness EI' is replaced in the evaluation of flexure by a reduced value equal to

$$EI'\left(1 - \frac{P}{P_*}\right),$$

where P/P_* is the ratio of the actual force of axial compression to the critical force.

5. Torus shell

Circular or nearly circular torus shells loaded by normal pressure or edge forces (axisymmetric or of wind type) are considered using the linear and a nonlinear approximation.

5.1. Axial extension of toroidal expansion joint

Consider an expansion joint of circular toroidal form, connected at the edges $\eta = \pm\pi$ to tubes which transmit to the shell axial forces P (Fig. 67(b)). The stress state periodic with respect to the coordinate η and symmetric to the plane $\eta = 0$ of the shell can be described in terms of the sine series

$$\vartheta = P^\circ \vartheta_P = P^\circ \sum_j a_j \sin j\eta ; \quad \psi = P^\circ \psi_P = P^\circ \sum_j b_j \sin j\eta . \tag{5.55}$$

This solution satisfies the edge conditions

$$\eta = \pm\pi : \quad \vartheta = 0, \quad \psi = 0. \tag{5.56}$$

This means, in fact, that on both edges, the angle of rotation and the radial force $T_r = Q_2$ vanish.

The condition $\vartheta = 0$ is satisfied when the adjoining tubes are thick enough.

The tubes do cause radial reactive forces on the edges of the shell $(T_r(\pm\pi) \neq 0)$. These are, however, of no substantial influence on the design stresses in the toroidal shell, since in most cases the maximum stresses occur away from the edges (Fig. 67(b)).

In the following we are concerned with the periodic solution (5.55) and (5.56). Other conditions at the edges can be taken into account, for instance, in the way described in present chapter, §3.

The Fourier coefficients of the solution (5.55) are determined by (5.3) where for the circular torus

$$\alpha = \eta: \qquad r = R/R_m = 1 + \lambda \cos\eta \quad (\lambda = b/R_m)$$

and thus according to Chapter 2, §7.2,

$$r_+ = E + \lambda c_+ . \tag{5.57}$$

Inserting into the system (5.3) also the matrices c_-, s_+, s and c according to (3.23) and (2.148) gives

$$\tag{5.58}$$

Solution of this system determines the series (5.55) for ϑ, ψ and, by means of formulas of Chapter 2, §2, the stresses and displacements.

The strength of the toroidal expansion-joint under axial extension is determined practically by the wall-bending stress σ_2 alone. The design stress is the $|\sigma_2|_{max}$ occurring at a line ($\eta = \eta_\sigma$) where the value of $|\vartheta|$ has its maximum. Formula (5.5) gives

$$|\sigma_2|_{max} \cong \frac{Eh}{1-\nu^2} \frac{|\vartheta(\eta_\sigma)|}{2b} = \sigma_2^\circ \frac{P_z b}{2\pi R_m} \frac{6}{h^2} \tag{5.59}$$

with the stress factor determined by the solution of (5.58)

$$\sigma_2^\circ = \left|\sum_j ja_j \cos j\eta\right|_{max}. \tag{5.60}$$

The extension of the shell in the axial direction is evaluated with the aid of the formula (2.25) as

$$\Delta_z = -\int_{-\pi}^{\pi} \vartheta sb \, d\eta = \Delta^\circ \frac{P_z b^3}{2R_m D_m}, \quad \Delta^\circ = -a_1. \tag{5.61}$$

The values of the stress and displacements factors σ_2° and Δ° are determined by the graphs of Fig. 80.

Consider the limiting case of a shell with comparatively thick wall and $b \ll R$, when

$$\mu = \sqrt{12(1-\nu^2)} \frac{b^2}{R_m h} < 1, \quad \frac{b}{R_m} = \mu h^\circ \ll 1; \quad h \approx h_m, \quad R \approx R_m. \tag{5.62}$$

For such shells, the system (5.58) gives all the Fourier coefficients a_n and b_n equal to zero except a_1,

$$\Delta^\circ = -a_1 = 1. \tag{5.63}$$

In this limiting case, formulas (5.59) and (5.61) become

$$\sigma_{2max} = \frac{6P_z b}{2\pi R h^2}, \quad \Delta_z = \frac{P_z b^3}{2RD}. \tag{5.64}$$

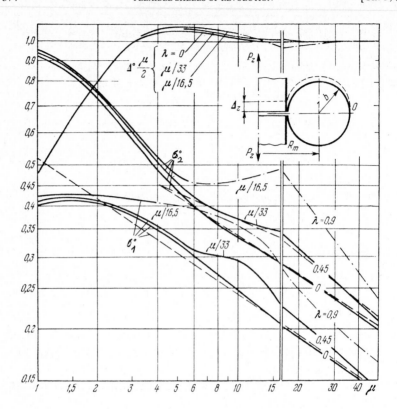

FIG. 80. Graphs for stress and stiffness factors.

The maximum stress (5.64) occurs at $\eta = 0$, π (Fig. 67(b)). Equations (5.63) and (5.64) indicate that by $\mu < 1$ the stressed state of the shell is a purely bending one. The shell resists the loading as would a beam with a cross-section of breadth $2\pi R$ and height h.

Return now to the more general case of toroidal shell. A comparison of formulas (5.60) and (5.61) with (5.64) indicates that the value of μ determines the contribution of the stress resultants T_1 to increasing the stiffness of the shell (reducing $\Delta°$) and to diminishing the stress factor $\sigma_2°$.

The larger the μ-value the more terms must be retained in the series of ϑ, ψ and, correspondingly, more equations in the system (5.58). For a correct evaluation of the stress, $\mu^{1/2} + 2$ terms are required in each of the series. (For the determination of Δ_z a smaller number of terms is sufficient.)

A substantial simplification is possible by neglecting the influence of the parameter λ. The value of λ is restricted by conditions

$$\lambda \equiv \frac{b}{R_m} < 1, \qquad \lambda \equiv \mu h^\circ < \mu \, \frac{1-\lambda}{16}. \qquad (5.65)$$

The second condition corresponds to a wall-thickness, h, as large as $1/5$ of the curvature radius $R_2 = b$ or R_1.

Figure 80 shows the influence of λ on the design stress to be relatively small. It is of the order of magnitude of h/b, and the stiffness factor Δ° is in fact independent of λ.

Neglecting the λ terms in (5.58) sets half of the unknowns equal to zero

$$b_1 = b_3 = \cdots = 0, \qquad a_2 = a_4 = \cdots = 0 \qquad (5.66)$$

reducing one half of the equations to identities. Of the remaining unknowns one half (either $b_2, b_4, \ldots$ or $a_1, a_3, \ldots$) are easily eliminated. This reduces the system to one similar to (3.25) or (3.26).

For $\mu \leqslant 28$, Δ° may be evaluated from the formula dual to (3.31)

$$\Delta^\circ = -a_1 = \frac{1 + (\mu/12)^2 - \delta}{1 + 10(\mu/12)^2 - \delta}, \qquad \delta = \frac{(\mu/12)^4}{4 + 17\mu^2/1800}. \qquad (5.67)$$

For shells with larger values of μ the series solution becomes less effective. But for $\mu > 10$ it can be replaced by the *asymptotic solution* set forth in (2.79) and Fig. 15. This renders formulas

$$\Delta^\circ = -a_1 = \frac{2}{\mu}, \qquad (5.68)$$

$$\sigma_2^\circ = \frac{0.754}{\mu^{1/3}}, \qquad \eta_\sigma = \frac{1.23}{\mu^{1/3}} - \frac{\pi}{2}. \qquad (5.69)$$

Formulas (5.68) and (5.69) are sufficiently accurate for $\mu > 10$; this is evident from Fig. 80, where this solution is displayed by dashed straight lines.

5.2. Torus under normal pressure

It appears intuitively plausible that a circular toroidal shell can support a normal-pressure load without the shell wall being resistant to bending. Assuming the stressed state to be a *membrane* one (Chapter 1, §9.3), simple formulas for the stress resultants T_1 and T_2 can be derived directly from the equilibrium equations. Indeed, the equilibrium of forces acting on the part of shell $0 < \eta < \pi/2$ (Fig. 81, when $Q_2 = 0$, $P_z(\pi/2) = 0$)

$$q\pi R^2 - q\pi R_m^2 - 2\pi R T_2 \cos \eta = 0$$

yields an expression of T_2 in terms of q. Inserting T_2 into the equilibrium equation (1.136) with $R_1 = R/\cos \eta$, $R_2 = b$ and $1/R_{12} = 1/R_{21} = 0$ we obtain

$$T_1 = \frac{qb}{2} , \qquad T_2 = qb \, \frac{1 + \frac{1}{2}\lambda \cos \eta}{1 + \lambda \cos \eta} . \tag{5.70}$$

(These well-known formulas are named after A. Föppl.)

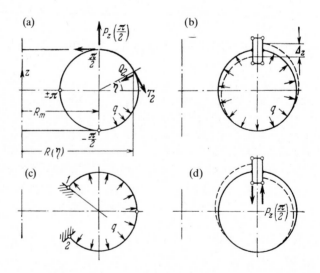

Fig. 81. Circular toroidal shell and its model.

Consider now a solution taking into account also the *bending* of the shell wall and, thus, the lateral stress resultant Q_2.

To simplify the analysis we assume the angle of rotation $\vartheta(\eta)$ to remain small in comparison to unity. This is justified by the smooth variation of the curvatures $1/R_i$ and by the uniformity of the load q, making the deformation of the shell a smooth one. Under these conditions large angles of rotation $\vartheta \sim 1$ would be possible only for components of strain of the order of magnitude of unity.

We employ equations (2.18) simplified by the restriction $|\vartheta| \ll 1$, but *retain* those *load* terms, which depend on ϑ linearly. Introducing the simplifications based on dropping the terms of the order of magnitude of the quantities (2.27), equations (2.18) for uniform wall-thickness may be written in the form

$$(r\dot{\psi})^{\cdot} + \mu c \vartheta = -h^{\circ}\lambda \frac{s}{r} F_1^{\circ} + \nu h^{\circ}(F_1^{\circ})^{\cdot} \,,$$

$$(r\dot{\vartheta})^{\cdot} - \mu c \psi = P^{\circ}s + q^{\circ}p^0 + q^{\circ}p^1(\vartheta) \,,$$

$$p^0 = \frac{\lambda}{2} \eta \cos \eta \,, \quad P^{\circ} = P_z\left(\frac{\pi}{2}\right)\frac{b^2}{2\pi R_m D} \,, \quad q^{\circ} = qb^3/D \,, \tag{5.71}$$

$$p^1(\vartheta) = c \int_0^{\eta} \vartheta sr \, d\eta + \vartheta s \int_0^{\eta} cr \, d\eta - s \int_{\pi/2}^{\eta} \vartheta cr \, d\eta - \vartheta c \int_{\pi/2}^{\eta} sr \, d\eta \,.$$

Compared to (2.30) these equations retain the loading terms with $h^{\circ}F_1^{\circ}$. For the circular torus, when the terms $q^{\circ}p^0$ and $q^{\circ}p^1$ are small, the terms with $h^{\circ}F_1^{\circ}$ are of the same order of magnitude.

In the symmetric case under consideration, the series solution has the form (2.170) with $N^{\circ} = q^{\circ}h^{\circ}/2$ (as determined in Chapter 3, §10.2),

$$\psi = \tfrac{1}{2}q^{\circ}h^{\circ}\eta + \psi_s \,, \qquad \psi_s = \sum_j b_j \sin j\eta \,,$$

$$\vartheta = \sum_j a_j \sin j\eta \,. \tag{5.72}$$

The stress resultants are expressed in terms of ψ and ϑ with the aid of formulas (2.24),

$$T_1 = \tfrac{1}{2} qb + Ehh^\circ \dot{\psi}_s ,$$

$$T_2 = \frac{1 + \tfrac{1}{2}\lambda \cos\eta}{1 + \lambda \cos\eta} + \frac{1}{2\pi R} P_z\!\left(\frac{\pi}{2}\right) \cos\eta - \frac{b}{R} Ehh^\circ \psi_s \sin\eta .$$

(5.73)

The underlined terms represent the T_1 and T_2 of the *membrane* state according to (5.70). The other terms turn out to be negligible for the case under consideration. To ascertain this let us look at closer range at the problems of determining ψ_s, P_z and ϑ.

After introducing (5.72) and dropping those of the P° terms of relative order of magnitude of $h^\circ\lambda$, equations (5.71) become

$$(r\dot{\psi}_s)^{\cdot} + \mu\vartheta \cos\eta = -q^\circ h^\circ \lambda \frac{\sin\eta}{r} \left(1 + \frac{\lambda}{2}\cos\eta\right),$$

$$(r\dot{\vartheta})^{\cdot} - \mu\psi_s \cos\eta = P^\circ \sin\eta + q^\circ p^1(\vartheta) .$$

(5.74)

The system of algebraic equations determining the Fourier coefficients a_j and b_j of the series (5.72) can be obtained from (5.74) in the form

$$\begin{bmatrix} \Lambda_+ r_+ \Lambda & -\mu c_- \\ \mu c_- & \Lambda_+ r_+ \Lambda + q^\circ F \end{bmatrix} \begin{bmatrix} \psi \\ \vartheta \end{bmatrix} = \begin{bmatrix} q^\circ h^\circ \lambda\, r_-^{-1}\!\left(s + \dfrac{\lambda}{2} c_- s\right) \\ -P^\circ s \end{bmatrix},$$

(5.75)

where $r_\pm = E + \lambda c_\pm$; matrices ψ, ϑ, c_+, c_- and s are written out in Chapter 2, §8.1, (3.23), (2.148) and

$$F = \begin{bmatrix} 1 & 0 & 1/8 & 0 & \cdots \\ 0 & 4/3 & 0 & 0 & \cdots \\ 0 & 0 & 9/8 & 0 & \cdots \\ 0 & 0 & 0 & 16/15 & \cdots \\ \cdots & \cdots & \cdots & \cdots & \cdots \end{bmatrix}$$

$$+ \lambda \begin{bmatrix} 0 & 7/16 & 0 & 5/48 & \cdots \\ 13/24 & 0 & 29/48 & 0 & \cdots \\ 0 & 1/2 & 0 & 1/2 & \cdots \\ 1/16 & 0 & 113/240 & 0 & \cdots \\ \cdots & \cdots & \cdots & \cdots & \cdots \end{bmatrix} .$$

(5.76)

For an open shell (such as the expansion joint of the Fig. 80 or Fig. 81(c)) either the axial force P_z on the edges $\eta = \pm\pi$ or the displacement Δ_z of one of the edges with respect to the other is given. This allows us to determine the inner axial force P_z by means of one of the two relations (cf. Chapter 2, §2.3)

$$2\pi R T_2 = P_z \quad (\eta = \pm\pi) , \tag{5.77}$$

$$u_z(\pi) - u_z(-\pi) = -a_1 \pi b = \Delta_z , \quad u_z = -\int \vartheta \sin \eta \, b \, d\eta . \tag{5.78}$$

The condition (5.78) is stated with the aid of (2.25) where the ε_2 term has been neglected.

For a *closed* torus shell we have, of course, $\Delta_z = 0$, and the relation (5.78) expresses the condition of continuity.

Let us assess the *bending* deformation of a closed torus shell loaded by the uniform pressure.

We begin by setting in (5.75) the dependent variables as a sum of two parts

$$[\psi_s \ \vartheta] = q^\circ[\psi_q \ \vartheta_q] + P^\circ[\psi_P \ \vartheta_P] . \tag{5.79}$$

The system (5.75) renders for ψ_q, ϑ_q and ψ_P, ϑ_P, respectively, separate algebraic systems. These being solved, we have P°/q° from (5.78), i.e. from

$$\Delta_z = -\int_{-\pi}^{\pi} \vartheta \sin \eta \, d\eta = 0 . \tag{5.80}$$

The functions ψ_s and ϑ determining the bending effects tend to zero with λ and, as $\lambda \equiv \mu h^\circ$, with μ. It is evident from (5.75). Thus, if this effect is negligible for a particular value of λ it should also be so for the smaller values of λ.

Consider a limiting-case example: a shell with $\mu = 8$ and $\lambda = 0.25$ —the highest value of λ corresponding to the largest wall-thickness ($h = b/10$) still consistent with the use of the thin-shell theory. Calculating the stress with formula (5.59)

$$\sigma_{M2} \approx \frac{E}{1 - \nu^2} \frac{h}{2} \frac{\dot{\vartheta}}{b} \tag{5.81}$$

and $\vartheta(\eta)$ found from (5.75) and (5.80) renders for the example $\sigma_{M2} = 0.07qb/h$. In relation to the membrane stress $T_2/h = qb/h$, the σ_{M2} contribution is less than h/b and thus negligible within the accuracy of the thin-shell theory. The contribution represented by the terms with ψ_s and P_z in (5.73) is even smaller.

For the shells with *larger values of* μ we can make the assessment in an explicit form with the aid of the asymptotic solution (2.79). Inserting into formulas (2.79) the right-hand sides of equations (5.74) in the modified form

$$ g_{ki} = q^\circ h^\circ \lambda \sin \eta \left(1 + \frac{\lambda}{2} \cos \eta \right), \qquad g_{kr} = 0, $$

i.e. with $P^\circ = 0$, $r = 1$, and without the second-order term $q^\circ p^1(\vartheta)$, we obtain $\vartheta(\eta)$ which satisfies the continuity condition (5.80) thus confirming that $P^\circ = 0$. The maximum bending stress and the maximum of the ψ_s term in (5.73), both occurring in the area of $\cos \alpha \sim 0$, are

$$ \sigma_{M2} \approx \frac{qb}{h} \sqrt{\frac{3}{1-\nu^2}} \, \lambda \, \frac{0.939}{\mu^{1/3}}, \qquad Ehh^\circ \frac{b}{R} \, |\psi_s|_{\max} \approx qb\lambda^2 \frac{1.288}{\mu^{2/3}}. $$

$$(5.82)$$

This asymptotic assessment of the wall-bending contribution (being based on (2.73) assuming $r = 1$) is somewhat approximate[3] for large λ. But compared to $|T_2|/h$ the entire contribution is small.

Moreover, the actual stressed state is nonlinear, which makes it in the case of inner pressure, nearer to the membrane state.[4] Even *small* rotations can produce the normal component ϑT_2 of the *large* resultant T_2, which is of the same order of magnitude as Q_2.

We note, finally, that the stresses σ_{M2} indicated in (5.82) occur at $\eta \sim \pi/2$ and thus away from the zone where the design stress T_2/h occurs. The design stress is determined by (5.70) for $\eta = \pm\pi$,

$$ \sigma_{2\max} = \frac{qb}{h} \frac{1 - \lambda/2}{1 - \lambda}. $$

$$(5.83)$$

---

[3] Cf. comprehensive results of KALNINS (*J. Appl. Mech.* **31** (1964) 467–476) for shells with $\lambda = 2/3$, $b/h = 20$, 50 and 200.

[4] An interesting investigation of this problem is due to JORDAN [154].

The simple approximate formulas (5.70) for the stress resultants T_1 and T_2 lead to equally simple expressions for the radial displacement $u_r = R\varepsilon_1$,

$$u_r = \frac{R}{Eh}(T_1 - \nu T_2) = \frac{qbR_m}{2Eh}[1 - 2\nu + \lambda(1 - \nu)\cos\eta].\quad (5.84)$$

An expansion joint causes in the two adjoint tubes resisting its axial expansion ($\Delta_z = 0$, in Fig. 80) the axial forces determined by the condition (5.77)

$$P_z(\pi) = -qb2\pi R_m\left(1 - \frac{\lambda}{2}\right).\quad\quad\quad (5.85)$$

The minus sign means that under internal pressure when, $q > 0$, the forces $P_z(\pi)$ compress the tubes.

The foregoing is applicable for any internal pressure. The picture is different for the external pressure: when q approaches q_{cr}, the wall-bending may become substantial. In this case, however, the critical pressure q_{cr} is the one for *axisymmetric* buckling form. This upper-bound value of the critical pressure is indicated in Fig. 44 by the curves (3.225) and (3.222). It is considerably higher than the buckling pressure of an unstiffened torus shell. For such shells, formulas (5.70) and (5.82) are valid in the entire prebuckling range.

5.3. Noncircular torus

Deviations of the torus meridian from the circular form are in fact unavoidable. Axisymmetric deviations, even small ones—of the order of magnitude of the wall-thickness—substantially influence the stressed state due to the uniform pressure q.

The stresses in a nearly circular torus depend on the pressure nonlinearly. Even comparatively small displacements can change the small noncircularity relatively much. Internal pressure causes the deformation that smooths out the noncircularity. On the contrary, the external pressure causes a more intensive wall-bending than predicted by the linear approximation.

But similarly to the basic case of the circular profile, the normal pressure q causes only small angles of rotation, that is

$$|\vartheta| \ll 1, \tag{5.86}$$

when the pressure is not close to the critical value q_{cr} and the deviations of the meridian form are small enough,

$$(q/q_{cr})^2 \ll 1, \qquad |\alpha(\eta) - \eta| \ll 1. \tag{5.87}$$

Conditions (5.86) and (5.87) justify substantial simplifications of the solution. The governing equations can be used in the form similar to (5.71)

$$(r\dot{\psi})^{\cdot} + \mu c\vartheta = 0,$$
$$(r\vartheta)^{\cdot} - \mu c\psi = P^{\circ}s + q^{\circ}p^{0} + q^{\circ}p^{1}(\vartheta). \tag{5.88}$$

The left-hand side of the equations (the functions $r(\eta)$ and $c(\eta)$) can be assumed the same as for the circular tube in linear approximation. That is,

$$c = \cos \eta, \qquad s = \sin \eta, \qquad r = 1 + \lambda \cos \eta.$$

The noncircularity of the profile and the nonlinearity of the dependence of ϑ and ψ on the normal pressure are accounted for by the load terms only.

Deformation of the toroidal shell by the normal pressure q and the axial force P_z extending the expansion joint is determined by series (5.72) of ϑ and ψ, formulas (5.73) and equations

$$\begin{bmatrix} \Lambda r_+ \Lambda & -\mu c_- \\ \mu c & \Lambda r_+ \Lambda + q^{\circ}F \end{bmatrix} \begin{bmatrix} \psi \\ \vartheta \end{bmatrix} = \begin{bmatrix} 0 \\ -P^{\circ}s + q^{\circ}p^0 \end{bmatrix}. \tag{5.89}$$

The left-hand sides of these equations can be assumed equal to those of a circular tube. On the right-hand sides this applies for $s = [1\ 0\ \cdots]$. Only the p^0 must be determined taking into account the noncircular form of the torus according to (3.22). For a closed shell the value of P° is determined by the condition $\Delta_z = 0$ as described in present chapter, §5.2.

The results of this evaluation are presented in Fig. 82 for an example. It concerns a toroidal shell with $\mu = 7.5$; $\lambda = 0.1$ and the meridian in the form of an ellipse with the axis ratio $A/B = 1.035$.

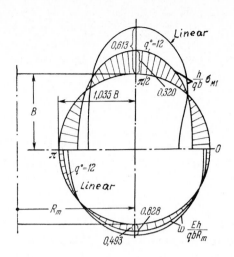

Fig. 82. Linear approximation overestimates the stresses for internal pressure.

In contrast to the circular torus, the wall-bending stress σ_{M2} is for the elliptical profile substantial. In the linear approximation we have $\sigma_{2\max} = 1.654qb/h$. (For the circular profile it would be $1.055qb/h$.)

The linear result is correct for $|q| \ll |q_{cr}|$. In this case ($\mu = 7.5$) according to the curve (3.222) of Fig. 44 we have the critical external pressure $q° \approx -12$. For *internal* pressure $q° = 12$ we find $\sigma_{2\max} = 1.34\ qb/h$, i.e. 81% of the result of the linear solution. This decrease concerns in fact only the *flexural* part σ_{M2} of the stress. The nonlinear value of σ_{M2} for $q° = 12$ is about a *half* of the value given by a linear solution (Fig. 82).

5.4. Nonsymmetric deformation of the toroidal joint

Besides the axisymmetric deformation the expansion joints can be loaded with tilting moment L_y as shown in Fig. 67. The stressed state is determined in this case similarly to the solution described in present chapter, §4. We need only take into account the (now circular, $\alpha = \eta$) form of the meridian. Therefore, we restrict the discussion to remarks on the formulas for the flexure angle θ, the design stress $\sigma_{2\max}$ and on the contribution of the normal pressure to the nonsymmetric deformation.

The governing equations (5.29) differ from those of the axisymmetric problem (5.71) only by terms of the order of magnitude λ (besides the terms with $q°$). With this accuracy their solutions ϑ and ψ coincide. Introducing into the relations (5.32), (5.81), (2.57) and (1.89) the $\vartheta(\eta)$ function of the axisymmetric deformation with $P° = L°$ we obtain formulas

$$\theta \approx \Delta° L_y \frac{b^3}{R_m^3 D}, \tag{5.90}$$

$$\sigma_{2max} = \sigma_2° L_y \frac{b6}{\pi R_m^2 h^2}, \tag{5.91}$$

where the values of $\Delta°$ and $\sigma_2°$ may be taken from axisymmetric case—from the graphs of Fig. 80. The so-calculated θ and σ_{2max} have the relative error under $\lambda/2$ for $\lambda \leqslant 0.2$. When $\lambda > 0.2$ a more exact evaluation—with the aid of the full Schwerin–Chernina equations (5.29)—may become necessary.

The *influence of* the uniform *normal pressure* q on the flexure of the expansion joint can be taken into account by supplementing the equations (5.29) with the load terms (3.85). Thus, for the case of cross-bending (Fig. 67) the series solution can be found with the aid of the equations

$$\begin{bmatrix} Ar_+ t_+^{-1} \Lambda + L_{(-1)} & -\mu c_- \\ \mu c_- & Ar_+ t_+^3 \Lambda + L_{(3)} + q° F \end{bmatrix} \begin{bmatrix} \psi \\ \vartheta \end{bmatrix}$$

$$= \begin{bmatrix} 0 \\ -P_x° c_- + L_-° r_-^{-1} s \end{bmatrix},$$

$$\psi = \{b_1 \ b_2 \ \cdots\}, \quad \vartheta = \{a_1 \ a_2 \ \cdots\},$$

$$\psi = \sum_j b_j \sin j\eta, \quad \vartheta = \sum_j a_j \sin j\eta.$$

This system is a direct extension of (5.31). In most cases the λ^2 terms $L_{(n)}$ can be disregarded and the wall-thickness considered constant ($t = 1$). The so-simplified left-hand sides of equations differ from (5.75) only by the term $q° F$. The influence of normal pressure is represented here by the matrix F with the accuracy of formulas (3.83) and (3.85) (where by setting $r = 1$ the λ terms are neglected). The F is determined by formula (3.22) or by (5.76).

REFERENCES[1]

[1] ARON, H. "Das Gleichgewicht und die Bewegung einer unendlich duennen beliebig gekruemmten elastischen Schale", *J. Reine Angew. Math.* **78** (1874) 1–2, 136–174.

[2] LOVE, A.E.H. "On the small free vibrations and deformation of thin elastic shell", *Phil. Trans. Roy. Soc.* **A179** (1888) 491–546.

[3] BASSET, A. "On the extension and flexure of thin elastic shells", *Phil. Trans. Roy. Soc.* **A181** (1890) 433–480.

[4] KÁRMÁN, TH. VON. "Ueber die Formaenderung duennwandiger Rohre—insbesondere federnder Ausgleichsrohre", *VDI-2* **55** (1911) 1889–1895.

[5] REISSNER, H. "Spannungen in Kugelschalen", Festschrift H. Mueller-Breslau; Leipzig (1912).

[6] MEISSNER, E. "Das Elastizitaetsproblem fuer duenne Schalen von Ringflaechen, Kugel- oder Kegelform", *Physikalische Zeitschrift* **14** (1913) 41–52.

[7] SCHWERIN, E. "Ueber Spannungen in symmetrisch und unsymmetrisch belasteten Kugelschalen—insbesondere bei Belastung durch Winddruck", Dissertation, Berlin (1918).

[8] TIMOSHENKO, S.P. "Analysis of bi-metal thermostats", *J. Opt. Soc. Amer.* **11** (1925) 233–255.

[9] BRAZIER, L.G. "On the flexure of thin cylindrical shells and other "thin" sections", *Proc. Roy. Soc.* **116** (1927) 104–114.

[10] THULOUP, M.A. "Essai sur la fatigue des tuyaux minces a fibre moyenne plane ou gauche", *Bull. l'Assoc. Tech. Maritime Aeron.* **32** (1928) 643–680; **36** (1932) 443–463; **41** (1937) 317–325.

[11] MISES, R. VON "Der Kritische Aussendruck fuer allseits belastete zylindrische Rohre", Festschrift zum 70, Geburtstag von Prof. A. Stodola; Zurich (1929).

[12] FLÜGGE, W. "Die Stabilitaet der Kreiszylinderschale", *Ing.-Arch.* **3** (1932) 463–506.

[13] DONNELL, L.H. "Stability of thin-walled tubes under torsion", *NACA Report* **479** (1933).

[14] MUSHTARI, KH.M. "On stability of circular thin cylindrical shells under torsion", *Trudy Kasansk. Aviats. Instituta* **2** (1934) [in Russian].

[15] DONNELL, L.H. "A new theory for the buckling of thin cylinders and columns under axial compression and bending", *Trans. ASME* (1934) 795–806.

[1] Ordered chronologically; for a publication year, alphabetically.

[16] TUEDA, M. "Mathematical theories of Bourdon pressure tubes and bending of curved pipes", *Mem. Coll. Eng.*; *Kyoto Imp. Univ.* **8** (1934) 102–115; **10** (1936) 132–152.

[17] TREFFTZ, E. "Ableitung der Schalenbiegungsgleichungen mit dem Castigliano' schen Prinzip", *ZAMM* **15** (1935) H1/2.

[18] STAERMAN, I.J. "Stability of shells", *Trudy Kievskogo Aviatsionnogo Instituta* **1** (1936) [in Russian].

[19] HECK, O.S. "The stability of orthotropic elliptic cylinders in pure bending", *NACA TN* **834** (1937).

[20] GOL'DENVEIZER, A.L. "Additions and corrections to the Love's thin-shell theory", *Plastinki i obolotchki—Gosstrojizdat* (1939) [in Russian].

[21] LUR'E, A.I. "General theory of elastic thin shells", *Prikl. Mat. Mekh.* (*PMM*) **4** (1940) 7–34 [in Russian].

[22] PANOV, D. JU. "On large deflections of circular diaphragms with shallow corrugation", *Prikl. Mat. Meh.* (*PMM*) **5** (1941) 303 [in Russian].

[23] REISSNER, E. "A new derivation of the equations for the deformation of elastic shells", *Amer. J. Math.* **63** (1941) 1.

[24] REISSNER, E. "Note on expression for strains in bent thin shells", *Amer. J. Math.* **64** (1942) 768–772.

[25] KARL, H. "Biegung gekruemmter duennwandiger Rohre", *ZAMM* **23** (1943) 331–345.

[26] VIGNESS, I. "Elastic properties of curved circular thin-walled tubes", *Trans. ASME* **65** (1943) 2.

[27] CHIEN, W.Z. "The intrinsic theory of shells and plates: Part, 1 and 2", *Quart. Appl. Math.* **1** (1944) 297–327; **2** (1945) 120–135.

[28] LOVE, A.E.H. *A Treatise on the Mathematical Theory of Elasticity*, 4th ed., Dover Publications, New York (1944).

[29] BESKIN, L. "Bending of curved thin tubes", *J. Appl. Mech.* **12** (1945) A1–A7.

[30] NOVOZHILOV, V.V. "Analysis of shells of revolution", *Izv. AN SSSR OTN* (1946) no. 7 [in Russian].

[31] LUR'E, A.I. *Statics of Thin-Walled Elastic Shells.* AEC-TR-3798 (1947) [Translated from Russian edition of 1947].

[32] FEODOS'EV, V.I. *Uprugie elementy tochnogo priborostroenija*, Oborongiz, Moscow (1949) [in Russian].

[33] LAGALLY, M. *Vorlesungen ueber Vektorrechnung*, Leipzig (1949).

[34] REISSNER, E. "On bending of curved thin-walled tubes", *Proc. Nat. Acad. Sci. USA* **35** (1949) 204–208.

[35] REISSNER, E. *On the Theory of Thin Elastic Shells.* Hans Reissner Anniversary Volume. J.W. Edwards, Eds., Ann Arbor (1949) 231–247.

[36] WLASSOW, W.S. *Allgemeine Schalentheorie und ihre Anwendung in der Technik.* [Translated from Russian (1949)]; Akademie-Verlag, Berlin (1958).

[37] LUR'E, A.I. "On equations of general theory of elastic shells", *Prikl. Mat. Meh.* **14** (1950) no. 5 [in Russian].

[38] REISSNER, E. "On axi-symmetrical deformations of thin shells of revolution", *Proc. Sympos. Appl. Math.* **3** (1950) 27–52.

[39] CLARK, R.A. and REISSNER, E. "Bending of curved tubes", *Adv. in Appl. Mech.* **2** (1951) 93–122.

[40] NOVOZHILOV, V.V. *The Theory of Thin Shells.* [Translated from Russian by P.G. Lowe (1951)]; Wolters-Noordhoff, Groningen (1970).

[41] PARDUE, T.E. and VIGNESS, I. "Properties of thin-walled curved tubes of short-bend radius", *Trans. ASME* **73** (1951) 77–87.

[42] ASHWELL, D.G. "A characteristic type of instability in the large deflections of elastic plates", *Proc. Roy. Soc. Ser. A* **214** (1952) 116.

[43] ASHWELL, D.G. "The stability in bending of slightly corrugated plates", *J. Roy. Aeron. Soc.* **56** (1952) 502.

[44] CHEN, CHU. "The effect of initial twist on the torisional rigidity of thin prismatical bars and tubular members", and "A theory of twisted Bourdon tubes", *Proc. 1st US Nat. Congr. Appl. Mech.*, J.W. Edwards, Ed., Ann Arbor/Michigan (1952) 271–280.

[45] CLARK, R., GILROY, T. and REISSNER, E. "Stresses and deformations of toroidal shells of elliptical cross-section", *J. Appl. Mech.* **19** (1952) 37–48.

[46] HARINGX, J.A. "Instability of bellows subjected to internal pressure", *Phillips Res. Rep.* **7** (1952) no. 3.

[47] REISSNER, E. "Stress strain relations in the theory of thin elastic shells", *J. Math. Phys.* **31** (1952) 109–119.

[48] GOL'DENVEIZER, A.L. *Theory of Elastic Thin Shells.* [Translation from the Russian edition (1953)]; G. Hermann, Ed., Pergamon Press, Oxford (1961).

[49] REISSNER, E. "On a variational theorem for finite elastic deformations", *J. Math. Phys.* **32** (1953) 129–135.

[50] ASHWELL, D.G. "Curved rectangular plates in axial compression", *Proc. Roy. Soc. Ser. A* **222** (1954) 44.

[51] WUEST, W. "Die quergewoelbte Biegefeder", *VDI-Z* **34** (1954) 179–182.

[52] WUEST, W. "Einige Anwendungen der Theorie der Zylinderschale", *ZAMM* **12** (1954) 444–454.

[53] SCHNELL, W. "Zur Krafteinleitung in die versteifte Zylinderschale", *Zeitschrift fuer Flugwissenschaften* **3** (1955); **4** (1957).

[54] CRANDALL, S.H. and DAHL, N.C. "The influence of pressure on bending of curved tubes", *Proc. of the 9th Int. Congr. of Appl. Mech.*; Brussels (1956) 101–111.

[55] JENNINGS, F. "Theories on Bourdon tubes", *Trans. ASME* **78** (1956) 55–64.

[56] KAFKA, P.G. and DUNN, M.B. "Stiffness of curved circular tubes with internal pressure", *J. Appl. Mech.* **23** (1956) no. 2; *Trans. ASME* **78** (1956) 247–254.

[57] MASON, H. "Sensitivity and life data on Bourdon tubes", *Trans. ASME* **78** (1956) no. 1.

[58] WUEST, W. "Theorie der Hochdruckrohrenfeder", *Ing.-Arch.* **24** (1956).

[59] MUSHTARI, K.M. and GALIMOV, K.Z. *Nonlinear Theory of Thin Elastic Shells.* [Translated from Russian by J. Morgenstern and J.J. Schorr-Kon (Kasan, 1957)]; NASA-TT-F62 (1961).

[60] RODABAUGH, E.C. and GEORGE, H.H. "Effect of internal pressure on flexibility and stress-intensification factors of curved pipe or welding elbows", *Trans. ASME* **79** (1957) 939–948.

[61] AXELRAD, E.L. "On the theory of nonhomogeneous isotropic shells. On temperature deformation of nonhomogeneous shells", *Izv. AN SSSR OTN* (1958) nos. 6/8
[in Russian].

[62] AXELRAD, E.L. "Deformation of a cantilever bi-metallic plate under heating", *Isv.
Vuz'ov, Priborostrojenie* (1958) no. 4 [in Russian].

[63] WOOD, J.D. "The flexure of a uniformly pressurized circular cylindrical shell", *J.
Appl. Mech.* **25**; *Trans. ASME* **80** (1958) 453–458.

[64] AXELRAD, E.L. "Bending of thin-walled beams with shallow open section under
large elastic displacements", *Izv. AN SSSR OTN, Mek. i Mash.* (1959) [in
Russian].

[65] AXELRAD, E.L. "Analysis of thermobimetallic strip and spiral", *Priborostrojenie
Mashgiz* (1959) no. 9 [in Russian].

[66] CHERNINA, V.S. "On a system of differential equations of equilibrium of rotationally symmetric shell under bending load", *Prikl. Mat. Meh. (PMM)* **23** (1959) no.
2 [in Russian].

[67] CHERNYKH, K.F. "Meissner equations for the skew-symmetric load", *Izv. AN
SSSR OTN, Mekh. i. Mash.* (1959) no. 6 [in Russian].

[68] KARDOS, G. "Tests on deflections of flat-oval Bourdon tubes", *Trans. ASME, J.
Basic Engng.* (1959) Dec.

[69] KOSTOVETSKI, D.L. "On bending of curved thin-walled tubes with nearly circular
profile by inside or outside pressure", *Izv. AN SSSR OTN, Mekh. i Mash.* (1959)
no. 6 [in Russian].

[70] REISSNER, E. "On finite bending of pressurized tubes", *J. Appl. Mech.* (1959)
386–392.

[71] TUMARKIN, S.A. "Asymptotic solution of a linear non-homogeneous differential
equation and its applications", *Prikl. Mat. Meh. (PMM)* **23** (1959) no. 6 [in
Russian].

[72] TURNER, C.E. "Stress and deflection studies of flat plate and toroidal expansion
bellows, subjected to axial, excentric or internal pressure loading", *J. Mech. Eng.
Sci.* **2** (1959).

[73] AXELRAD, E.L. "Nonlinear equations of shells of revolution and of bending of
thin-walled beams", *Izv. AN SSSR OTN, Mekh. i Mash.* (1960) no. 4, 84–92 [in
Russian]; [Translated in *Amer. Rocket Soc. J. (Supplement)* **32** (1962) 1147].

[74] CHERNYKH, K.F. "St. Venant problems for thin-walled tubes with circular axis",
Prikl. Mat. Meh. (PMM) **24** (1960) no. 3 [in Russian].

[75] COHEN, J.W. "The inadequacy of the classical stress-strain relations for the right
helicoidal shell", *Proc. IUTAM Symp. on the Theory of Thin Elastic Shells*, W.T.
Koiter, Ed., North-Holland, Amsterdam (1960) 415–433.

[76] KOSTOVETSKY, D.L. "Bending of thin-walled curved tubes by large elastic displacements", *Izv. AN SSSR OTN, Mekh. i Mash.* (1960) no. 3 [in Russian].

[77] KOITER, W.T. "A consistent first approximation in the general theory of thin elastic
shells", *Proc. IUTAM Symp. on the Theory of Thin Elastic Shells*, W.T. Koiter,
Ed., North-Holland, Amsterdam (1960) 12–33.

[78] AXELRAD, E.L. "Flexure of thin-walled beams under large elastic displacements",
Izv. AN SSSR OTN, Mekh. i Mash. (1961) no. 3 [in Russian].

[79] AXELRAD, E.L. "On the theory of non-homogeneous anisotropic shells", *Izv. AN SSSR OTN, Mekh. i Mash* (1961) no. 2 [in Russian].

[80] GÜNTHER, W. "Analoge Systeme von Schalen-Gleichungen", *Ing.-Arch.* **30** (1961) 160–186.

[81] KOSTOVETSKY, D.L. "On stability of curved thin-walled tube under external pressure", *Izv. AN SSSR OTN, Mekh. i Mash* (1961) no. 1, 111 [in Russian].

[82] REISSNER, E. "On finite pure bending of cylindrical tubes", *Oesterr. Ing.-Arch.* **15** (1961) 165–172.

[83] SEIDE, P. and WEINGARTEN, V.I. "On the buckling of circular cylindrical shells under pure bending", *J. Appl. Mech.* **28** (1961) 112–116.

[84] AXELRAD, E.L. "Flexure and stability of thin-walled tubes under hydrostatic pressure", *Izv. AN SSSR OTN, Mekh. i Mash.* (1962) no. 1 [in Russian].

[85] ANDREEVA, L.E. *Uprugie elementy priborov*, Mashgiz, Moscow (1962) [in Russian].

[86] CHERNYKH, K.F. "Linear theory of shells: Parts 1 and 2", *NASA Techn. Trans.* **F441**; [Translated from Russian edition, Leningrad University (1962)]; (1964).

[87] WUEST, W. "Die Berechnung von Bourdonfedern", *VDI Forschungsheft* **489** (1962).

[88] BUDIANSKY, B. and SANDERS, J.L. "On the 'best' first-order linear shell theory", *Progress in Appl. Mech.*, *Prager Anniversary Volume*, MacMillan Co. (1963) 129–140.

[89] MACNAUGHTON, J.D. "Unfurable metal structures for spacecraft", *Canad. Air and Space J.* **9** (1963) 103–116.

[90] REISSNER, E. and WEINITSCHKE, H.J. "Finite pure bending of circular cylindrical tubes", *Quart. Appl. Math.* **21** (1963) no. 2, 305–319.

[91] REISSNER, E. "On stresses and deformations in toroidal shells of circular cross section, which are acted upon by uniform normal pressure", *Quart. Appl. Math.* **21** (1963) no. 3, 177–188.

[92] SANDERS, J.L. "Nonlinear theory for thin shells", *Quart. Appl. Math.* **21** (1963) 21–36.

[93] HOFF, N.J., CHAO, C.C. and MADSEN, W.A. "Buckling of a thin-walled circular cylindrical shell heated along an axial strip", *J. Appl. Mech.* **31** (1964) 253.

[94] REISSNER, E. "On the form of variationally derived shell equations", *J. Appl. Mech.* **31** (1964) 233–238.

[95] AXELRAD, E.L. "Refinement of critical load analysis for tube flexure by way of considering precritical deformation", *Izv. AN SSSR OTN, Mekh. i Mash.* (1965) no. 4, 133 [in Russian].

[96] HAMADA, M. and TAKEZONO, S. "Strength of U-shaped bellows", *Bull. JSME* **8** (1965) no. 32; **9** (1966) no. 35; **10** (1967) nos. 40/41.

[97] LARDNER, T.J. and SIMMONDS, J.G. "On the lateral deformation of shallow shells of revolution", *Internat. J. Solids and Structures* **1** (1965) 337–384.

[98] VASIL'EV, B.N. "Stressed state and deformation of Bourdon spring", *Izv. AN SSSR OTN, Mekh. i Mash.* (1965) no. 4 [in Russian].

[99] AXELRAD, E.L. "Periodic solutions of the axisymmetric problem of the shell theory", *Inzhenern. Zhurnal Mekhanika Tverdogo Tela (MTT)* (1966) no. 2 [in Russian].

[100] JAHNKE, E., EMDE, F. and LOESCH, F. *Tafeln hoeherer Funktionen.* Stuttgart (1966).

[101] KOITER, W.T. "On the nonlinear equations of thin elastic shells", *Proc. Kon. Ned. Akad. Wetenschappen* **B69** (1966) nos. 1/2.

[102] RIMROTT, F.P.J. "Two secondary effects in bending of slit thin-walled tubes", *J. Appl. Mech.* (1966) 75–78.

[103] VASIL'EV, B.N. *On Analysis of Bourdon Tubes.* Sbornik; Trudy Leningr. Instituta inzhenerov zhel.-dor. transporta Nr. 249 (1966) [in Russian].

[104] AXELRAD, E.L. "The constrained torsion of thin-walled beams", *Mech. Solids* **2** (1967) 142–146 [translated].

[105] AXELRAD, E.L. "Stability of a curved pipe of circular cross section under external pressure", *Mech. Solids* **2** (1967) 170–176 [translated].

[106] AXELRAD, E.L. "On different definitions of parameters of shell-curvature change and compatibility equations", *Mech. Solids* (1967) no. 2 [translated].

[107] FERNANDEZ SINTEZ, I. "Un elemento de connexion tubular plegable", *Ingegneria Aeronaut.* **19** (1967) no. 97, 23–33.

[108] HAMADA, M., KATSUHISA, F. and KIYOSHI, A. "Deformations and stresses in flat-oval tubes", *Bull. JSME* **10** (1967) no. 10, 618–625.

[109] JONES, H. "In-plane bending of a short-radius curved pipe bend", *J. Engng. for Industry; Trans. ASME Ser. B* **39** (1967) 271.

[110] KOITER, W.T. "General equations of elastic stability for thin shells", *Proc. Symp. on the Theory of Shells to Honour L.H. Donnell*, Houston (1967) 189–227.

[111] KRAUS, H. *Thin Elastic Shells.* Wiley, New York (1967).

[112] LANCZOS, C. *Applied Analysis.* Prentice-Hall, Englewood Cliffs, N.J. (1967).

[113] SOBEL, L.H. and FLÜGGE, W. "Stability of toroidal shells under uniform external pressure", *AIAA J.* **5** (1967) 425–431.

[114] CHERNINA, V.S. *Statika Tonkostennykh Obolotchek Vrashtchenija.* Nauka, Moscow (1968) [in Russian].

[115] HAMADA, M. and SEGUCHI, S. "Numerical method for nonlinear axisymmetrical bending of arbitrary shells of revolution and large deflection analysis of corrugated diaphragm and bellows", *Bull. JSME* **11** (1968) no. 43.

[116] HUTCHINSON, J.W. "Buckling and initial postbuckling behaviour of oval cylindrical shells under axial compression", *J. Appl. Mech.* **35** (1968) 66–72.

[117] IL'IN, V.P. "Stability of curved thin-walled tubes with stiffened edges under bending", *Mech. Solids* (1968) no. 2 [translated].

[118] IL'IN, V.P. "Experimental investigation of prebuckling deformation and buckling of cylinder shells under pure bending", *Stroitelnoje prjektirovanie promyslennych predprijatij* (1968) no. 1 [in Russian].

[119] REISSNER, E. "Finite inextensional pure bending and twisting of thin shells of revolution", *Quart. J. Mech. Appl. Math.* **21** (1968) 3.

[120] RIMROTT, F.P.J. "Entwurf und Berechnung von Lapprohren", *Luftfahrttechnik und Raumfahrttechnik* **14** (1968) 1.

[121] ALMROTH, B.O., SOBEL, L.H. and HUNTER, A.R. "An experimental investigation of the buckling of toroidal shells", *AIAA J.* **7** (1969) 2185–2186.

[122] DANIELSON, D.A. and SIMMONDS, J.G. "Accurate buckling equations for arbitrary and cylindrical elastic shells", *Internat. J. Engng. Sci.* **7** (1969) 459–468.

[123] KALNINS, A. "Stress analysis of curved tubes", *Proc. 1st Int. Conf. on Pressure Vessel Technology*; Delft (1969) 1–19, 223–235.

[124] KOITER, W.T. "Foundations and basic equations of shell theory. A survey of recent progress", *Theory of Thin Shells, 2nd Symp.*, F. Niordsen, Ed., Springer, Berlin (1969) 93–105.

[125] REISSNER, E. and WAN, F.Y.M. "Rotationally symmetric stress and strain in shells of revolution", *Stud. Appl. Math.* **48** (1969) no. 1.

[126] RIMROTT, F.P.J. "Large uniform torsion of a thin-walled open section of circular cross section; *C.A.S.I. Trans.* **2** (1969) no. 1.

[127] SANDERS, J.L., JR. "On the shell equations in complex form", *Theory of Thin Shells, 2nd Symp.*, F. Niordsen, Ed., Springer, Berlin (1969).

[128] SAVKIN, N.M. "Analysis of bellows for axisymmetric loading", *Izv. Vuz'ov, Mashinostroenie* **8** (1969) [in Russian].

[129] TENNYSON, R.C. "Buckling modes of circular cylindrical shells", *AIAA J.* **7** (1969) 1476.

[130] AXELRAD, E.L. and SAVKIN, N.M. "Graphoanalytical method for calculation of bellows", *Pribory i sistemy upravlenija* (1970) no. 8 [in Russian].

[131] CHENG, E.H. and THAILER, H.T. "In-plane bending of a U-shaped circular tube with end constraints", *Trans. ASME* **B92** (1970) no. 4.

[132] REISSNER, E. "On the derivation of two-dimensional shell theory from three-dimensional elasticity theory", *Stud. Appl. Math.* **49** (1970) 205–224.

[133] SIMMONDS, J.G. and DANIELSON, D.A. "Nonlinear shell theory with a finite rotation', *Proc. Kon. Ned. Akad. Wetenschappen* **B73** (1970) 460–478.

[134] SINTO, SEGUCHI and JOKODA. "Stresses and deformations of welded bellows", *Trans. ASME* **36** (1970) no. 283 [in Japanese].

[135] WAN, F.Y.M. "Rotationally symmetric shearing and bending of helicoidal shells", *Stud. Appl. Math.* **49** (1970) 351–369.

[136] WAN, F.Y.M. "Circumferentially sinusoidal variable stress and strain in shells of revolution", *Internat. J. Solids and Structures* **6** (1970) 959–973.

[137] WEINITSCHKE, H.J. "Die Stabilitaet elliptischer Zylinderschalen bei reiner Biegung", *ZAMM* **50** (1970) 411–422.

[138] HAMADA, M. et al. "Flexural deformation of U-shaped bellows", *Bull. JSME* **14** (1971) no. 71.

[139] IOKHELSON, J.J. "Bending of bellows by moments applied on the ends", *Izv. Vuz'ov, Mashinostroenie* (1971) no. 10.

[140] RIMROTT, F.P.J. "Das Waermeflattern von Lapprohren", *Ing.-Arch.* **40** (1971) 40–54.

[141] JAIN, V.K. and RIMROTT, F.P.J. "The ploy region of a slit tube", *C.A.S.I. Trans.* **4** (1971) no. 2.

[142] VASIL'EV, V.V. "Graphoanalytical analysis for bellows loaded by forces and moments on the ends", *Izv. Vuz'ov, Priborostroenie* (1971) no. 6 [in Russian].

[143] AXELRAD, E.L. and IL'IN, V.P. *Pipes Analysis.* (Rastchet truboprovodov) Mashgiz, Leningrad (1972) [in Russian].

[144] AXELRAD, E.L. and VASIL'EV, V.V. "Analysis of bellows loaded by bending moments", *Izv. Vuz'ov, Priborostroenie* (1972) no. 5 [in Russian].

[145] DODGE, W.G. and MOORE, S.E. "Stress indices and flexibility factors for moment loadings on elbows and curved pipe", *Welding Research Council Bulletin* (1972) no. 179.

[146] KRUGLJAKOVA, V.I. "Analysis of thin-walled tubes with curvilinear axis", *Izv. Akad. Nauk SSSR Meh. Tverd. Tela (Mech. Solids)* (1972) no. 6.

[147] LISOVSKIJ, A.S., OKISHEV, V.K. and USMANOV, J.A. "Ploskij izgib i rastjazhenie...", *Plane Flexure and Extension of Curved Thin-Walled Beams*, Mashinostroenie, Moscow (1972) [in Russian].

[148] NAGHDI, P.M. "The theory of shells and plates", *Handbuch der Physik*, 2nd ed., Vol. 6-2. W. Flügge, Ed., Springer, Berlin (1972).

[149] REISSNER, E. "On finite symmetrical strain in thin shells of revolution", *J. Appl. Mech.* **39** (1972) 1137.

[150] SIMMONDS, J.G. and DANIELSON, D.A. "Nonlinear shell theory with finite rotation and stress-function vectors", *J. Appl. Mech.* **39** (1972) 1085–1090.

[151] AXELRAD, E.L. *Statics of Elastic Beams*, A.P. Filin, Ed., Rastshet prostranstvennykh konstruktsij na protshnost i zhestkost. Strojizdat, Leningrad (1973) 38–52.

[152] FLÜGGE, W, *Stresses in Shells*, 2nd ed., Springer, Berlin (1973).

[153] IL'IN, V.P. "On analysis of curved bi-metallic tubes", *Mech. Solids* (1973) no. 5.

[154] JORDAN, P.F. "Buckling of toroidal shells under hydrostatic pressure", *AIAA J.* **11** (1973) 1439–1441.

[155] KOITER, W.T. and SIMMONDS, J.G. "Foundations of shell theory", *Proc. 13th ICTAM*, E. Becker and G.K. Michailov, Eds., Springer, Berlin (1973).

[156] NORDELL, W.J. and CRAWFORD, J.E. "Analysis of behaviour of unstiffened toroidal shells", *IASS Paper 4-4*; *Pacif. Symp. Hydromech. Loaded Shells*; University of Hawaii, Honolulu (1973) 304–313.

[157] WAN, F.Y.M. "Laterally loaded shells of revolution", *Ing.-Arch.* **42** (1973) 245–258.

[158] WÖBBECKE, W. "Allgemeine Differentialgleichungen duenner elastischer Schalen in Matrizendarstellung mit einer Loesung fuer verwundene Roehre", *Forschungsbericht der Deutschen Gesellschaft fuer Luft und Raumfahrt DGLR-RB* (1973).

[159] AXELRAD, E.L. and KVASNIKOV, B.N. "Semi-membrane theory of curved beam-shells", *Mech. Solids* **9** (1974) 125–132 [translated from Russian].

[160] BABCOCK, CH.D. "Experiments in shell buckling", *Proc. of the Symp. on Thin Shell Structures*, Y.C. Fung and E.S. Sechler, Eds., Prentice-Hall, Englewood Cliffs, N.J. (1974).

[161] BEGUN, P.I. "Analysis of twisted tubes", *Izv. Vus'ov, Priborostroenie* (1974) no. 6 [in Russian].

[162] KEMPNER, J. and CHEN, Y.N. "Buckling of oval cylindrical shells under combined axial compression and bending", *Trans. New York Acad. Sci.* **36**, Series 2 (1974) no. 2, 171–191.

[163] REISSNER, E. "Linear and nonlinear theory of shells", *Proc. of the Symp. on Thin Shell Structures*, Y.C. Fung and E.E. Sechler, Eds., Prentice-Hall, Englewood Cliffs, N.J. (1974).

[164] SEAMAN, W.J. and WAN, F.Y.M. "Lateral bending and twisting of thin-walled curved tubes", *Stud. Appl. Math.* **8** (1974) no. 1.

[165] ANDREEVA, L.E. et al. *Silfony*. [*Bellows*.] Mashinostronenie, Moscow (1975) [in Russian].

[166] AXELRAD, E.L. "Semi-membrane theory of flexible shells", *Proc. 10th USSR Conf. on Shell Theory*, Tbilisi (1975) no. 1. 487–496 [in Russian]; (Trudy 10 Vsesojuznoj konferenzii teorii obolotchek i plastin).

[167] BRUSH, D.O. and ALMROTH, B.O. *Buckling of Bars, Plates and Shells*. McGraw-Hill, New York (1975).

[168] ESSLINGER, M. and GEIER, B. *Postbuckling Behaviour of Structures*. Springer, Wien (1975).

[169] LIBRESCU, L. *Elastostatics and Kinetics of Anisotropic and Heterogeneous Shell-Type Structures*. Noordhoof Internat. Publ., Leyden (1975).

[170] NATARAJAN R. and BLOMFIELD, J.A. "Stress analysis of curved tubes with end constraints", *Computers and Structures* 5 (1975) 187–196.

[171] REISSNER, E. "Note on the equations of finite-strain force and moment stress elasticity", *Stud. Appl. Math.* 14 (1975) 1–8.

[172] RIMROTT, F.P.J. "On pure torsion of a cantilevered open section", *Trans. CSME* 3 (1975) 111.

[173] SEIDE, P. *Small Elastic Deformations of Thin Shells*, Noordhoff, Leiden (1975).

[174] SIMMONDS, J.G. "Rigorous expunction of Poisson's ratio from the Reissner–Meissner equations", *Internat. J. Solids and Structures* 11 (1975) 1051–1056.

[175] STEPHENS, W.B., STARNESS, J.H. and ALMROTH, B.O. "Collapse of long cylindrical shells under combined bending and pressure loads", *AIAA J.* 13 (1975) 20–25.

[176] AXELRAD, E.L. *Flexible Shells*. [Gibkie obolotshki.] Nauka, Moscow (1976) [in Russian].

[177] CHEN, J.N. and KEMPNER, J. "Buckling of oval cylindrical shells under compression and axisymmetrical bending", *AIAA J.* 14 (1976) 1235.

[178] GOL'DENVEIZER, A.L. *Theory of Elastic Thin Shells*. Nauka, Moscow (1976) [in Russian].

[179] GRIGOLJUK, E.I. and KABANOV, V.V. *Stability of Cylinder Shells* [in Russian]. Itogi Nauki, Mekhanika tverd. deform. tel, Moskva (1976).

[180] KABANOV, V.V. and KURTSEVICH, G.I. "Investigation of the stability of a cylindrical shell under axial compression nonuniform along the length", *Mech. Solids* (1976) no. 11, 171.

[181] EPSTEIN, M. and GLOCKNER, P. "Nonlinear analysis of mulilayered shells", *Internat. J. Solids and Structures* 13 (1977) 1081–1089.

[182] FABIAN, OLE. "Collapse of cylindrical elastic tubes", *Internat. J. Solids and Structures* 13 (1977) 1257–1270.

[183] LIBAI, A. and DURBAN, D. "Buckling of cylindrical shells subjected to nonuniform axial loads", *J. Appl. Mech.* 44 (1977) 714.

[184] THURSTON, G.A. "Critical bending moment of circular cylindrical tubes", *J. Appl. Mech.* 44 (1977) 173.

[185] AXELRAD, E.L. "Flexible shell theory and buckling of toroidal shells and tubes", *Ing.-Arch.* 47 (1978) 95–104.

[186] ODEN, J.T. and BATHE, K.J. "A commentary on computational mechanics", *Appl. Mech. Rev.* **31** (1978) 1053–1058.

[187] RODABAUGH, E.O., ISKANDER, S.K. and MOORE, S.E. "End effects on elbows subjected to moment loadings", *Battelle Lab. Rep. ORNL Sub-2913/7* (1978).

[188] PIETRASKIEWICZ, W. *Finite Rotations and Lagrangian Description in the Non-Linear Theory of Shells*, Polish Scientific Publishers, Warszawa-Poznań (1979).

[189] SAAL, H., KAHMER, H. and HEIN, J.L. "Experimentelle und theoretische Untersuchungen an beulgefaehrdeten Strukturen", *Der Stahlbau* **98** (1979) 353–359.

[190] SPENCE, J. and TOH, S.L. "Collapse of thin orthotropic elliptical cylindrical shells under combined bending and pressure loads", *J. Appl. Mech.* **46** (1979) 363–371.

[191] SPENCE, J. and BOYLE, J.T. "The influence of shape imperfections on stresses in piping components", *Paper C14/79*; *Proc. Conf. Significance of Deviations from Design Shapes*; Institute of Mechanical Engineers, London (1979).

[192] WHATHAM, J.F. and THOMPSON, J.J. "The bending and pressurization of pipe bends with flanged tangents", *Nuclear Engrg. Design* **54** (1979) 17–28.

[193] AXELRAD, E.L. "Flexible shells", *Theoretical and Applied Mechanics*; *Proc. of the 15th Int. Cong. Toronto* (1980), F.P.J. Rimrott and B. Tabarrok, Eds., North-Holland, Amsterdam (1981).

[194] BUSHNELL, D. "Buckling of shells—pitfall for designers", *AIAA 80-0665CP, 21st Structures Conference* (1980).

[195] KOITER, W.T. "The intrinsic equations of shell theory with some applications", *Mechanics Today* (E. Reissner Volume), S. Nemat-Nasser, Ed., Pergamon Press, Oxford (1980) 139–154.

[196] VOLPE, V., CHEN, Y.N. and KEMPNER, J. "Buckling of orthogonally stiffened finite oval cylindrical shells under axial compression", *AIAA J.* **18** (1980) 571–580.

[197] AXELRAD, E.L. "On vector description of arbitrary deformation of shells", *Internat. J. Solids and Structures* **17** (1981) 301–304.

[198] EMMERLING, F.A. "Nichtlineare Biegung eines schwach gekrümmten Rohres", *ZAMM* **61** (1981) T86–T89.

[199] REISSNER, E. "On finite pure bending of curved tubes", *Internat. J. Solids and Structures* **17** (1981) 839–844.

[200] EMMERLING, F.A. "Nichtlineare Biegung von geschlitzten dünnwandigen Rohren", *ZAMM* **62** (1982) 343–345.

[201] EMMERLING, F.A. "Nichtlineare Biegung und Beulen von Zylindern und krummen Rohren bei Normaldruck", *Ing.-Arch.* **52** (1982) 1–16.

[202] KOITER, W.T. "The application of the initial postbuckling analysis to shells", *Buckling of Shells*, E. Ramm, Ed., Proc. of a State-of-the-Art Coll., Springer, Berlin (1982).

[203] AXELRAD, E.L. and EMMERLING, F.A. "Finite bending and collapse of elastic pressurized tubes", *Ing.-Arch.* **53** (1983) 41–52.

[204] THOMSON, G. and SPENCE, J. "The influence of flanged end constraints on smooth curved tubes under in-plane bending", *Int. J. Press. Ves. & Piping* **13** (1983) 65–83.

[205] AXELRAD, E.L. *Schalentheorie*, B.G. Teubner, Stuttgart (1983).

[206] LIBAI, A. and SIMMONDS, J.G. "Nonlinear elastic shell theory", *Advances in Appl. Mech.* **23** (1983) 271–371.

[207] ÖRY, H. and WILCZEK, E. "Stress and stiffness calculation of thin-walled curved pipes with realistic boundary conditions being loaded in the plane of curvature", *Int. J. Press. Ves. & Piping* **12** (1983) 167–189.

[208] AXELRAD, E.L. "Flexible shells", *Flexible Shells, Theory and Applications*, E.L. Axelrad and F.A. Emmerling, Eds., Springer-Verlag, Berlin (1984) 44–63.

[209] BUFLER, H. "The principle of virtual displacements and the principle of virtual forces in the case of large deformations", *Acta Mech.* **53** (1984) 15–26.

[210] EMMERLING, F.A. "Nonlinear bending of curved tubes", *Flexible Shells, Theory and Applications*, E.L. Axelrad and F.A. Emmerling, Eds., Springer-Verlag, Berlin (1984) 175–191.

[211] HÜBNER, W. "Large deformations of elastic conical shells", *Flexible Shells, Theory and Applications*, E.L. Axelrad and F.A. Emmerling, Eds., Springer-Verlag, Berlin (1984) 257–270.

[212] RIMROTT, F. and DRAISEY, S.H. "Critical bending moment of double-slit tubing", *J. Spacecraft* **21** (1984) 316–318.

[213] SIMMONDS, J.G. "General helicoidal shells undergoing large one-dimensional strains or large inextensional deformations", *Internat. J. Solids and Structures* **20** (1984) 13–30.

[214] WHATHAM, J.F. "Results of pipe bend analysis", *Australian Atomic Energy Commission Reports, Parts I–X, AAEC/E551–554, E576–577, E585–587* (1982–1984).

[215] YAMAKI, N. *Elastic Stability of Circular Cylindrical Shells*, North-Holland, Amsterdam (1984).

[216] AXELRAD, E.L. "Elastic tubes—assumptions, equations, edge conditions", *Thin-Walled Structures* **3** (1985) 193–215.

[217] AXELRAD, E.J. "On local buckling of thin shells", *Internat. J. Non-Linear Mech.* **20** (1985) 249–259.

[218] AXELRAD, E.L. and EMMERLING, F.A. "Collapse load of elastic tubes under bending", *Israel J. Technology* **22** (1984/5) 89–94.

[219] BENSON, R.C. "Postbuckling analysis for the bending of a long beam with a thin, open, circular cross section", *J. Appl. Mech.* **52** (1985) 129–132.

[220] BUFLER, H. "Zur Potenzialeigenschaft der von einer Flüssigkeit herrührenden Druckbelastung", *ZAMM* **65** (1985) 130–132.

[221] FRIED, I. "Nonlinear finite element analysis of the thin elastic shell of revolution", *Comput. Methods Appl. Mech. Engrg.* **48** (1985) 283–299.

[222] NAGHDI, P.M. and YONGSARPIGOON, L. "Some general results in the kinematics of axisymmetrical deformation of shells of revolution", *Quart. Appl. Math.* **43** (1985) 23–26.

[223] REISSNER, E. "On mixed variational formulations in finite elasticity", *Acta Mech.* **56** (1985) 117–125.

[224] SCHMIDT, R. "A current trend in shell theory: constrained geometrically nonlinear Kirchhoff–Love type theories based on polar decomposition of strains and rotations", *Computers & Structures* **20** (1985) 265–275.

[225] AXELRAD, E.L. and EMMERLING, F.A., "Intrinsic shell-theory formulation, effective for large elastic rotations, and an application", *Finite Rotations in Structural Mechanics*, W. Pietraskiewicz, Ed., Springer-Verlag, Berlin (1986) 1–18.

[226] Bielski, J., "Postcritical deformations of meridional cross-section of elastic toroidal shells subject to external pressure", *Finite Rotations in Structural Mechanics*, W. Pietraskiewicz, Ed., Springer-Verlag, Berlin (1986) 47–61.

[227] Bufler, H., "Finite rotations and complementary extremum principles", *Finite Rotations in Structural Mechanics*, W. Pietraskiewicz, Ed., Springer-Verlag, Berlin (1986) 82–100.

[228] DUMIR, P.C. "Nonlinear axisymmetric response of orthotropic thin truncated conical and spherical caps", *Acta Mech.* **60** (1986) 121–132.

[229] HÜBNER, W. "Curved tubes with flanges under bending", *Applied Solid Mechanics–1*, A.S. Tooth and J. Spence, Eds., Elsevier Applied Science Publications, London (1986) 255–271.

[230] MURAKAMI, H., "Laminated composite plate theory with improved in-plane responses", *J. Appl. Mech.* **53** (1986) 661–666.

[231] RECKE, L. and WUNDERLICH, W., "Rotations as primary unknowns in the nonlinear theory of shells and corresponding finite elements models", *Finite Rotations in Structural Mechanics*, W. Pietraskiewicz, Ed., Springer-Verlag, Berlin (1986) 239–258.

[232] STUMPF, H. "General concept of the analysis of thin elastic shells", *ZAMM* **66** (1986) 337–350.

[233] SUHIR, E., "Stresses in bi-metal thermostats", *J. Appl. Mech.* **53** (1986) 657–660.

AUTHOR INDEX